Harmonisation of Regulatory Oversight in Biotechnology

Safety Assessment of Transgenic Organisms in the Environment, Volume 5

OECD CONSENSUS DOCUMENTS

OECD
BETTER POLICIES FOR BETTER LIVES

Please cite this publication as:
OECD (2016), *Safety Assessment of Transgenic Organisms in the Environment, Volume 5: OECD Consensus Documents*, Harmonisation of Regulatory Oversight in Biotechnology, OECD Publishing, Paris.
http://dx.doi.org/10.1787/9789264253018-en

ISBN 978-92-64-25044-4 (print)
ISBN 978-92-64-25301-8 (PDF)

Series: Harmonisation of Regulatory Oversight in Biotechnology
ISSN 2414-6854 (print)
ISSN 2311-4622 (online)

Foreword

From their first commercialisation in the mid-1990s, genetically engineered crops (also known as "transgenic" or "genetically modified" plants) have been approved for commercial release in an increasing number of countries, for planting or for entering in the composition of foods and feeds, or use in industrial processing. Up to now, the large majority of these agricultural productions remain for soybean, maize, cotton and rapeseed (canola), as outlined in *The Bioeconomy to 2030: Designing a Policy Agenda* (OECD, 2009). Despite some differences in total estimates, all analyses and statistics concur in underlining the general increasing trend in volumes produced and traded, number of countries involved and growth potential. For instance, James reports in the *Global Status of Commercialized Biotech/GM Crops: 2014, ISAAA Brief No. 49* that the surface area of transgenic crops worldwide constantly increased over the 19-year-period from 1996 to 2014, to reach 181.5 million hectares grown in 28 countries. To date, genetically engineered varieties of over 25 different plant species (including crops, flowers and trees) have received regulatory approval in OECD and non-OECD countries from all regions of the world. Such approvals for release in the environment usually follow a science-based risk/safety assessment before being granted.

The five main producers of genetically engineered crops in 2014 were the United States, followed by Brazil, Argentina, India and Canada, covering together almost 90% of the total area. Interestingly, developing countries grew more of global transgenic crops (53%) than industrial countries, at 47%. Among the 28 countries having planted those crops in 2014, only 9 of them were OECD countries, listed by decreasing area as follows: the United States, Canada, Australia, Mexico, Spain, Chile, Portugal, the Czech Republic and the Slovak Republic. In addition, some countries do not grow genetically engineered plants but import the produced commodities, for use in their feed industry in particular, as it is the case in several jurisdictions of Europe as well as some other economies worldwide.

Information on the transgenic crops which have been approved for commercial release in at least one country (for use in agriculture and/or foods and feeds processing) can be found in the OECD *Biotrack Product Database* (*www2.oecd.org/biotech*). Each transgenic product and its Unique Identifier are described, with information on approvals in countries. To date, this database covers about 240 approved genetically engineered plant varieties, and will be extended in future years to include additional species and information from a larger group of countries.

Modern biotechnologies are applied to plants, and also trees, animals and micro-organisms. The safety of the resulting genetically engineered organisms when released in the environment for their use in agriculture, food and feed industry, as biofuel or for other applications represents a challenging issue.

This is already true nowdays with the increasing cultivation of transgenic crops. It will be even more critical in the future as applications of biotechnologies widen to new species and new areas: a growing number of novel organisms will have to be assessed before their possible use and market release. Among the ongoing developments of modern biotechnology, crop varieties modified for gaining adaptation features such as the resistance to certain biotic/abiotic stresses, result in better resilience to climate change. "Bio-fortification" (applied to rice, tuber crops and other species) develop varieties with enhanced content in some constituents, e.g. vitamins or minerals. Plants with reduced lignine or with increased oil content are examples of products sought to facilitate industrial uses of the commodities and decrease the production costs. As highlighted in the proceedings of the OECD Conference "*Biosafety and the Environmental Uses of Micro-organisms*" held in 2012, a range of new species are contemplated as potential biofuels to provide renewable energy; among them algaes, with photosynthetic cyanobacteria, are of special interest as they can be cultivated year round on non-arable land, alleviating the pressure on agricultural land and freshwater resources that would be exerted by crops growng for biofuel purposes. Less anticipated, genetically engineered mosquitos are used in few places since 2014 to control the insect population and fight tropical diseases transmitted by them. Other biotechnology developments, and in particular applied to micro-organisms, might lead to other products such biofertilizer organisms living in symbiosis in crop roots and optimising the nitrogen fixation, or biocontrol agents acting as plant protection products to control disease and attack by insects. Other exploratory fields may comprise bioremediation by using of living organisms for removing contaminants from the environment such as polluted land, or the development of detergents containing micro-organisms.

Even if it is difficult to predict which of these new biotechnology developments would lead to large applications in a medium term, it is expected that some of the products will have important impacts in their respective economic sectors. A scientifically sound approach to their risk assessment should inform biosafety regulators and support the national decisions regarding their potential release. Genetically engineered products are rigorously assessed by their developers during their elaboration, and by governments when ready for commercial use, to ensure high safety standards for the environment, human food and animal feed. Such assessments are felt essential for a healthy and sustainable agriculture, industry and trade.

An environmental safety/risk assessment of transgenic organisms is normally based on the information on the characteristics of the host organism, the introduced traits, the environment into which the organism is introduced, the interaction between these and the intended application. The OECD's Working Group on Harmonisation of Regulatory Oversight in Biotechnology (the "Working Group") decided, at its first session in June 1995, to focus its work on identifying parts of this information which could be commonly used in countries for environmental safety/risk assessment, to encourage information sharing and prevent duplication of efforts. The biosafety consensus documents are one of the major outputs of its work.

The biosafety consensus documents constitute a "snapshot" of current information on a specific host organism or trait, for use during regulatory assessments. They are not intended to be a comprehensive source of information on everything that is known about a specific host or trait, but they do address the key or core set of issues that OECD member countries believe are relevant to risk/safety assessment. Several non-member economies, as well as other international organisations, are associated with the work and share their expertise. The information collated in the consensus documents is said to be mutually acceptable among the OECD community and beyond in other juridictions wishing to use them during their assessment process.

As of December 2015, a total of 53 consensus and guidance documents on biosafety have been published by the Working Group. They include documents which address the biology of plants, trees and micro-organisms as well as those dealing with specific traits that are used in genetically engineered crops. In addition, documents of broader nature aiming to facilitate harmonisation have been developed.

The volumes of this publication published in 2016 contain a compilation of those biosafety consensus documents issued in 2011 and 2012 (Volume 5), and from 2013 to 2015 (Volume 6). Both of them contain the "Introduction to the biosafety consensus documents" published earlier (and slightly updated since Volumes 3 and 4 of 2010). The introduction explains the purpose of the documents and how they are relevant to risk/safety assessment. It also describes the process by which the documents are drafted, using a "lead country" approach.

Along with previous Volumes 1-4 (OECD, 2006a; 2006b; 2010a; 2010b) the present publication offers ready access to those consensus documents published on the OECD BioTrack website thus far. As such, Volumes 5 and 6 should be of value to applicants for commercial uses of transgenic organisms, regulators in national authorities, breeders, risk assessors as well as the wider scientific community.

This biosafety work is complementary of the activities of the OECD programme on novel food and feed safety, in particular to the consensus documents developed on the composition of foods and feeds derived from transgenic organisms, which detail the key nutrients, anti-nutrients, toxicants and other constituents that can be used in a comparative approach. More information on this programme can be found in the introduction.

As each of the consensus documents may be updated in the future when new knowledge becomes available, users of this book are encouraged to provide any information or opinions regarding the contents of the consensus documents or indeed, the OECD's other harmonisation activities. Comments can be provided to: ehscont@oecd.org.

The published consensus documents are also freely available individually, in their original form, from the OECD's Biotrack website (*www.oecd.org/biotrack*). Some updates have been made to data and citations in this edition.

Acknowledgements

This book is the result of the common effort of the participants in the OECD's Working Group on Harmonisation of Regulatory Oversight in Biotechnology. Each chapter is composed of a "consensus document" which was prepared under the leadership of one or several countries and observer delegations, as listed at the end of this volume. During their successive draftings, valuable inputs and suggestions for the documents were provided by a number of delegates and experts in the Working Group, whether from OECD member countries, non-member economies or observer organisations.

Each consensus document was issued individually, as soon as it was finalised and agreed on declassification, by the OECD Environment, Health and Safety (EHS) Division in the Series on Harmonisation of Regulatory Oversight in Biotechnology. Volumes 5 and 6 of this publication, containing the 2011-15 consensus documents, were prepared by Jennifer Allain and edited by Bertrand Dagallier, under the supervision of Peter Kearns, at the EHS Division, OECD Environment Directorate.

Table of contents

Tables

Figures

Executive summary

This document constitutes the fifth volume of the OECD Series on Harmonisation of Regulatory Oversight in Biotechnology, which relates to the environmental risk/safety assessment of transgenic organisms, also called "biosafety". It is a compendium collating in a single volume the individual "consensus documents" published by the Working Group on the Harmonisation of Regulatory Oversight in Biotechnology. The four previous volumes of the series covered the documents issued during the 1996-2010 period. This volume contains the consensus documents issued in 2011 and 2012, while Volume 6 will collate those published in 2013, 2014 and 2015.

Modern biotechnologies are applied to plants, and also trees, animals and micro-organisms. The safety of the resulting transgenic organisms when released in the environment for their use in agriculture, food and feed industry, or for other applications, represents a challenging issue. This is true nowdays with the increasing cultivation of genetically engineered crops, and might become more crucial with future biotechnology developments widening to new species (e.g. insects, algae) and new targets such as crops adapted to climate change, plants of improved composition (biofortification), products for easier processing, renewable biofuels, insects modified to prevent diseases, biofertilisers and other applications. Genetically engineered products are rigorously assessed by their developers during their elaboration, and by governments when ready for release, to ensure high safety standards for the environment, human food and animal feed. Such assessments are felt essential for a healthy and sustainable agriculture, industry and trade. The growing number of novel organisms will also need to be assessed through a scientifically sound approach to risk assessment that will inform biosafety regulators and support the decision concerning their release.

The OECD Working Group on Harmonisation of Regulatory Oversight in Biotechnology was established in 1995. It gathers national authorities responsible for the environmental risk/safety assessment of products of modern biotechnology in OECD countries and in other economies which are key stakeholders in their production and use. Observer international organisations and experts involved in biosafety are associated to this work. The Working Group's primary goals are to promote international regulatory harmonisation, to ensure that methods used is the risk assessment of genetically engineered products are as similar as possible, therefore opening the way to possible recognition and even acceptance of information from other countries' assessments. The benefits of harmonisation are multiple: it strengthens mutual understanding among countries, avoids duplication and saves resources, increases the efficiency of the risk assessment process. Overall, it improves safety, while reducing unnecessary barriers to trade.

The consensus documents constitute the main output of the Working Group. They offer practical tools which compile science-based information relevant to the risk/safety assessment of transgenic organisms intended for release in the environment. They are publicly available and considered worldwide as solid references for biosafety.

In this volume, the introduction to the biosafety consensus documents presents the OECD Working Group, the key background concepts, principles and common approach prevailing in risk/safety assessment of transgenic organisms. The purpose of the consensus documents and how they are developed, is also described.

Chapter 1 provides guidance on issues relevant to the risk/safety assessment of commercial environmental applications involving genetically engineered micro-organisms, especially bacteria. It explores the important aspects in bacteria for causing adverse human health effects, and how this knowledge can be used in biosafety regulatory assessment. It contains information on bacterial pathogenicity (general considerations, factors and determinants, molecular aspects), and other elements on assessing potential for bacteria-mediated adverse human health effects.

Chapter 2 deals with the biology of squashes, pumkins, zucchinis and gourds (*Cucurbita* species). This information can be used as a useful tool for the biosafety assessment. It contains elements of taxonomy, centres of origin and distribution, morphological characters, reproductive biology, genetics, hybridisation and introgression, crop production, interactions with other organisms, pests and pathogens, and biotechnological developments.

Chapter 3 relates to the biology of *Brassica* species which include oilseed rape, turnip rape, mustards, cabbages and other oilseed crops. Taxonomy for a range of *Brassica* species is described, their centres of origin and distribution, reproductive biology, genetics, hybridisation and introgression, crop production, interactions with other organisms, pests and pathogens, breeding methods and biotechnological developments, common pathogens and pests.

The set of science-based information and data contained in this volume, previously agreed by consensus and published by the OECD, constitute a solid reference recognised internationally. It is already widely used as part of the biosafety assessments. As such, this publication should be of value to applicants for commercial uses of transgenic organisms, to risk assessors and regulators in national authorities in charge of granting approvals to their release in the environment, as well as the wider scientific community.

Introduction to the biosafety consensus documents

About the OECD's Working Group for biosafety

The OECD's Working Group on Harmonisation of Regulatory Oversight in Biotechnology (the "Working Group") comprises delegates from the 34 member countries of the OECD and the European Commission. Typically, delegates are from those government ministries and agencies which have responsibility for the environmental risk/safety assessment of products of modern biotechnology. The Working Group also includes a number of observer delegations and invited experts who participate in its work, such as Argentina, the Russian Federation, the United Nations Environment Programme (UNEP), the Secretariat of the Convention on Biological Diversity (SCBD), the Food and Agriculture Organization of the United Nations (FAO), the United Nations Industrial Development Organisation (UNIDO) and the Business and Industry Advisory Committee to the OECD (BIAC).

In recent years, with the increasing use of biotech products in many regions of the world, together with the development of activities relating to tropical and subtropical species, participation was enlarged to other non-member economies including Brazil, Bangladesh, the People's Republic of China, Colombia, India, Indonesia, Kenya, Moldova, Paraguay, the Philippines and South Africa, as well as the African Biosafety Network of Expertise from the New Partnership for Africa's Development, a body from the African Union (AU-NEPAD-ABNE). From July 2011 to December 2014, a programme was jointly implemented by the World Bank, the ILSI Research Foundation–Center for Environmental Risk Assessment (ILSI-CERA) and the OECD in the framework of the "Partnership for Biosafety Risk Assessment and Regulation", which developed new links, enhanced collaboration and supported the participation of four non-member economies in the activities of the Working Group.

Regulatory harmonisation

The Working Group was established in 1995[1] at a time when the first commercial transgenic crops were being considered for regulatory approval in a number of OECD member countries. From the beginning, one of the group's primary goals was to promote international regulatory harmonisation in biotechnology among members. Regulatory harmonisation is the attempt to ensure that the information used in risk/safety assessments, as well as the methods used to collect such information, are as similar as possible. It could lead to countries recognising or even accepting information from one anothers' assessments. The benefits of harmonisation are clear. It increases mutual understanding among countries, which avoids duplication, saves on scarce resources and increases the efficiency of the risk/safety assessment process. This, in turn, improves safety while reducing unnecessary barriers to trade (OECD, 2000).

The need for harmonisation activities at the OECD

The establishment of the Working Group and its programme of work followed a detailed analysis by member countries of whether there was a need to continue work on harmonisation in biotechnology at the OECD, and if so, what it should entail. This analysis was undertaken by the Ad Hoc Group for Environmental Aspects of Biotechnology (established by the Joint Meeting),[2] in 1994 mainly.

The Ad Hoc Group for Environmental Aspects of Biotechnology took into consideration, and built upon, the earlier work at the OECD which began in the mid-1980s. Initially, these OECD activities focused on the environmental and agricultural implications of field trials of transgenic organisms, but this was soon followed by a consideration of their large-scale use and commercialisation. (A summary of this extensive body of work is found in the annex to this introduction.)

Key background concepts and principles

The Ad Hoc Group for Environmental Aspects of Biotechnology took into account previous work on risk analysis that is summarised in *Safety Considerations for Biotechnology: Scale-up of Crop Plants* (OECD, 1993a). The following quote gives the flavour: "Risk/safety analysis is based on the characteristics of the organism, the introduced trait, the environment into which the organism is introduced, the interaction between these, and the intended application." This body of work has formed the basis for environmental risk/safety assessment that is now globally accepted. In considering the possibilities for harmonisation, the Ad Hoc Group paid attention to these characteristics and the information used by risk/safety assessors to address them.

This was reinforced by the concept of familiarity, also elaborated in the above-mentioned document (OECD, 1993a). This concept "is based on the fact that most genetically engineered organisms are developed from organisms such as crop plants whose biology is well understood... Familiarity allows the risk assessor to draw on previous knowledge and experience with the introduction of plants and micro-organisms into the environment." For plants, familiarity takes account of a wide-range of attributes including, for example, knowledge and experience with "the crop plant, including its flowering/reproductive characteristics, ecological requirements, and past breeding experiences" (OECD, 1993a – see also the annex for a more detailed description). This illustrates the role of information related to the biology of the host organism as a part of an environmental risk/safety assessment.

The Ad Hoc Group for Environmental Aspects of Biotechnology also considered the document *Traditional Crop Breeding Practices: An Historical Review to Serve as a Baseline for Assessing the Role of Modern Biotechnology* (OECD, 1993b), which focuses on host organisms. It presents information on an initial group of 17 different crop plants, which are used (or are likely to be used) in modern biotechnology. It includes sections on phytosanitary considerations in the movement of germplasm and on current uses of these crop plants. There is also a detailed section on current breeding practices.

A common approach to risk/safety assessment

An important aspect for the Ad Hoc Group for Environmental Aspects of Biotechnology was to identify the extent to which member countries address the same questions and issues during risk/safety assessment. Big differences would mean difficulties in working towards harmonisation, while a high level of similarity would suggest it is more feasible.

This point was resolved by two studies considered by the Ad Hoc Group: one covered crop plants (OECD, 1995a; 1995b) while the other concerned micro-organisms (OECD, 1995c; 1995d). Both studies involved a survey with national authorities responsible for risk/safety assessment. The aim was to identify the questions they address during the assessment process (as outlined in national laws/regulations/guidance texts) in order to establish the extent of similarity among national authorities. The studies used the information provided in the OECD's "Blue Book" on Recombinant DNA Safety Considerations (OECD, 1986) as a reference point, in particular, the sections covering: 1) general scientific considerations; 2) human health considerations; and 3) environmental and agricultural considerations (Appendices B, C and D). Both studies showed a remarkably high degree of similarity among countries in the questions/issues addressed in risk/safety assessment.

The emergence of the concept of consensus documents

The Working Group was therefore established in the knowledge that national authorities have much in common in terms of the questions/issues addressed when undertaking risk/safety assessment. It also took into account those characteristics identified as part of the assessment (i.e. the organism, the introduced trait and the environment) around which harmonisation activities could focus.

It was further recognised that much of the information used in risk/safety assessment relating to the biology of host organisms (crop plants, trees, animals or micro-organisms) would be similar or virtually the same in all assessments involving the same organism. In other words, the questions addressed during risk/safety assessment which relate to the biology of the organism, for example the potential for gene transfer within the crop plant species, and among related species, as well as the potential for weediness remain the same for each application involving the same host species. This also applies to some extent to information related to introduced traits.

Consequently, the Working Group evolved the idea of compiling information common to the risk/safety assessment of a number of transgenic products, and decided to focus on two specific categories: the biology of the host species and traits used in genetic modifications. The aim was to encourage information sharing and prevent duplication of effort among countries by avoiding the need to address the same common issues in applications involving the same organism or trait. It was recognised that biology and trait consensus documents could be agreed upon relatively quickly by member countries (within a few years). This compilation process was quickly formalised in the drafting of consensus documents.

The purpose of consensus documents

The consensus documents are not intended to be a substitute for a risk/safety assessment, because they address only a part of the necessary information. Nevertheless, they should make an important contribution to environmental risk/safety assessment.

Consensus documents are intended to be a "snapshot" of current information, for use during the regulatory assessment of products of biotechnology. They are not intended to be a comprehensive source of information covering the full knowledge about a specific host organism or trait; but they address – on a consensual basis – the key or core set of issues that countries believe to be relevant to risk/safety assessment.

The aim of the documents is to share information on these key components of an environmental safety review in order to prevent duplication of effort among countries. The documents are envisaged to be used: 1) by applicants as information to be given in applications to regulatory authorities; 2) by regulators as a general guide and reference source in their reviews; and 3) by governments for information sharing, research reference and public information.

Originally, it was said that the information in the consensus documents is intended to be mutually recognised or mutually acceptable among OECD member countries, though the precise meaning of these terms is still open for discussion. During the period of the Ad Hoc Group for Environmental Aspects of Biotechnology and the early days of the Working Group (1993-95), the phrase "mutual acceptance of data" was discussed. This concept, borrowed from OECD's Chemicals Programme, involves OECD Council decisions that have legally binding implications for member countries. In the case of the consensus documents, there has never been a legally binding commitment to use the information they contain, though the Working Group is interested in enhancing the commitment of countries to make use of the documents. Participation in the development of documents, and the intention by countries to use the information, is done in "good faith." It is expected, therefore, that reference will be made to relevant consensus documents during risk/safety assessments. As these documents are publicly available tools, they can be of interest for any country wishing to use them in national assesments.

The process through which consensus documents are initiated and brought to publication

There are a number of steps in the drafting of a specific consensus document. The first step occurs when a delegation, in a formal meeting of the Working Group, makes a proposal to draft a document on a new topic, typically a crop species or a trait. If the Working Group agrees to the proposal, a provisional draft is prepared by either a single country or two or more countries working together ("lead country approach"). Typically, the lead country(ies) has had experience with the concerned crop or trait and is able to draw on experts to prepare a provisional draft.

The provisional draft is first reviewed by the Bureau of the Working Group[3] to ensure that it addresses the range of issues normally covered by consensus documents and is of sufficiently high quality to merit consideration by the Working Group as a whole.

Based on the comments of the Bureau, a first draft is prepared for consideration by the full Working Group. This is the opportunity for each delegation to review the text and provide comments based on their national experiences. Inputs are incorporated in a second draft, which is again circulated to the Working Group. At this point, the Working Group may be asked to recommend that the document be declassified. Such a recommendation is only forthcoming when all delegations have come to a consensus that the document is complete and ready for publication. Sometimes, however, the text may need a third or even more discussions in the Working Group before a declassification can be contemplated.

When the Working Group has agreed to recommend a document for declassification, it is forwarded to the supervisory committee – the Joint Meeting – which is invited to declassify the document. Following the agreement of the Joint Meeting, the document is then published.

It is important to note that the review of consensus documents is not limited to formal meetings of the Working Group. Much discussion also occurs through electronic means, especially via the protected website dedicated to the Working Group. This enables a range of experts to have input into drafts.

For a number of documents, it has also been necessary to include information from non-member countries. This wider share of expertise has become increasingly important in recent years with the development of activities relating to tropical and subtropical species. This has been particularly true in the case of crop plants where the centre of origin and diversity occurs in a non-member country(ies). In these cases, UNEP, UNIDO and the FAO have assisted in the preparation of documents by identifying experts from concerned countries. For example, this occurred with the consensus document on the biology of *Oryza sativa* (rice) published in 1999.

The full series of consensus documents developed by the Working Group is also published in compendium documents, as it is the case for these volumes 5 and 6 covering 2011-15. Previous volumes 3 and 4 were published in 2010 (covering 2007-10), while volumes 1 and 2 were issued in 2006 (covering1996-2006) (OECD, 2010b; 2010c; 2006a; 2006b).

Current and future trends in the Working Group

The Working Group continues its work on the preparation of specific consensus documents, and on the efficiency of the process by which they are developed. An increasingly large number of crops and other host species (trees, animals, micro-organisms) are being modified, for an increasing number of traits, and the Working Group aims to fulfil the current needs and be prepared for emerging topics.

At the OECD Workshop on Consensus Documents and Future Work in Harmonisation, held in Washington, DC in October 2003, the Working Group considered how to set priorities for drafting future consensus documents among the large number of possibilities. The workshop also recognised that published consensus documents may be in need of review and updating from time to time, to ensure that they include the most recent information. The Working Group considers these aspects on a regular basis when planning future work. For the preparation of future documents, the workshop identified the usefulness of developing a standardised structure of consensus documents. The Working Group contemplated to develop, firstly, a guidance document on "Points to consider" for consensus documents on the biology of cultivated plants that was published in 2006, and then that of the trait documents. The "Points to consider' document, included in Volumes 3 and 4 of the compendia series, is currently under review by the Working Group to update it with the latest developments.

Within the important ongoing activities of the Working Group, a new document is being developed on the "Environmental considerations for the risk/safety assessment for the release of transgenic plants". Focused on the core of the biosafety work that is applied to crops and trees, and taking into account the most recent views from countries of all regions of the world, this document will constitute a key guidance tool for developers, assessors and regulatory authorities. It is expected to be published around 2017.

Other projects are implemented to prepare consensus documents on the biology of animals, to date on the Atlantic salmon (*Salmo salar*), and on the mosquito *Aedes aegypti*, for which some genetically engineered strains are used since 2014 in limited areas to control the virus-vector insect population and participate in the fight against the tropical diseases such as dengue fever and chikungunia that have been dramatically extending in many regions of the world over the last decade.

The Working Group is also considering projects on micro-organisms, therefore opening to new areas, for instance, bioenergy, with the preparation of a document on eukaryotic micro-algae having started recently. The photosynthetic cyanobacteria are potential providers of renewable energy and are of special interest as they can be cultivated year round on non-arable land, alleviating the pressure on farmland and freshwater resources that would be exerted by crops grown for biofuel purposes, as stated in the proceedings of the OECD Conference on Biosafety and the Environmental Uses of Micro-Organisms set up by the Working Group in 2012 (OECD, 2015a). Other biotechnology developments applied to micro-organisms might be considered to prepare future documents: updated review of biofertilizer organisms living in symbiosis in crop roots and optimising the nitrogen fixation, or biocontrol agents acting as plant protection products to control disease and attack by insects and other herbivores. Other exploratory fields may comprise bioremediation by using living organisms for removing contaminants from the environment such as polluted land, or the development of detergents containing micro-organisms.

In recent years, the Working Group started to exchange knowledge and promote discussion on the new plant-breeding techniques and their potential impact of risk/safety assessment. An OECD workshop was organised on these matters by the Working Group in 2014, and the report will be published soon.

The OECD Task Force for the Safety of Novel Foods and Feeds

The OECD Task Force for the Safety of Novel Foods and Feeds ("Task Force"), established in 1999, addresses aspects of the assessment of human food and animal feed derived from genetically engineered crops. As with the Working Group, the main focus of the Task Force work is to ensure that the types of information used in risk/safety assessement, as well as the methods to collect such information, are as similar as possible amongst countries. The approach is to compare transgenic crops and derived products with similar conventional ones that are already known and considered safe because of recognised experience in their use. Harmonised methods and the sharing of information are facilitated through the Task Force's activities.

Similarly to the biosafety programme, the main outcome of the foods and feeds programme is the set of consensus socuments on compositional considerations of new varieties of specific crops. The Task Force documents compile a common base of scientific information on the major components of crop plants, such as key nutrients, toxicants, anti-nutrients and allergens. These documents constitute practical tools for regulators and risk/safety assessors dealing with these new varieties, with respect to foods and feeds. To date, 26 consensus documents have been published on major crops and on general considerations for facilitating harmonisation. They constitute the Series on the Safety of Novel Foods and Feeds which is also available on the OECD's website (*www.oecd.org/env/ehs/biotrack*).

The full series of consensus documents developed by the Task Force was published in 2015 in two compendium documents, Volume 1 covering 2002-08 and Volume 2 covering 2009-14 (OECD, 2015b; 2015c).

The Working Group and the Task Force are implementing closely related and complementary programmes, focused on environmental aspects for the first and on food and feed aspects for the second. Their co-operation on issues of common interest resulted in the first document developed jointly by the two bodies, the "Consensus document on molecular characterisation of plants derived from modern biotechnology", published in 2010 (included in Volume 3 of the current series).

Notes

1. The original title of the Working Group was the "Expert Group for the Harmonisation of Regulatory Oversight in Biotechnology". It became an OECD working group in 1998.
2. The Joint Meeting was the supervisory body of the Ad Hoc Group for Environmental Aspects of Biotechnology and, as a result of its findings, established the Working Group as a subsidiary body. Today, its full title is the Joint Meeting of the Chemicals Committee and the Working Party on Chemical, Pesticides and Biotechnology.
3. The Bureau comprises the Chair and Vice-Chairs of the Working Group. The Bureau is elected by the Working Group once per year. At the time of preparing this publication – Volumes 5 and 6 – the Chair is from the United States, and the Vice-Chairs from Australia, Belgium, Finland, Japan and Mexico.

Annex: OECD biosafety principles and concepts developed prior to the Working Group on Harmonisation of Regulatory Oversight in Biotechnology (1986-94)

Since the mid-1980s the OECD has been developing harmonised approaches to the risk/safety assessment of products of modern biotechnology. Prior to the establishment of the Working Group on Harmonisation of Regulatory Oversight in Biotechnology, the OECD published a number of reports on safety considerations, concepts and principles for risk/safety assessment as well as information on field releases of transgenic crops, and a consideration of traditional crop breeding practices. This annex notes some of the highlights of these achievements that were background considerations in the establishment of the Working Group and its development of consensus documents.

Underlying scientific principles

In 1986, the OECD published its first safety considerations for genetically engineered organisms (OECD, 1986). These included the issues relevant to human health, the environment and agriculture that might be considered in a risk/safety assessment. In its recommendations for agricultural and environmental applications, it suggested that risk/safety assessors:

- "Use the considerable data on the environmental and human health effects of living organisms to guide risk assessments.
- Ensure that recombinant DNA organisms are evaluated for potential risk, prior to application in agriculture and the environment by means of an independent review of potential risks on a case-by-case basis.
- Conduct the development of recombinant DNA organisms for agricultural and environmental applications in a stepwise fashion, moving, where appropriate, from the laboratory to the growth chamber and greenhouse, to limited field testing and finally to large-scale field testing. And,
- Encourage further research to improve the prediction, evaluation, and monitoring of the outcome of applications of recombinant DNA organisms."

The role of confinement in small-scale testing

In 1992, OECD published its *Good Developmental Principles* (OECD, 1992) for the design of small-scale field research involving transgenic plants and micro-organisms. This document describes the use of confinement in field tests. Confinement includes measures to avoid the dissemination or establishment of organisms from a field trial, for example, the use of physical, temporal or biological isolation (such as the use of sterility).

Scale-up of crop-plants – "risk/safety analysis"

By 1993, the focus of attention had switched to the scale-up of crop plants as plant breeders began to move to larger scale production and commercialisation of transgenic plants. The OECD published general principles for scale-up (OECD, 1993a), which reaffirmed that, "safety in biotechnology is achieved by the appropriate application of risk/safety analysis and risk management. Risk/safety analysis comprises hazard identification and, if a hazard has been identified, risk assessment. Risk/safety analysis is based on the characteristics of the organism, the introduced trait, the environment into which the organism is introduced, the interaction between these and the intended application. Risk/safety analysis is conducted prior to an intended action and is typically a routine component of research, development and testing of new organisms, whether performed in a laboratory or a field setting. Risk/safety analysis is a scientific procedure which does not imply or exclude regulatory oversight or imply that every case will necessarily be reviewed by a national or other authority" (OECD, 1993a).

The role of familiarity in risk/safety assessment

The issue of scale-up also led to an important concept, familiarity, which is one key approach that has been used subsequently to address the environmental safety of transgenic plants.

The concept of familiarity is based on the fact that most genetically engineered organisms are developed from organisms such as crop plants, whose biology is well understood. It is not a risk/safety assessment in itself (US-NAS, 1989). However, the concept facilitates risk/safety assessments, because to be familiar means having enough information to be able to make a judgement of safety or risk (US-NAS, 1989). Familiarity can also be used to indicate appropriate management practices, including whether standard agricultural practices are adequate or whether other management practices are needed to manage the risk (OECD, 1993a). Familiarity allows the risk assessor to draw on previous knowledge and experience with the introduction of plants and micro-organisms into the environment and this indicates appropriate management practices. As familiarity depends also on the knowledge about the environment and its interaction with introduced organisms, the risk/safety assessment in one country may not be applicable in another country. However, as field tests are performed, information will accumulate about the organisms involved, and their interactions with a number of environments.

Familiarity comes from the knowledge and experience available for conducting a risk/safety analysis prior to scale-up of any new plant line or crop cultivar in a particular environment. For plants, for example, familiarity takes account of, but need not be restricted to, knowledge and experience with the following (OECD, 1993a):

- "The crop plant, including its flowering/reproductive characteristics, ecological requirements, and past breeding experiences
- the agricultural and surrounding environment of the trial site
- specific trait(s) transferred to the plant line(s)
- results from previous basic research including greenhouse/glasshouse and small-scale field research with the new plant line or with other plant lines having the same trait
- the scale-up of lines of the plant crop varieties developed by more traditional techniques of plant breeding
- the scale-up of other plant lines developed by the same technique

- the presence of related (and sexually compatible) plants in the surrounding natural environment, and knowledge of the potential for gene transfer between crop plant and the relative, and
- interactions between/among the crop plant, environment and trait."

Risk/safety assessment and risk management

Risk/safety assessment involves the identification of potential environmental adverse effects or hazards, and determining, when a hazard is identified, the probability of it occurring. If a potential hazard or adverse affect is identified, measures may be taken to minimise or mitigate it. This is risk management. Absolute certainty, or "zero risk", in a safety assessment is not achievable, so uncertainty is an inescapable aspect of all risk assessment and risk management (OECD, 1993a). For example, there is uncertainty in extrapolating the results of testing in one species to identify potential effects in another. Risk assessors and risk managers thus spend considerable effort to address uncertainty. Many of the activities in intergovernmental organisations, such as the OECD, address ways to handle uncertainty (OECD, 2000).

References and additional reading

Agrawala, S. et al. (2012), "Adaptation and innovation: An analysis of crop biotechnology patent data", *OECD Environment Working Papers*, No. 40, OECD Publishing, Paris, *http://dx.doi.org/10.1787/5k9csvvntt8p-en*.

James, C. (2015), "Global status of commercialized biotech/GM crops: 2014", ISAAA Brief No. 49, International Service for the Acquisition of Agri-Biotech Applications (ISAAA), Ithaca, New York, at: *www.isaaa.org/resources/publications/briefs/49* (executive summary accessed 28 January 2015).

OECD (2015a), *Biosafety and the Environmental Uses of Micro-Organisms: Conference Proceedings,* OECD Publishing, Paris, *http://dx.doi.org/10.1787/9789264213562-en*.

OECD (2015b), *Safety Assessment of Foods and Feeds Derived from Transgenic Crops, Volume 2 (2009-2014)*, OECD Publishing, Paris, *http://dx.doi.org/10.1787/9789264180147-en*.

OECD (2015c), *Safety Assessment of Foods and Feeds Derived from Transgenic Crops, Volume 1 (2002-2008)*, OECD Publishing, Paris, *http://dx.doi.org/10.1787/9789264180338-en*.

OECD (2010a), *Safety Assessment of Transgenic Organisms: OECD Consensus Documents: Volume 4*, OECD Publishing, Paris, *http://dx.doi.org/10.1787/9789264096158-en*.

OECD (2010b), *Safety Assessment of Transgenic Organisms: OECD Consensus Documents: Volume 3*, OECD Publishing, Paris, *http://dx.doi.org/10.1787/9789264095434-en*.

OECD (2009), *The Bioeconomy to 2030: Designing a Policy Agenda*, OECD Publishing, Paris, *http://dx.doi.org/10.1787/9789264056886-en*.

OECD (2006a), *Safety Assessment of Transgenic Organisms: OECD Consensus Documents: Volume 2*, OECD Publishing, Paris, *http://dx.doi.org/10.1787/9789264095403-en*.

OECD (2006b), *Safety Assessment of Transgenic Organisms: OECD Consensus Documents: Volume 1*, OECD Publishing, Paris, *http://dx.doi.org/10.1787/9789264095380-en*.

OECD (2000), "Report of the Working Group on Harmonisation of Regulatory Oversight in Biotechnology", prepared for the G8 Summit held in Okinawa, Japan on 21-23 July 2000, C(2000)86/ADD2, OECD, Paris, at: *www.oecd.org/chemicalsafety/biotrack/Report-of-the-Working-Group-on-Harmonisation-of-Regulatory.pdf* (accessed 18 November 2015).

OECD (1995a), "Commercialisation of agricultural products derived through modern biotechnology: Survey results", *OECD Environment Monograph: Series No. 99*, OECD, Paris, at: *www.oecd.org/science/biotrack/1876950.pdf* (accessed 20 November 2015).

OECD (1995b), "Report of the OECD Workshop on the Commercialisation of Agricultural Products Derived through Modern Biotechnology", *OECD Environment Monograph: Series No. 107*, OECD, Paris, at: *www.oecd.org/officialdocuments/publicdisplaydocumentpdf/?cote=OCDE/GD(95)72&docLanguage=En* (accessed 20 November 2015).

OECD (1995c), "Analysis of information elements used in the assessment of certain products of modern biotechnology", *OECD Environment Monograph: Series No. 100*, OECD, Paris, at: *www.oecd.org/officialdocuments/publicdisplaydocumentpdf/?cote=OCDE/GD(95)11&docLanguage=En* (accessed 20 November 2015).

OECD (1995d), *Safety Considerations for Biotechnology: Scale-up of Micro-organisms as Biofertizers*, OECD, Paris, at: *www.oecd.org/env/ehs/biotrack/Safety-considerations-scale-up-of-micro-organisms-as-biofertilizers.pdf* (accessed 20 November 2015).

OECD (1993a), *Safety Considerations for Biotechnology: Scale-up of Crop Plants*, OECD, Paris, at: *www.oecd.org/env/ehs/biotrack/1958527.pdf* (accessed 20 November 2015).

OECD (1993b), *Traditional Crop Breeding Practices: An Historical Review to Serve as a Baseline for Assessing the Role of Modern Biotechnology*, OECD, Paris, at: *www.oecd.org/env/ehs/biotrack/1946204.pdf* (accessed 20 November 2015).

OECD (1992), *Safety Considerations for Biotechnology – Part Two: Good Developmental Principles (GDP)*, OECD, Paris, at: *www.oecd.org/sti/biotech/2375496.pdf* (accessed 20 November 2015).

OECD (1986), *Recombinant DNA Safety Considerations. Safety Considerations for Industrial, Agricultural and Environmental Applications of Organisms Derived by Recombinant DNA Techniques* ("The Blue Book"), OECD, Paris, at: *www.oecd.org/env/ehs/biotrack/Recombinant-DNA-Safety-Considerations.pdf* (accessed 20 November 2015).

US-NAS (1989), *Field Testing of Genetically Modified Organisms: Framework for Decisions*, National Research Council, Committee on Scientific Evaluation of the Introduction of Genetically Modified Microorganisms and Plants into the Environment, National Academy Press, Washington, DC, at: *www.nap.edu/catalog/1431/field-testing-genetically-modified-organisms-framework-for-decisions*.

Part I:

Micro-organisms

Chapter 1.

Bacteria: Pathogenicity factors

This chapter provides guidance on topics and issues relevant to the risk/safety assessment of commercial environmental applications involving genetically engineered micro-organisms, especially bacteria. It explores the important aspects in bacteria for causing adverse human health effects, and how this knowledge can be used in biosafety regulatory assessment. It contains information on bacterial pathogenicity (general considerations, factors and determinants, genetics and molecular biology), and also elements on assessing potential for bacteria-mediated adverse human health effects.

The chapter was prepared by the OECD Working Group on the Harmonisation of Regulatory Oversight in Biotechnology, Sub-working Group on Micro-organisms, with the Netherlands and Canada having served as lead countries. It was initially issued in September 2011.

General considerations for bacterial pathogenicity

This chapter provides guidance on the concept of bacterial pathogenicity in the context of risk/safety assessment of deliberate release of "genetically engineered", or "genetically modified",[1] micro-organisms intended for commercial environmental applications (e.g. bioremediation, biosensors, biofertilisers, biopesticides, biomining, biomass conversion or oil recovery). It is limited in scope to bacteria that may exhibit properties pathogenic to human beings. Not included in the scope are environmental releases of known (potential) pathogens, e.g. vaccine strains. The chapter explores the factors that are important in bacteria for causing adverse human health effects and assesses how this knowledge can be used in risk/safety assessment of environmental applications of bacteria. Where appropriate, the chapter also refers to certain aspects of mamalian bacterial pathogens. For specific aspects of plant and/or other animal (e.g. fish, insects and other invertebrates) pathogens, separate documents on these issues would be needed.

Genetically engineered bacteria applied for environmental purposes, including field trials, should be evaluated to determine whether they may pose hazards to human health, which this chapter addresses. The analysis from the OECD "Blue Book" on recombinant DNA safety (OECD, 1986) appears to be still valid: Agricultural applications may result in release of large quantities of modified [micro]-organisms into terrestrial or aquatic ecosystems. Recombinant DNA-derived vaccines for animals and humans, as well as certain plant-associated micro-organisms, may in some cases have a limited pattern of environmental exposure because of biological specificity to the host, but incidental release to the environment certainly occurs in sewage and feed-lot or run-off waters, and may be significant. Environmental applications (e.g. metal extraction, pollutant and toxic waste degradation) may be confined initially to a specific location or may result in broad ecosystem exposure. The scientific considerations for assessing risk/safety will vary with each particular environmental application, depending on the organism, the physical and biological proximity to man and/or other significant biota. Local quarantine regulations, confinement measures and monitoring methodologies utilised during research and development will also be relevant.

In general, prior to their release, bacterial strains should be submitted to an assessment of their potential health effects, including their pathogenicity. As "virulence" is the quantitative measure of the pathogenicity of a micro-organism, the virulence factors of a bacterial strain are its traits that will be taken into account in the risk/safety assessment. For the special case of genetically engineered micro-organisms, the risk/safety assessment should take into account any characteristics of the engineered micro-organism related to pathogenicity, and whether any introduced traits are associated with pathogenicity.

When performing a regulatory review of the role of a donor gene as a virulence factor in the recipient micro-organism, regulators need a good understanding of the significance of a given virulence gene in the physiological background of the donor organism, as well as of the constitution of the recipient micro-organism. A large number of interacting factors affect the ability of a micro-organism to become pathogenic, and acquisition of a single gene in the absence of other genes necessary for pathogenicity will not likely convert a non-pathogen to a pathogen. Only if the newly acquired gene can have a role in the pathogenicity of the recipient micro-organism can an interaction be expected between the newly acquired gene and the resident genes contributing to a pathogenic lifestyle.

Pathogenicity is a multifactorial process which depends on the immune status of the host, the nature of the bacterial species or strain, and the number of organisms in the exposure. Therefore, the risk/safety assessment for human health can only be done on a case-by-case basis, taking into account the activity(s) of the introduced gene(s), the (potential) health hazards of the bacterial strain depending on the route of exposure (e.g. ingestion, inhalation, dermal contact) and the actual way that exposure to the strain is expected to occur under the conditions of the release. Exposure can depend on a number of factors, including the pattern of release (e.g. aerial spray, ground application, deep well injection, application into water bodies or effluent streams, shedding from inoculated humans or animals) and the scale of use (e.g. pilot, field trial, commercial use).

Because this chapter is intended as an aid to general risk/safety assessment tool, its nature is generic, i.e. not organism specific, and refers to specific bacteria and characteristics only to illustrate specific concepts. In addition to describing potential adverse health effects, and the bacterial factors that can contribute to these effects, the chapter describes general considerations in assessing the potential hazard of unmodified bacteria, e.g. a description of some tools available for predicting pathogenicity. Lastly, the chapter addresses considerations for the potential to introduce or alter pathogenicity as a result of genetic modifications to the micro-organism.

General considerations in assessing the hazardous potential of bacteria: The concept of bacterial pathogenicity

This section and the following two sections deal with the concept of bacterial pathogenicity in general, as it is discussed for unmodified bacteria; the concept also applies to genetically modified bacteria. Pathogenic bacteria have the ability to invade their hosts and produce disease. In this chapter, "pathogenicity" is referred to as the property of a micro-organism to cause disease. The great majority of bacteria that are encountered in the environment usually do not present problems to human health, in the sense that no record exists of them behaving as pathogens. Many bacteria are even beneficial, e.g. because of their role in essential processes in the environment such as mineralization, or their function as human symbionts. There are many bacteria that may act as opportunistic pathogens, i.e. organisms that are normally present in the environment or as part of the commensal bacterial population of a host, but that may cause disease when defense systems of the host become debilitated, or when the equilibrium within the existing bacterial population is disrupted. In general, given the interplay between members of microbial communities and the interplay between micro-organisms and potential hosts, it is unrealistic to say that a bacterium can never be a pathogen, and probably "non-pathogenic" bacteria can best be seen as bacteria that have not yet proven to have pathogenic potential.

Although "pathogenicity" can be defined in terms of properties of a micro-organism, it is important to keep in mind that the concept of pathogenicity is highly anthropomorphic, as it implies that a micro-organism would cause disease "on purpose". A more realistic view is that the body is a habitat for micro-organisms to adapt to and use as a favourable environment for survival and growth. Some bacteria have developed a "lifestyle" that enables them to colonise this niche in symbiotic as well as in pathogenic ways (Wassenaar and Gaastra, 2001). Each body surface – skin, conjunctiva, mucous membranes of the upper and lower respiratory tract, intestinal tract, genital tract and so forth – harbors a characteristic commensal bacterial population which differs qualitatively from the population of other areas of the body. Bacteria with pathogenic behaviour may establish a foothold in this microbial ecosystem. Once established, other pathogenic properties allow the pathogen to penetrate into deeper tissues, to avoid or counteract host defense mechanisms, and to multiply. As they pursue this strategy, pathogenic bacteria produce damage to the host. Virulence-associated factors may be defined as all factors that are essential for expressing pathogenicity.

Whether a host will develop disease is, however, not just determined by the pathogenic potential of the bacterium, but also by host factors. There is a formidable array of specific and non-specific host factors that affect the outcome of an encounter between a host and a pathogenic bacterium. For example, the normal commensal population plays an important role in protecting the host from invasion by pathogenic organisms. They do this by mechanisms such as: 1) competition for the same nutrients; 2) competition for the same receptors on the host cells (tropism); 3) production of bacteriocins or other antimicrobial agents (interference); and 4) stimulation of cross-protective immune factors. The commensal population of the host may be affected by a number of activities (e.g. use of antibiotics). Additional host factors that can affect pathogenicity include the production of antimicrobial substances (e.g. lysozyme in bronchial secretions; or the pancreatic enzymes, bile or intestinal secretions; or secretion of acid [HCl] for low pH of the stomach). Also, humans have an innate immune system that protects against invasion. When this system breaks down, e.g. in advanced stages of acquired immunodeficiency syndrome (AIDS) (Gradon, Timpone and Schnittman, 1992), bacteria that are normally not able to cause disease in humans may become opportunistic pathogens that cause conditions that clinically mimic the more commonly encountered "frank" pathogens. The potential of bacteria that normally occur in the environment to cause opportunistic infections in hosts with debilitated defense systems is recognised as an important human health hazard. The case of the *Burkholderia cepacia* complex (Bcc) is an example (Mahenthiralingham, Urban and Goldberg, 2005). Bacteria of the Bcc are found throughout the environment, some as plant pathogens.

General considerations in assessing the hazardous potential of bacteria: Classification of risk groups of bacteria

Pathogenic bacteria are commonly classified in risk groups, according to their pathogenic potential. The classification of the World Health Organization (WHO), as found in its *Laboratory Biosafety Manual* (WHO, 2004), is generally accepted. It should be noted, though, that these risk groups are primarily concerned with laboratory applications, where exposure may be high. They are valid for persons that are not immunocompromised. According to this classification, risk group 1 ("no or low individual or community risk") comprises micro-organisms that are unlikely to cause human or animal disease. Risk group 2 ("moderate individual risk, low community risk") comprises pathogens that can cause human or animal disease but that are unlikely to be a serious hazard to laboratory workers, the community, livestock or the environment; laboratory exposures may cause serious infection, but effective treatment and preventive measures are available and the risk of the spread of infection is limited. Risk group 3 ("high individual risk, low community risk") comprises pathogens that usually cause serious human or animal disease but do not ordinarily spread from one infected individual to another; effective treatment and preventive measures are available. Risk group 4 ("high individual and community risk") comprises pathogens that usually cause serious human or animal disease and that can be readily transmitted from one individual to another, directly or indirectly; effective treatment and preventive measures are not usually available.

For practical reasons, also in regulatory practice, a distinction is drawn between bacteria that are pathogenic to humans and bacteria that are pathogenic to other animals. Host specificity of bacteria is the result of differences between the environment that bacteria encounter in different hosts, i.e. in the human body and the bodies of other animals. If there are similarities between these environments, it may be expected that pathogenic organisms frequently "jump the species barrier". Indeed, there are a number of bacteria that are primarily pathogenic to other vertebrates that are also pathogenic to humans, e.g. *Bacillus anthracis*, *Brucella abortus*, *Yersinia pestis*, *Leptospira* spp. and a number of *Salmonella* species. Human diseases caused by these bacteria are called zoonoses (see also Blancou et al., 2005, for a review). In some cases insect vectors play a specific role in passing the pathogenic bacteria from the animal to the human host. Zoonotic diseases are "animal borne": animals, or animal products, act as a source of the disease. Consequently, exposure to the disease may change with changing social, behavioral and consumer practices. The risk class of a zoonotic bacterial species may differ depending on the host. For environmental risk/safety evaluations of activities with these bacterial species, the highest risk class has to be taken into consideration.

As pointed out previously, it is difficult to definitively state that a bacterial strain is non-pathogenic. The evidence given for non-pathogenicity can only be tentative. The determination of whether a bacterial strain may be considered non-pathogenic is usually made in a stepwise fashion. The strain may be considered non-pathogenic if it belongs to a species or taxonomic group for which no pathogenic strains are known. If it has direct relatives that are pathogenic, or if it is derived as an attenuated pathogenic strain, it should be shown that the strain effectively lacks the virulence determinants of its pathogenic relatives. If this fails, evidence for non-pathogenicity can be obtained through appropriate animal testing. This requires, however, a validated animal model. If none of this evidence is available or can be obtained, the strain may be considered non-pathogenic because it has a long history of safe use under conditions where no specific physical containment, like a closed fermentor system, has been applied to reduce worker exposure.

Although there is a clear value in using risk groups in practice (e.g. refer to WHO, 2004, Chapters 1 and 2), the concept of "opportunistic pathogenicity" implies that there is a continuum from non-pathogens to full frank pathogens. Some bacteria complete their life cycle independent of a human or animal host. Others that lack the ability to cause disease may still be able to recognise, adhere to and multiply in or on the host, as commensals. Opportunistic pathogens have some limited ability to cause disease, but are normally kept under control by the host immune response and defense systems and the competitive, harmless micro-organisms with which they compete in the host's habitat. However, they may acquire a toehold, with adverse consequences for the host, generally under circumstances where the host's defense mechanisms are compromised (e.g. weakening of the immune system through age or HIV infection) or destroyed (e.g. through skin lesions or burns). Some opportunistic pathogens are acquired from the environment while others may constitute part of the host's normal bacterial population. Some bacterial species causing infections at hospitals are used in bioremediation and/or bioaugmentation processes that may involve inoculation of soil with large amounts of bacteria. For instance, *Pseudomonas aeruginosa* and *Stenotrophomonas maltophilia* are organisms used industrially that cause nosocomial infections in cystic fibrosis and burn patients. *Serratia marcescens*, a common soil bacterium, causes pneumonia, urinary tract infections and bacteremia in compromised human hosts and is lethal to certain insect species with commercial use as a biopesiticde while commensal on the rhizoplane of many plant species. Other bacteria, such as *Lactobacillus acidophilus*, may be considered to be non-pathogens, because they rarely or never cause human disease. However, it should be noted that categorisation as non-pathogens may change due to the inherent variability and adaptability of bacteria and the potential for detrimental effects on host defense systems caused, for example, by radiation therapy, chemotherapy and immunotherapy; genetic defects (cystic fibrosis); or immunosuppressive infection (HIV).

General considerations in assessing the hazardous potential of bacteria: Approaches to bacterial virulence

In 1890 Robert Koch established his "postulates", a standard for the evidence of causation in infectious disease. The evidence should show that: 1) the micro-organism occurs in every case of the disease in question and under circumstances which can account for the pathological changes and clinical course of the disease; 2) after being isolated from the body and grown in pure culture, 3) the micro-organism can be inoculated into a healthy host and induce the disease anew; and 4) the micro-organism can be re-isolated after this experimental infection.

Virulence factors can be defined in terms of Koch's postulates as phenotypic properties of a micro-organism that are present in pathogenic strains that fulfill Koch's postulates but that are not observed in related strains that are not pathogenic. Although the postulates have been generally accepted for over 100 years (Fredricks and Relman, 1996), Koch himself already recognised the limitations of these guidelines. For instance, the ability to cause disease as an invariant virulence trait has been challenged. In recent years, a more integrated view of microbial pathogenesis has been developed which recognises that the contributions of both the pathogen and its host are required. The lack of experimental models for human-specific pathogens limits testing of the third postulate, and consequently also the rigorous testing of the role of a human-specific virulence factor.

Still, based on the notions of Koch's postulates, a number of virulence factors have been identified because of their clear role in the pathogenesis or their clear-cut coincidence with pathogenic strains, (e.g adhesins, invasins, haemolysins or, in general, cytolysins). With the development of molecular biological techniques, it became possible to identify the genes encoding these known virulence factors and to identify genes of unknown function for which a possible role in virulence could be determined. This resulted in a new approach of research on bacterial pathogenicity, in which the role of specific genes in bacterial virulence was the key point.

Virulence of a micro-organism is usually considered as the "degree" of pathogenicity of the micro-organism in a susceptible host. Finlay and Falkow (1997) discussed the various definitions of microbial pathogenicity, and the idea that pathogens can be distinguished from their non-virulent counterparts by the presence of such virulence genes. A virulence factor is a phenotypic trait associated with the virulence level of a micro-organism. The term is also used for a gene product (or group of gene products) that is responsible for the phenotypic trait. Virulence factors add to the pathogenicity, by enhancing one or more of the processes involved in the stages of pathogenicity: 1) the ability of the bacterial pathogen to gain access to the individual by surviving on or penetrating skin and mucous membranes; 2) the *in vivo* multiplication of the pathogen; 3) the inhibition or avoidance of host protective mechanisms; and 4) the production of disease or damage to the host. In this chapter microbial toxins are regarded as virulence factors even though these toxins are defined as gene products[2] produced by a bacterium that can cause harmful effects in the absence of the active living bacterium because in most cases the bacterium producing the toxin has to be established within the host in order to deliver the toxin most effectively. Therefore, the phenotypic trait of toxin production may be seen as increasing the pathogenic potential of a bacterium, while the full-blown effects of a toxin may be dependent on other virulence factors of the producing micro-organism, (e.g. the ability to colonise the host). It should, however, be noted that some bacteria that are not regarded as pathogenic (e.g. neurotoxin producing cyanobacteria) may also produce toxins, and that some bacteria producing toxins that can act at a distance (e.g. *Clostridium botulinum* causing foodborne disease) are characterised as pathogens.

Bacterial factors and determinants for pathogenicity

"Virulence" is a quantitative measure of the pathogenicity of a micro-organism that may be expressed by the ratio of the number of individuals developing clinical illness to the number of individuals exposed to the micro-organism, or in a comparative manner, by the number of individuals that develop clinical illness if the same dose of different micro-organisms is applied to each of them.

Pathogenic bacteria have evolved a number of different mechanisms, which result in disease in the host. The virulence factors and determinants used by bacteria to interact with the host can be unique to specific pathogens or conserved across several different species or even genera. For instance, common mechanisms for adherence, invasion, evasion of host defenses and damage to host cells are shared by profoundly different microbial pathogens. However, a virulence factor can only contribute to the pathogenic potential of a bacterium in and as far as the micro-organism possesses the constellation of traits conducive to pathogenicity. This section examines bacterial factors/determinants that contribute to pathogenicity in bacteria. While these are the determinants that would generally be considered in a risk/safety assessment, it should be noted that the same factor/determinant will not necessarily have a similar effect on the virulence of two different bacteria, and thus simple possession of a trait is not an indicator that the micro-organism is pathogenic. The concept of the "pathogenicity" of bacteria is further discussed in the next section.

Host recognition/adherence

Bacterial adherence to host surfaces is an essential first step in colonisation, infection and disease production. Colonisation establishes the organism at the portal of entry. Whereas intact outer skin is generally impervious to invasion by organisms, surface penetration of the urogenital, digestive and respiratory tracts as well as the mucosal barrier is more easily accomplished. Much of the body that is usually regarded as internal is topologically connected to the exterior. For example, the surfaces of the intestinal lumen, the lung alveoli, the bile cannaliculi and the kidney tubules are continuous with the outside skin. Organisms infecting these regions usually have elaborate adherence mechanisms and some ability to overcome or withstand the constant pressure of the host defenses on the surface. Bacterial adherence to host cells is usually a prerequisite to invasion. Consequently, a great deal of research has focused on elucidating bacterial mechanisms of adherence to host cells (adhesin biosynthesis, regulation of adhesins, identification of host receptors).

Adhesion can be defined as the coupling of a bacterium with a substratum. For molecules on the surface of the bacterium to interact with molecules on the surface of a host cell or the extracellular matrix, the two molecules must come into contact, an action that leads to the creation of intermolecular bonds requiring a certain amount of energy or effort to break. Bacterial adherence to a eukaryotic cell or tissue surface requires the participation of two factors: a receptor and an adhesin. The receptors so far defined are usually specific carbohydrate or peptide residues on the eukaryotic cell surface. Many bacterial adhesins are a macromolecular component of the bacterial cell surface which interacts with the host cell receptor. This interaction is usually complementary and specific, although most receptors can bind several ligands. It is this specificity which determines the tropism of the bacteria for a particular tissue (or a specific animal).

Bacterial adherence to cells or tissue surfaces may be specific or non-specific. Non-specific adherence or "docking" involves attractive forces and allows for the approach and reversible attachment of the bacterium to the eukaryotic surface (Kachlany et al., 2000). Possible interactions and forces involved include: hydrophobic interactions, electrostatic attractions, Brownian movement, recruitment and trapping by biofilm polymers interacting with the bacterial glycocalyx or capsule (Gilbert, Das and Foley, 1997; An, Dickinson and Doyle, 2000; Ukuku and Fett, 2002; Foong and Dickson, 2004). Specific adherence occurs when the bacterium forms a more permanent, yet still reversible, attachment with the eukaryotic surface and may proceed as one or more steps. Many specific lock-and-key bonds between complementary molecules on each cell surface are formed. Complementary receptor and adhesin molecules must be accessible and arranged in such a way that many bonds form over the area of contact between the two cells. Once the bonds are formed, separation under physiological conditions requires significant energy input. Some Gram positive bacteria with microbial surface components recognising adhesive matrix molecules (MSCRAMMs) employ a dock, lock and latch mode of ligand binding (Ponnuraj et al., 2003). Generally, reversible attachment precedes irreversible attachment, but in some cases specific adherence is not observed.

Mammalian cells communicate with each other through cell surface receptors. Once a receptor is bound with its ligand, a cellular response is triggered. Bacterial recognition of and interaction with host cell ligands facilitates the initial adherence to, and subsequent invasion of, host cells (Table 1.1). Through host receptor binding, bacteria exploit normal cellular processes to invade host cells.

Many micro-organisms have elaborate properties that can be used for industrial purposes in extensive biotechnological applications. For example, *Rhodococcus* spp. have elaborated adhesive properties for attachment to environmental surfaces or for biofilm formation that are particularly useful for adherence to heavy metals and hydrocarbons (Shabtai and Fleminger, 1994; Stratton et al., 2002). Although *Rhodococcus* spp. are not generally considered to be human pathogens, some species have emerged as rare opportunistic human pathogens. *Rhodococcus equi* infection is characterised by bronchiopneumonia following adherence and entry into alveolar macrophages. Garton et al. (2002) postulated that a novel lipoarbinomannan (LAM) variant may contribute to pathogenesis of disease caused by *R. equi*., similar to Manosylated LAM of *Mycobacterium tuberculosis* which facilitates adherence to alveolar macrophages via mannose receptors. Evaluators must always be cognisant that those factors which have extensive industrial applications (for instance, adhesive properties) may also confer one of the properties that allow a micro-organism to cause disease in susceptible individuals.

Table 1.1. Examples of specific bacterial adherence to host cell surfaces

Bacterium/disease	Adherence factors	Cellular receptors	Attachment rites	Reference(s)
Bordetella pertussis/ whooping cough	fimD; FHA; pertactin; pertussis toxin	VLA5; β2 integrin via LRI/IAP	Monocytes/macrophages respiratory epithelium	Hazenbos et al. (1995); Mattoo et al. (2001); Ishibashi et al. (2002); McGuirk, McCann and Mills (2002)
Burkholderia cepacia/ opportunistic infection	Cable (cbl) type II pili	Mucus glycoproteins	Respiratory epithelium	Sajjan et al. (1995)
Enterococcus faecium/ opportunistic bacterimia	Collagen adhesine gene (*acm*)	Collagen	Various tissue	Nallapareddy, Singh and Murray (2008)
Escherichia coli – ETEC/diarrhoea	FaeG (F4 or K88 fimbriae, pigs); FanC (F5 or K99 fimbriae, calves, lambs); GAG (humans)	Specific glycoconjugates	Brush borders of intestinal enterocytes	Nagy and Fekete1(999); Van den Broeck et al. (2000); Grange et al. (2002)
Escherichia coli – EPEC/diarrhoea	Bfp; intimin	Phosphatidylethanolamine; Tir	Intestinal epithelium	Hicks et al. (1998); Nougayrède et al. (2006); Touze et al. (2004)
Escherichia coli – EHEC/haemolytic uremic syndrome	Intimin	Tir	Colonic epithelium	Li et al. (2000); Goosney, DeVinney and Finlay (2001); Liu et al. (2002)
Escherichia coli – UPEC/ pyelonephritis	P pili [PapG (I, II, III)]; FimH, fimbriae	Gb03, Gb04, Gb05; CD55, Gal(α1-4)Gal containing isoreceptors, mannosylated glycoproteins	Kidney epithelial cells, erythrocytes; urinary tract epithelium	Dodson et al. (2001); Johnson et al. (2001); Ishikawa et al. (2004), Nowicki, Selvarangan and Nowicki (2002)
Escherichia coli – NMEC/neonatal meningitis	SfaII (S fimbriae)	Sialyl-α2-3 β-galactose-containing receptor molecules	Endothelial and epithelial cells	Tullus et al. (1992); Saren et al. (1999); Bonacorsi et al. (2000)
Escherichia coli O157:H7	Flagellin	TLR-5	Human colonic epithelium	Miyamoto et al. (2006)
Non-typeable *Haemophilus influenzae*/otitis media, sinusitis, conjunctivitis	HifE (pilus adhesin); HMW1, HMW2; Hap; Hia	Fibronectin; α2-3 linked sialic acid glycoprotein, unknown; fibronectin, laminin, collagen IV; unknown	Respiratory epithelium	McCrea et al. (1997); Laarmann et al. (2002); St. Geme III (2002); O'Neill et al. (2003)
Haemophilus influenzae	Fimbriae LKP family	Sialic acid-containing lactosyl ceramides and AnWJ antigen	Oropharyngal epithelial cells and erythrocytes	van Alphen et al. (1991)
Helicobacter pylori/peptic ulcer disease	BabA, SabAB adhesins	MUC5AC, MUC5B and MUC57	Oral cavity and stomach epithelium	Goodwin et al. (2008;) Lindén et al. (2008)
Legionella pneumophila/ Legionnaires disease	pilE; pilBCD (type IV pili); MOMP; enhD	β2 integrin (CR3), C1q, FcR	Macrophage; monocytes, epithelial cells	Cirillo et al. (2001); Samrakandi et al. (2002)
Listeria monocytogenes/ listeriosis	InlA; InlB	E-cadherin; gC1q-R	Epithelial cells	Braun, Ghebrehiwet and Cossart (2000); Kathariou (2002)
Moraxella catarrhalis	UspA1 and UspA2	Fibronectin	Epithelial cells	Tan, Forsgren and Riesbeck (2006)
Mycobacterium leprae/ leprosy	PGL-1 glycolipid	α2-laminin-dystroglycan complex	Schwann cells	Marques et al., 2001; Brophy (2002)
Mycobacterium tuberculosis/tuberculosis	HbhA protein	$\alpha_M\beta_2$ integrin (CR3 or CD11b/CD18)	Macrophage	Mueller-Ortiz, Wanger and Norris (2001); Velasco-Velázquez et al. (2003)
Mycobacterium avium/ pulmonary disease	FAP	Fibronectin	Extracellular matrix (damaged epithelial cells)	Schorey et al. (1996); Middleton et al. (2000)
Mycoplasma pneumoniae/atypical "walking" pneumonia	P1; P30	Sulfated glycolipids, sialylated compounds	Respiratory epithelium, alveolar macrophages	Athamna, Kramer and Kahane (1996); Seto et al., 2001; Balish et al. (2003); Seto and Miyata (2003)
Mycoplasma genitalium, M. pneumoniae	MG: P140 and P110; MP: P1 and P30 (specific adhesins of attachment organelles)			Burgos et al. (2006)
Neisseria meningitidis/carrier state	pilC (type IV pili;) Opa; Opc	CD46; HSPGs, fibronectin, vitronectin	Nasopharyngeal epithelium, endothelium	Merz and So (2000); Dehio, Gray-Owen and Meyer (2000); Hauck and Meyer (2003)
Porphyromonas gingivalis	Type II fimbriae	α5ß1-integrin		Nakagawa et al. (2002)
Pseudomonas aeruginosa	Type IV pili; OprF; PA-IL, PA-IIL	asialo-GM1 -GM2; galactose- and fucose/mannose-containing glycoconjugates	Epithelium	Craig, Pique and Tainer (2004); Azghani et al. (2002); Winzer et al. (2000); Imberty et al. (2004)
Rickettsia sp.	rOmpA: Crystalline rpoteic layer (S-layer) made of surface protein antigen (SPA)		Endothelial cells	Li and Walker (1998)

Table 1.1. **Examples of specific bacterial adherence to host cell surfaces** (*cont.*)

Bacterium/disease	Adherence factors	Cellular receptors	Attachment rites	Reference(s)
Salmonella enterica serovar Typhimurium/ gastroenteritis	fim (type I fimbriae); pef (PE fimbrae); lpf (LP fimbrae); agf (curli fimbriae); ShdA	oligomannose motifs; Hep-2 domain of fibronectin	Intestinal epithelium	Bäumler, Tsolis and Heffron (1997); Thankavel et al. (1999); Kingsley et al. (2004)
Shigella flexneri/dysentery	Invasion plasmid antigen BCD	$\alpha_5\beta_1$ integrin; carbohydrate moieties associated with mucin layer	Colonic epithelial cells	Rajkumar, Devaraj and Niranjali (1998); Kerr (1999); Kohler, Rodrigues and McCormick (2002)
Staphylococcus aureus/boils, furuncles, impetigo, septic shock	MSCRAMMs (FnBP, Protein A, PNSG, Cna, coagulase, Clf)	$\alpha_5\beta_1$ integrin; other unknown receptors on collagen, fibrinogen, IgG, prothrombin	Extracellular matrix	Shuter, Hatcher and Lowy (1996); Salyers and Whitt (2002); Roche et al. (2004)
Streptococcus pneumoniae/ sepsis, meningitis, otitis media, pneumonia	PavA	Fibronectin	Nasopharynx and alveolar epithelium	Holmes et al. (2001)
Streptococcus pyogenes/throat infections, other serious infections	MSCRAMMs (SfbI/F1, Fpb54, SfbII/SOF, F2)	$\alpha_5\beta_1$ integrin (fibronectin receptor)	Pharyngeal epithelium	Cue et al. (2000); Towers et al. (2003); Kreikemeyer et al. (2004)
Group A Streptococcus (*S. pyogenes*)	Protein M	CD46	Keratinocytes,	Rezcallah et al. (2005)
Group B Streptococcus	BibA	hC4bp	Epithelial cells	Santi et al. (2007)
Group G Streptococcus (*S. dysgalactiae*)	Surface protein FOG	Collagens I fibrils		Nitsche et al. (2006)
Treponema Pallidum/syphilis	MSCRAMMs	fibronectin receptor containing α_5;laminin receptor	Mucosal epithelium	Cameron (2003); Lee et al. (2003)
Vibrio cholerae/cholera	Tcp pili; others (O Ag of LPS, MSHA, MFRHA)	Specific carbohydrate and glycoprotein receptors	Intestinal epithelium	Franzon, Barker and Manning (1993); Häse and Mekalanos (1998); Sasmal et al. (2002)
Yersinia enterocolitica/diarrhoea	Invasin (OMP); YadA	β_1 integrins ($\alpha_3\beta_1$, $\alpha_4\beta_1$ $\alpha_5\beta_1$ $\alpha_6\beta_1$ and $\alpha_v\beta_1$); collagen, laminin, fibronectin	Intestinal epithelium and submucosa	Schulte et al. (2000); El Tahir and Skurnik (2001); Isberg and Barnes (2001)

Notes:

asialo-GM1 -GM2: glycolipids
Bfp: bundle forming pili
C4bp: complement component 4 binding protein
CD46: membrane cofactor protein, member of superfamily of complement resistant proteins
CD55: decay accelerating factor for complement
Clf: clumping factor
Cna: collagen binding protein
CR1: complement receptor type 1
CR3: complement receptor type 3
Curli fimbriae: thin aggregative fimbriae
EHEC: Enterohemorrhagic *Escherichia coli*
EPEC: Entoropathogenic *Escherichia coli*
ETEC: Enterotoxigenic *Escherichia coli*
FAP: fibronectin attachment protein
FcR: Fc receptor
FHA: filamentous hemagglutinin
Fim: fimbriae
FnBP: fibronectin binding protein
FOG: Friend of GATA
HbhA: heparin-binding hemagglutinin
HSPGs: heparansulphate proteoglycans
InlA: internalin A
InlB: internalin B
LKP: long-thick fimbriae
LP fimbrae: long polar fimbriae
LPS: lipopolysaccharide
LRI/IAP: leukocyte response integrin ($\alpha V\beta 3$, CD61)/integrin associated protein (CD47)
MFRHA: mannose fucose resistant hemagglutinin
MOMP: major outer membrane protein
MSCRAMMS: microbial surface components recognizing adhesive matrix molecules
MSHA: mannose sensitive hemagglutinin
MUC: mucin gene
NMEC: Neonatal Meningitidis *Escherichia coli*
OMP: outer membrane protein
Opa: opacity associated
Opc: class 5 outer memebrane protein
OprF: porin F
PavA: Adherence and virulence protein A
PE fimbrae: plasmid-encoded fimbriae
PGL-1: phenolic glycolipid 1
Pil: pili (fimbriae)
PNSG: Poly-n-succinyl-β-1,6 glucosamine
rOmpA: 190-kDa cell surface antigen
SfbI: streptococcal fibronectin-binding protein I
SfbII: streptococcal fibronectin-binding protein II
ShdA: host colonisation factor
Tcp: toxin co-regulated pili (demonstrably important in humans)
Tir: Translocated intimim receptor
TLR-5: Toll-like receptor 5
UspA: ubiquitous surface protein
UPEC: Uropathogenic *Escherichia coli*
VLA-5: very late antigen-5

While not an all-inclusive list, Table 1.1 gives examples of specific attachments of micro-organisms to host cell surfaces. It should be noted that many, but not all, adherence factors also play a role in invasion. For a more comprehensive review of adhesins, receptors and related structures, the reader is directed to articles by Connell et al. (1997), Soto and Hultgren (1999), Klemm and Schembri (2000), and Nougayrede et al. (2006).

In addition to determining pathogen location, adhesins affect important aspects of the biology of infection. Many pathogens have evolved the ability to bind to cell adhesion molecules (CAMs), which are eukaryotic cell-surface receptors that facilitate cell interaction and communication with other cells and the extracellular matrix. In these cases, cell signaling processes involving actin rearrangements are affected by virtue of their contact with the cytoskeleton (Mims, Nash and Stephen, 2001). Host cell adhesion receptors can be subdivided into several groups, for example, integrins, cadherins, immunoglobulin superfamily cell adhesion molecules (IgCAMs), selectins, receptor protein tyrosine phosphatases, syndecans and hyaluronate receptors (Freemont, 1998; Hauck, 2002). Since multiple adhesion molecules are found on a single host cell, they are ideal targets for pathogens trying to anchor themselves. Often, bacteria are able to bind to cell adhesion molecules by mimicking or acting in place of host cell receptors or their ligands, and may allow bacteria to exploit several of these molecules to establish tight contact with eukaryotic cell surfaces and the extracellular matrix (Hauck, 2002; Boyle and Findlay, 2003).

Bacterial adhesins[3] have been divided into two major groups: 1) pili (fimbriae) and 2) non-pilus (afimbrial) adhesins. Pili and fimbriae are interchangeable terms to designate short hairlike structures on the surface of bacterial cells. For the purposes of this chapter, the terms are used interchangeably and depend upon the article referenced.

Many bacteria express adhesive pili, which are hairlike surface appendages extending out from the bacterial surface to establish contact with the surface of the host cell. Pili may be displayed circumferentially (Salyers and Whitt, 2002; Hardy, Tudor and St. Geme III, 2003) or preferentially located on one part of the bacterial cell (Nougayrède et al., 2006). Binding to the host cell target is specific and it is this specificity that determines the preferential site/host for adherence.

The P pilus operon serves as a useful model for the general study of different bacterial pilus systems since the concepts are similar and many of the components are interchangeable, even though the host receptors differ. For example, the pyelonephritis-associated pili-D (PapD) chaperone, in addition to mediating the assembly of P pili, can modulate the assembly of type 1 pili (Bonci et al., 1997). There is a family of periplasmic PapD-like chaperones needed for the assembly of several pili, including K88, K99 and *Haemophilus influenzae* pili. Additionally, since the molecular machinery required for pilus biogenesis and bacterial surface assembly is conserved among diverse pili (Hultgren et al., 1993) the operons of type 1 and P pili are very similar with alignment of functionally analogous sequences. Nevertheless, they are structurally distinct pili (type 1 are flexible, rod-like fibers, while P pili are rigid structures) and bind to different receptors (Finlay and Falkow, 1997). Many adhesins of *E. coli* include their common pili and many strains of *E. coli* are able to express a variety of pili encoded by distinct regions of the chromosome or plasmids (Johnson, 1991).

Type 1 pili produced by *E. coli* strains recognise mannose receptors on host cells (Schwan et al., 2002). The mannose binding site may be located at the tips or inserted along the length of the pilus. Different tip protein adhesins allow the bacterium to adhere to different host cell receptors. This is of specific interest for evaluators since changes to tip proteins can significantly alter the tropism of the bacteria for a specific receptor. For example, tip proteins on pyelonephritis-associated (pap) pili recognise a galactose-galactose disaccharide, while tip proteins on S-fimbriae recognise sialic acid. It is equally important to recognise that while a receptor may be cell- or host-specific, this specificity may also change during the developmental stages of the host. Thus, while *E. coli* has been associated with meningitis in the neonate, in the adult this association is lost. Animal studies have demonstrated that endothelial receptors for *E. coli* are only present in the brain of the newborn (Parkkinen et al., 1988).

Type 4 pili (Tfp) constitute a separate, unique class of pili expressed by diverse gram-negative organisms of medical, environmental and industrial importance including *Pseudomonas aeruginosa*, *Neisseria* spp., *Moraxella* spp., Enteropathogenic *E. coli* (EPEC) and *Vibrio cholera*. Tfp share structural, biochemical, antigenic and morphological features (Strom and Lory, 1993) and a biogenesis pathway that is highly conserved and resembles the type II protein secretion pathway (Wolfgang et al., 2000). It has been suggested that the pilin molecules located at the tip may function as adhesins since the sequences exposed differ from those packed into repeating structures within a pilus. For instance, Tfp-mediated adherence is strongly correlated with a separate tip protein, PilC for *N. gonorrhoeae*, rather than the more abundant pilin subunit protein PilE (Winther-Larsen et al., 2001). Alterations in the pilus subunit can also affect adherence levels. Whereas *P. aeruginosa* strains usually express only one pilus subunit, the considerable variation exhibited by this subunit by the various strains affects the proficiency of adherence of the strains.

Bacteria usually adhere to receptor molecules via protein structures on their cell surface (typically pili) with distinct surface-binding capacities (Soto and Hultgren, 1999). However, other important adhesins found in a number of gram-negative pathogens may, alternatively, be anchored directly to the outer membrane (OM), resulting in an intimate attachment with the target cell receptor (Veiga, de Lorenzo and Fernandez, 2003). Afimbrial adhesins are bacterial surface proteins, structurally distinct from the adhesins of fimbriae, that facilitate the tighter binding of bacteria to host cell that usually follows initial binding via fimbriae. These proteins are important components of the systems that allow bacteria to attach to and invade host cells. Some may recognise proteins on host cell surfaces while others recognise carbohydrates (Salyers and Whitt, 2002). *Legionella pneumophila* afimbrial adhesin seems to be involved in attachment to and invasion of amoebae. Adhesins require presentation on the bacterial surface in an active binding conformation for interaction with the host cell. In gram-negative bacteria, surface localisation requires the translocation of the protein through the cytoplasmic membrane (export into the periplasm) and through the OM (secretion). Generally, surface localisation occurs via one of six different secretion pathways distinguished at least in part by the mechanisms of translocation across the OM and designated types I-VI (Stathopoulos et al., 2000; Cascales, 2008; Pukatzki, McAuley and Miyata, 2009).

Proteins secreted by the type V pathway are referred to as autotransporters (AT; Henderson, Cappello and Nataro, 2000). For example, the *H. influenzae* Hap autotransporter is a non-pilus adhesin that influences adherence to epithelial cells and some extracellular matrix proteins and impacts bacterial aggregation and microcolony formation. Other autotransporter proteins that function as adhesins include: ShdA and MisL of *Salmonella enterica* (Kinsgley et al., 2002); Pertactin, Vag8 and TcfA of *Bordetella* spp. (Li et al., 1992; Finn and Stevens, 1995; Finn and Amsbuagh, 1998); AIDA-I, TibA and Ag43 of *E. coli* (Benz and Schmidt, 1989; Lindenthal and Elsinghorst, 1999; Kjaergaard et al., 2000; Henderson and Owen, 1999); Hap, Hia and Hsf of

H. influenzae (St. Geme III, de la Morena and Falkow, 1994; St. Geme III, Cutter and Barenkamp, 1996; Barenkamp and St. Geme III, 1996; Yeo et al., 2004); BabA of *H. pylori* (Ilver et al., 1998); UspA2, UspA2h of *Moraxella catarhalis* (Aebi et al., 1998; Lafontaine et al., 2000) and rOmpA of *Rickettsia* spp. (Crocquet-Valdes, Weiss and Walker, 1994).

The AT secretion system is a modular structure consisting of three domains. These include a C-terminal transporter or β domain, an internal passenger domain and an N-terminal signal sequence. The β-domain ends up being inserted as an oligomer in the OM while the passenger domain is the protein moiety eventually presented on and anchored to the cell surface (Henderson, Navarro-Garcia and Nataro, 1998; Veiga, de Lorenzo and Fernandez, 2003; Desvaux, Parham and Henderson, 2004). The AT secretion system tolerate a wide range of protein modules that become displayed with the same structure, which favours the emergence of novel adhesins with new specificities. Veiga, de Lorenzo and Fernandez (2003) have demonstrated this property by creating hybrid fusion proteins containing the β-AT domain of an AT protein of *Neisseria gonorrhoeae* and the partner leucine zippers of eukaryotic transcription factors Fos and Jun. When the hybrid proteins were expressed in *E. coli*, the cells acquired novel adherence traits resulting in the self-association and clumping of planktonic bacteria in liquid media, or in formation of stable consortia between cells of strains expressing the dimerisation domains.

Another type of adherence is bacterial attachment to a surface and each other to form a biofilm. In a biofilm the adherence is mediated by an extracellular polysaccharide slime that acts as a kind of non-specific (although the signal to produce the biofilm may be specific) glue to bind the bacteria to each other and to a surface (Watnick and Kolter, 2000; Salyers and Whitt, 2002).

Many microbes can occupy a variety of habitats whereas others are confined to a specific microenvironment. The range of hosts, tissues or cell types colonised by bacteria is determined, in part, by adhesin recognition of and affinity for host receptors. For example, most *Bordetella* spp. can cause a similar disease in the upper respiratory tract of many mammals but their host specificities can differ considerably. *B. pertussis* is human specific while *B. bronchisepta* is responsible for infecting a wide variety of mammals and birds but only rarely causes disease in humans. Strains of *B. parapertussis* can be divided into two groups, one which is human specific, the other ovine specific (Cummings et al., 2004).

Host invasion

Subsequent to attachment, the bacterium may or may not invade the host, depending upon the pathogen. In any case, the host-associated pathogen must now repel the host defenses. Infection is the invasion of the host by micro-organisms, which then multiply in close association with the host's tissues. Mechanisms that enable a bacterium to invade eukaryotic cells make entry possible at mucosal surfaces. Whereas some invasive bacteria are obligate intracellular pathogens, most are facultative intracellular pathogens. In many cases, the exact bacterial surface factors that mediate invasion are not known, and multiple gene products are frequently involved. Pathogens may have mechanisms to disguise or switch antigens on their surface, thus confusing humoral and cellular immunity. Defensive mechanisms include the expression of proteins and enzymes to destroy phagocytes and weaken surrounding host tissues, making it easier to spread to new areas. Many pathogens have also developed resistance to common antibiotics, allowing them to continue infection even when the host is treated with antibiotics.

Entry into tissues may take several forms. Micro-organisms may pass directly through the epithelia, especially mucous membranes that consist of a single cell layer. However, in the case of skin, which is tough and multilayered, access is usually via trauma, insect bites or other damage to the surface.

Invasion through mucosal surfaces requires that the bacteria first cross the mucus layer coating the epithelium and then adhere to and infect the underlying target tissue. Many micro-organisms must first interact with specific receptors on the surface of the host cell to penetrate through mucosal epithelia. Mucosal and submucosal glands secrete a protective network of carbohydrate-rich glycoproteins called mucin. Aside from the lubricative value of mucin, the primary function is to trap bacteria and prevent them from gaining access to mucosal cells. Most bacteria have mucin-binding surface molecules and are removed with the mucus flow, some establish residence within the mucus layer or penetrate the mucus and adhere to epithelial cells (Salyers and Whitt, 2002). Bacteria which lack mucin-binding surface proteins or carbohydrates may have the ability to transit the mucin layer. Since mucin is an extremely viscous material that is relatively resistant to enzymatic digestion (de Repentigny et al., 2000; Moncada et al., 2000) bacteria that are able to move through viscous material or degrade mucin can overcome the first major barrier to mucosal invasion. In risk/safety evaluation, attention should be given, in general, to any changes in surface proteins or carbohydrate moieties involved in binding to mucin or with an ability to degrade mucin.

In most cases, once a micro-organism crosses an epithelial barrier, it is recognised by macrophages (mononuclear phagocytes and neutrophils) resident in tissues. Binding to specific cell-surface receptors triggers phagocytosis. When internalised bacteria become enclosed in a membrane vesicle or phagosome, it becomes acidified by the lysosomes. Fusion with lysosomes mediates an intracellular antimicrobial response to kill the bacteria. Most bacteria are destroyed by this process; however, there are various bacterial strategies for coping with phagolysosome formation and evading destruction. One strategy prevents phagosome-lysosome fusion and is used by *Mycobacterium*, *Legionella* and *Chlamydia* spp. Another strategy exemplified by *Actinobacillus* spp., *Listeria* spp., *Rickettsia* spp. and *Shigella* spp. involves disruption of the vesicle membrane and entry into the cytoplasm (Gouin et al., 1999). Bacterial survival and evasion of host response are covered in more detail in the section "Evasion of host immune response and multiplication in host".

Host invasion may be aided by the production of invasins which act against the host by breaking down primary or secondary defenses of the body. Part of the pathology of a bacterial infection may be the result of invasive activity. One of the best-studied invasins is produced by *Yersinia* spp. Isberg and Leong (1990) demonstrated that invasin tightly adheres to β_1 integrins (host cell adhesion receptors) to mediate bacterial uptake by "zippering" the host cell membrane around the bacterium as it enters. The ability of various bacteria to induce internalisation following contact with eukaryotic cells appears to play a crucial role in pathogenesis (Finlay and Cossart, 1997). This uptake is directed into host cells that are not naturally phagocytic, including epithelial and endothelial cells lining mucosal surfaces and blood vessels, and is manipulated by the invading bacteria.

The two main mechanisms of induced uptake are zipper and trigger. Bacteria utilising the zipper mechanism of entry express a surface protein which binds to host surface receptors involved in cell-matrix or cell-cell adherence. This directed contact between bacterial ligands and cellular receptors proceeds sequentially, inducing host membrane extension and bacterial uptake through a "zippering" mechanism (Cossart and Sansonetti, 2004). Various pathogens such as *Helicobacter pylori* (Kwok et al., 2002), *Listeria monocytogenes* (Lecuit et al., 1997), *Neisseria* spp. (McCaw, Liao and Gray-Owen, 2004) and some streptococci (Dombek et al., 1999) use this type of mechanism. With the trigger mechanism of entry, bacteria bypass the first step of adhesion and interact directly with the cellular machinery. Effectors are injected through a type III secretory system and the bacterial signals sent to the host cell induce prominent membrane ruffling and cytoskeletal rearrangements resulting in macropinocytosis and almost passive entry of bacteria (Finlay and Cossart, 1997). This type of system is used by *Salmonella* spp. (Hayward et al., 2002) and *Shigella flexneri* (Van Der Goot et al., 2004). Generally, invasion into normally non-phagocytic cells establishes a protected cellular niche for bacterial replication, survival and persistence.

It must be stressed that a same single invasion strategy may not be shared by all members of a species. *Streptococcus pyogenes* strains have been shown to trigger different uptake events via distinct mechanisms. For instance, in *S. pyogenes* strain A40, the protein SfbI (Streptococcal fibronectin binding protein) has been shown to be the main factor for attachment and invasion and uptake is characterised by the lack of actin recruitment and the generation of large membrane invaginations (Molinari et al., 1997). Whereas in *S. pyogenes* strain A8, the SfbI gene is absent and uptake involves major rearrangements of cytoskeletal proteins leading to recruitment and fusion of microvilli and the generation of cellular leaflets (Molinari et al., 2000).

There is little distinction between the extracellular proteins which promote bacterial invasion and various extracellular protein toxins or exotoxins which damage the host. The action of an invasin is usually proximal to the site of bacterial growth and may not kill the cells, whereas exotoxins may act at sites distant to those of bacterial growth and are usually cytotoxic. In general, exotoxins are more targeted and result in greater pathology than invasins (Henderson, Poole and Wilson, 1996; Al-Shangiti et al., 2004). However, some exotoxins such as diphtheria toxin or anthrax toxin play a role in invasion while some invasins (e.g. staphylococcal leukocidin) have a relatively specific cytopathic effect. Table 1.2 lists some extracellular proteins which act as invasins. Host damage by exotoxins is more fully discussed in the section "Ability to damage or kill host".

Evasion of host immune response and multiplication in host

Microbial infections rarely cause disease without first multiplying within the host. Usually, multiplication is the main cause of disease associated with bacterial infection. Following entry into a host cell, most bacteria, including pathogens, are killed by macrophages and polymorphonuclear leukocytes. The incubation period reflects the time needed for the bacteria to overcome these early defenses and increase in number. The potential of a pathogen to cause a successful infection is reflected in the infective dose (ID). There can be wide variations in IDs, depending on the nature of the bacterial strain, the route of exposure (oral, inhalation, etc.), age (IDs would likely be lower for the very young and the very old) and the immune status of the host. Since the success of many pathogens relies on their ability to circumvent, resist or counteract host defense mechanisms, pathogens have developed numerous ways to avoid and manipulate host responses. This is reflected in the constant evolution of host defenses and bacterial pathogenic mechanisms.

Phagocytes are the first line of defense encountered by bacteria following tissue invasion. Phagocytosis has two main functions: 1) disposal of microbial pathogens; and 2) antigen processing and presentation for the induction of specific immune responses. Bacteria that readily attract phagocytes and are easily ingested and killed are generally unsuccessful pathogens. In contrast, most successful pathogens interfere to some extent with the activities of phagocytes or in some way avoid their attention. Bacterial pathogens have devised numerous and diverse strategies to avoid phagocytic engulfment and killing, with most strategies aimed at blocking one or more of the steps in phagocytosis, thereby halting the process. Other bacterial pathogens, exemplified by *Brucella* spp., *Mycobacterium* spp. and *Legionella* spp., survive and proliferate within "professional" phagocytes such as macrophages, neutrophils and dendritic cells. Survival inside of phagocytic cells, in either neutrophils or macrophages, protects the bacteria from antibodies, antibiotics, bacteriocides, etc. during the early stages of infection or until they develop a full complement of virulence factors.

Table 1.2. **Extracellular bacterial proteins that act as invasins**

Invasin	Bacteria	Action	Reference
C5a peptidase	Group A and B Streptococcus	Inactivates human C5a and promotes epithelial cells invasion leading to the dissemination of bacteria	Wexler, Chenoweth and Cleary (1985); Cheng et al. (2002)
Collagenase	*Clostridium* spp.	Dissolves collagen	Borriello (1998); Poilane et al. (1998)
Gingipain (cystein protease)	*Porphyromonas gingivalis*	Destruction of connective tissue, degradation of paxillin and focal adhesion kinase (FAK)	Nakagawa et al. (2006)
HAD superfamily member SerB653	*Porphyromonas gingivalis*	Secreted when in contact with gingival epithelial cells	Tribble et al. (2006)
Hyaluronidase (see also paragraph 'Spreading factor' under sub-section 'Ability to damage or kill host' below)	*Streptococcus* spp., *Staphylococcus* spp. and *Clostridium* spp.	Degrades hyaluronic acid of connective tissue	Paton et al. (1993); Borriello (1998); Hynes et al. (2000)
Hemolysins/cytolysins	*Edwardsiella tarda*, *Escherichia coli*, *Bordetella pertussis*, *Listeria monocytogenes*, *Streptococcus* spp., *Staphylococcus* spp. and *Clostridium* spp.	Destroy red blood cells and other cells by lysis	Paton et al. (1993); Strauss, Ghori and Falkow (1997); Bassinet et al. (2000); Cockeran, Anderson and Feldman (2002); Doran et al. (2002); Nizet (2002); Sierig et al. (2003)
Kinases (see also paragraph 'Kinases' under sub-section 'Ability to damage or kill host' below)	*Staphylococcus* spp. and *Streptococcus* spp.	Convert plasminogen to plasmin which digests fibrin	Ringdahl et al. (1998); Gladysheva et al. (2003)
Lecithinases	*Clostridium perfringens*, *Listeria monocytogenes*		Awad et al. (1995); Appelberg and Leal (2000)
Leukocidin	*Staphhylococcus aureus*	Disrupts neutrophil membranes and causes discharge of lysosymal granules	Rogolsky (1979); Dinges, Orwin and Schlievert (2000)
Phospholipases	*Clostridium perfringens*, *Neisseria gonorrhoeae*, *Shigella flexneri*, *Pseudomonas aeruginosa*	Hydrolytic enzymes involved in phospholipid cleavage	Vasil (1986); Awad et al. (1995); Meyer, Mintz and Fives-Taylor (1997); Guhathakurta et al. (1999); Edwards, Entz and Apicella (2003)
Sialidases/neuraminidases	*Vibrio cholerae*, *Shigella dysentariae*, *Streptococcus pneumonia*, *Trichomonas vaginalis*, *Bacteroides fragilis*, *Gardnerella vaginalis*, *Mycoplasma hominis*	Degradation of sialomucin on epithelial cell layer	Paton et al. (1993); Wiggins et al. (2001); Stewart-Tull, Lucas and Bleakley (2004)

Phagocytosis comprises several steps:

- Recognition and attachment of bacteria to professional (macrophages/neutrophils) or non-professional phagocytes (e.g. epithelial cells). The recognition is usually receptor-mediated (e.g. opsonisation – Fc receptors) but can be non-specific (bulk fluid pinocytosis).
- Endocytic entry of bacteria into the phagocytic cell with the generation of a phagocytic vacuole (endosome, phagosome).
- Generation of a phagolysosome via fusion of the phagosome with primary and secondary lysosomal granules.
- Degranulation and killing through the release of lysosomal or granular contents in direct apposition to the bacteria within the phagolysosome (maybe via oxygen-dependent and/or oxygen-independent mechanisms of killing).

The various strategies employed by bacteria to avoid destruction by phagocytes include: 1) adaptation to withstand the antimicrobial activity of the fused phagolysosome; 2) alteration of phagocytosis to target the bacterium to a novel phagosome; 3) escape from the phagosome into the cytosol by lysing the vacuolar membrane; 4) blocking lysosome/phagosome fusion or attenuating the acidification of phagolysosomes; 5) circumventing or resisting phagocytosis.

Adaptation to withstand the antimicrobial activity

With some intracellular bacteria, phagosome-lysosome fusion occurs, but the bacteria are resistant to inhibition and killing by the lysosomal constituents. Also, some extracellular pathogens can resist killing in phagocytes utilising similar resistance mechanisms. Resistance to phagocytic killing within the phagocytic vacuole is not completely understood, but it may be due to the surface components of the bacteria or due to extracellular substances produced which interfere with the mechanisms of phagocytic killing. *Brucella abortus* and *Staphylococcus aureus* are vigorous catalase and superoxide dismutase producers, which might neutralise the toxic oxygen radicals that are generated by the NADPH-oxidase and myeloperoxidase systems in phagocytes. *S. aureus* also produces cell-bound pigments (carotenoids) that "quench" singlet oxygen produced in the phagocytic vacuole. There are some micro-organsims, however, that are dependent upon phagosome-lysosome fusion for intracellular replication and persistence.

The pH that develops in the phagosome after engulfment induces bacterial gene products that are essential for their survival in macrophages. For instance, replication and synthesis of metabolic factors required for intracellular persistence of *Coxiella burnetti, Brucella suis* and *S. typhimurium* is induced by the acidic pH found within the phagolysosome (Hackstadt and Williams, 1981; Rathman, Sjaastad and Falkow, 1996; Porte, Liautard and Kohler, 1999; Ghigo et al., 2002).

Alteration of phagocytosis

Bacteria such as *Salmonella* spp. are able to induce phagocytosis in non-professional phagocytes. The Salmonella-containing vacuole (SCV), a unique cytoplasmic organelle formed following phagocytic induction, actually protects the bacterium; *Salmonella* spp. interfere with the ability of this phagosome to fully mature into a phagolysosome (Duclos and Desjardins, 2000).

Escape from the phagosome

Escape from the phagosome is a strategy employed by the Rickettsiae. *Rickettsia* spp. enter host cells in membrane-bound vacuoles (phagosomes) but are free in the cytoplasm a short time later, perhaps in as little as 30 seconds. A bacterial enzyme, phospholipase A, may be responsible for dissolution of the phagosome membrane. *Listeria monocytogenes* rely on several molecules for early lysis of the phagosome to ensure their release into the cytoplasm. These include listeriolysin O (LLO), a cholesterol-dependent cytolysin and two forms of phospholipase C. The low optimal pH activity of LLO allows the bacterium to escape from the phagosome into the host cytosol without damaging the plasma membrane of the infected cell.

Glomski et al. (2002) demonstrated that a single amino acid change from leucine 461 to threonine profoundly increased the hemolytic activity of LLO at a neutral pH and promoted premature permeabilisation of the infected cells. This discovery demonstrates how minor changes in proteins can be used by bacterial pathogens to establish and maintain the integrity of their specific niches or be exploited by researchers working with bacteria to produce a protein with novel properties. Once in the cytoplasm, *Listeria* spp. induce their own movement through a process of host cell actin polymerisation and formation of microfilaments within a comet-like tail. *Shigella* spp. also lyse the phagosomal vacuole and induce cytoskeletal actin polymerisation for the purpose of intracellular movement and cell-cell spread.

Blocking fusion or attenuating acidification

Some bacteria survive inside of phagosomes by blocking the fusion of phagocytic lysosomes (granules) with the phagosome thus preventing the discharge of lysosomal contents into the phagosome environment. This strategy is employed by *Salmonella* spp., *M. tuberculosis*, *Legionella* spp. and the chlamydiae. With *Legionella* spp., it is known that a single gene is responsible for the inhibition of phagolysosomal fusion. Attenuating the acidification of phagolysosomes is observed with *Rhodococcus* spp. Toyooka, Takai and Kirikae (2005) demonstrated that phagolysosomes did not acidify when they contained virulent *R. equi* organisms. Their research indicated that *R. equi* in phagolysosomes produced substance(s) to suppress acidification. Results by Tsukano et al. (1999) indicated that inhibition of phagosomal acidification by *Y. pseudotuberculosis* was due to attenuation of vacuolar-ATPase activity.

Phagocytic circumvention

Bacteria may avoid phagocytosis by simply penetrating areas inaccessible to phagocytes such as the lumens of glands and the urinary bladder and surface tissues such as the skin.

Other strategies for phagocyte evasion include suppression of the inflammatory response and inhibition of phagocyte chemotaxis. For example, pneumolysin (streptolysin) toxin produced by *Streptococcus pneumoniae* (Paton and Ferrante, 1983; Ernst, 2000) and components of *Mycobacterium* spp. inhibit polymorphonuclear leukocyte (PMN) migration. Also, studies involving pathogen-induced PMN alterations have suggested that *Anaplasma phagocytophilum* delays PMN apoptosis and lessens proinflammatory cytokine release (Yoshiie et al., 2000; Klein et al., 2000). Bacteria using host cell mimicry for phagocytic evasion cover their surface with a component which is recognised as "self" by the host phagocytes and immune system. This effectively hides the antigenic surface of the bacterial cell. Phagocytes are unable to recognise bacteria upon contact and thus opsonisation by antibodies to enhance phagocytosis is minimised. For example, *Staphylococcus aureus* produces cell-bound coagulase which clots fibrin on the bacterial surface, *Treponema pallidum* binds fibronectin to its surface, while Group A streptococci synthesise a capsule composed of hyaluronic acid which forms the ground substance of host connective tissue.

Resistance to phagocytic ingestion is usually due to a component of the bacterial cell surface (cell wall or fimbriae or a capsule). Examples of antiphagocytic substances on the bacterial surface include: Polysaccharide capsules (*S. pneumoniae*, *Haemophilus influenzae*, *Treponema pallidum* and *Klebsiella pneumoniae*); M protein and fimbriae of Group A streptococci; polysaccharide produced as biofilm by *Pseudomonas aeruginosa*; O polysaccharide associated with lipopolysaccharide (LPS) of *E. coli*; K or Vi antigens (acidic polysaccharides) of *E. coli* and *Salmonella typhi*, respectively; cell-bound or soluble Protein A produced by *Staphylococcus aureus* which attaches to the Fc region of IgG and blocks the cell-binding domain of the antibody.

Whereas phagocytic resistance and intracellular proliferation is accomplished via surface components, such as bacterial capsules and LPS, which effectively shield the bacteria, resistance to many bactericidal components of host tissues is usually a function of some structural property. For example, the poly-D-glutamate capsule of *Bacillus anthracis* protects the organisms against the action of cationic proteins (defensins) or by conventional proteases in sera or in phagocytes (Fouet and Mesnage, 2002). Similarly, the OM of gram-negative bacteria serves as a permeability barrier that is not easily traversed by hydrophobic compounds harmful to the bacteria, for example bile salts of the gastrointestinal tract. Intact LPS of gram-negative pathogens may protect the cells from complement-mediated lysis or the action of lysozyme. The OM and capsular components of gram-negative bacteria (e.g. *Salmonella* spp., *Yersinia* spp., *Brucella* spp., *E. coli*) can protect the peptidoglycan layer from the lytic activity of lysozyme (Hughey and Johnson, 1987; Martinez de Tejada et al., 1995). Mycobacteria (including *M. tuberculosis*) have waxy, hydrophobic cell wall and capsule components (mycolic acids), which are not easily attacked by lysosomal enzymes (Gao et al., 2003).

Other factors that enhance intracellular survival include bacterial enzymes which neutralise oxygen radicals and secreted proteolytic enzymes which degrade host lysosomal proteins. Another strategy in defense against phagocytosis is direct attack by the bacterium upon professional phagocytes. Most of these are extracellular enzymes or toxins that kill phagocytes either prior to or after ingestion and are discussed in the section "Ability to damage or kill host".

Multiplication in host

Multiplication in the host also requires that the micro-organism obtains the necessary nutrients and factors needed for growth and replication. Iron is an essential nutrient that is usually limited within eukaryotic hosts. Many pathogenic bacteria have developed regulated networks of genes important for iron uptake and storage. Also, available iron concentration may trigger the regulation of virulence gene expression (Merrell et al., 2003). *Salmonella* spp. and *E. coli* produce siderophores (extracellular iron-binding compounds) which extract Fe^{3+} from lactoferrin (or transferrin) and supply iron to bacterial cells for growth.

Successful intracellular lifestyle is conditional on the ability of the bacteria to obtain essential nutrients from the hostile phagosomal environment. For example, the virulence of both *M. tuberculosis* and *Salmonella enterica* (Hingley-Wilson, Sambandamurthy and

Jacobs Jr., 2003) is dependent upon their ability to acquire magnesium while inhabiting the phagosome.

Ability to damage or kill host

To counter infection the human host relies, initially, on the innate immune system. Prior to mounting an immune response, however, the host must first detect the pathogen. The innate immune system uses sets of recognition molecules, called pattern recognition receptors.

The toll-like receptors (TLRs) are one of the most important pattern recognition receptor families (Armant and Fenton, 2002). Pattern recognition receptors bind conserved molecular structures, unique to micro-organisms, termed pathogen-associated molecular patterns. Pathogen-associated molecular patterns such as peptidoglycan, teichoic acids, LPS, mycolic acid and mannose, bind to pattern recognition receptors on a variety of defense cells of the body causing them to synthesise and secrete a variety of cytokines. These cytokines can, in turn promote innate immune defenses such as opsonisation, activation of proinflammatory signaling cascades, phagocytosis, activation of the complement and coagulation cascades, and apoptosis (Wilson et al., 2002).

The host immune response plays a critical role in determining disease manifestations of chronic infections. Inadequate immune response may fail to control infection, although in other cases the specific immune response may be the cause of tissue damage and disease. Not infrequently, host defense mechanisms go overboard and it is this overaggressive immune response which contributes to the tissue damage observed with some infections.

A number of bacterial proteins that act as immune modulators are presented in Table 1.3. This chapter, however, focuses on specific bacterial factors directly responsible for tissue damage or host death.

Bacteria produce a large number of cell-associated or secreted proteins which play a role in colonisation, infection and subsequent tissue damage. The great majority of bacterial virulence factors are secreted products that augment the survival of the bacteria and/or damage the host (Jett, Huycke and Gilmore, 1994; Fournier and Philpott, 2005; Kuehn and Kesty, 2005). The following is a summary of activities of many bacterial proteins that contribute to host invasion, tissue damage or death.

Collagenase

Collagenase, produced by *Clostridium histolyticum* and *Clostridium perfringens* (Legat, Griesbacher and Lembeck, 1994; Rood, 1998), breaks down collagen, the single most abundant protein in mammals. Collagenases are thought to play a major role in the pathology of gas gangrene caused by clostridia because they can destroy the connective tissue barriers.

Spreading factor

Hyaluronidase, or more descriptively "spreading factor", affects the physical properties of tissue matrices and intercellular spaces. Hyaluronidase, an enzyme produced by streptococci, staphylococci and clostridia (Bergan, 1984; Li et al., 2000), is also a component of venom from snakes, spiders, jellyfish, etc. (Girish et al., 2004; Kuhn-Nentwig, Schaller and Nentwig, 2004). The enzyme attacks the ground substance of connective tissue by depolymerising hyaluronic acid thereby promoting the spread of the bacteria. Its activity also causes invasion, hence hyaluronidase is also seen as an invasin (Table 1.2).

Table 1.3. **Bacterial proteins that act as immunomodulator**

Bacteria/disease	Immunomodulator	Action	Reference
Borrelia burgdorferi/Lyme disease	OspE	Binds factor H	McDowell et al. (2004)
Enterococcus faecalis	Capsular polysaccharide	Resistance to opsonophagocytic killing	Hancock and Gilmore (2002)
Francisella tularensis/Tularemia	?	Survive and multiply inside macrophages	Maier et al. (2007)
Streptococcus pneumoniae	PspA	Inhibitor of factor B mediating complement activation and opsonisation	Tu et al. (1999)
Group A Streptococcus (*S. pyogenes*)	Fba	Binds factor H and fH-like protein, contribute to phagocytosis resistance	Wei et al. (2005)
Streptococci	Protein M	Alteration of opsonophagocytosis by recruitment of factor H	Jarva et al. (2003)
Group B Streptococcus (GBS)	Capsule	Protects from opsonisation by C3. ß-protein binds factorH	Rubens et al. (1987); Jarva et al. (2003)
Group B Streptococcus (GBS)	CspA (serine protease-like)	Evasion of opsonophagocytosis	Harris et al. (2003)
Lysteria monocytogenes	Lysterio-lysin O	Evasion of phagosome	Schnupf and Portnoy (2007)
Neisseria gonorrhoea	Por 1A	Binds factor H, C4 bp, mediates serum resistance	Ram et al. (1998; 2001)
Neisseria meningitides	Bind factor H	Avoids lysis by complement system	Schneider et al. (2006)
Staphylococcus aureus	Secretes extracellular adherence protein (EAP)	Binds to Inter-Cellular Adhesion Molecule (ICAM)-1, fibrinogen, vitronectin resulting in the disruption of the leukocyte recruitment	Athanasopoulos et al. (2006)
Yersinia spp.	Yop E, T A, H and Yop J	Block phagocytosis and suppress inflammatory mediators	Fällman and Gustavsson (2005)
Yersinia enterocolica	YadA	Alteration of opsonophagocytosis by recruitement of factor H	China et al. (1993)

Kinases

Streptokinase and staphylokinase are produced by streptococci and staphylococci, respectively. These enzymes convert inactive plasminogen to plasmin which digests fibrin and prevents clotting of the blood. The relative absence of fibrin in bacterial lesions allows more rapid diffusion of the bacteria (Gladysheva et al., 2003). Like hyaluronidase, kinases also cause invasion, and are seen as invasins (Table 1.2).

Sialidases/neuraminidase

Extracellular sialidases or neuraminidases, produced by various pathogens, have the ability to hydrolyse the sialic acid residues located on many mammalian cell membranes (Rood, 1998). The neuraminidase produced by *Mannheimia haemolytica* decreases the viscosity of respiratory mucus, thus providing the bacteria with greater access to the cell surface (Zecchinon, Fett and Desmecht, 2005).

Toxins

An overview of bacterial protein toxins may be found in Alouf (2000). Many toxins act on the animal cell membrane by insertion into the membrane (forming a pore that results in cell lysis), or by enzymatic attack on phospholipids, which destabilises the membrane. They may be referred to as lecithinases or phospholipases, and if they lyse red blood cells they may be called hemolysins. Hemolysins, notably produced by staphylococci (i.e. alpha toxin), streptococci (i.e. streptolysin/pneumolysin) and various clostridia, may be channel-forming membrane toxins capable of damaging a broad range of eukaryotic cell types (Awad et al., 1995; Menzies and Kourteva, 2000; Doran et al., 2002). Lecithinases destroy lecithin (phosphatidylcholine) in cell membranes (Awad et al., 1995; Appelberg and Leal, 2000). Phospholipases, for example alpha toxin produced by *Clostridium perfringens*, hydrolyse phospholipids in cell membranes by removal of polar head groups. Leukocidin, a bacterial exotoxin similar to streptolysin, is produced by staphylococci and specifically lyses phagocytes and their granules. Although leukocidin may be referred to as a bi-component leukotoxin (Morinaga, Kaihou and Noda, 2003; Futagawa-Saito et al., 2004), it should not be confused with the leukotoxins of the RTX family described below.

Exotoxins have sometimes been categorised according to the cells primarily affected by the toxin. For example, leukotoxins are a group of exotoxins that produce their primary effect on leukocytes, especially polymorphonuclear cells. *Mannheimia* (*Pasteurella*) *haemolytica*, one of the key pathogens associated with bovine respiratory disease complex produces a leukotoxin (LKT) that both activates and kills bovine leukocytes. Atapattu and Czuprynski (2005) have shown that LKT produced by *Mannheimia haemolytica* induces apoptosis of bovine lymphoblastoid cells (BL-3) via a caspase-9-dependent mitochondrial pathway. While LKT is able to bind leukocytes from various animal species, it is only cytotoxic for ruminant leukoctyes. This virulence factor is a member of the RTX (repeats in toxins) family of multidomain gram-negative bacterial toxins. RTX toxins fall into two categories: hemolysins which attack different cell types from a variety of species and leukotoxins which show a marked specificity for both cell type and host species (Lally et al., 1999). Other bacteria that produce RTX toxins include: *E. coli* (hemolysins), *Bordetella pertussis*, *Actinobacillus* spp. and *Actinobacillus actinomycetemcomitans* and various *Pasturella* spp. (leukotoxins), (Narayanan et al., 2002; Davies, Campbell and Whittam, 2002; Ward et al., 2002).

Toxins with short-range effects related to invasion

Bacterial protein toxins which have adenylate cyclase activity are thought to have immediate effects on host cells that promote bacterial invasion. One component of the anthrax toxin (EF or Edema Factor) is an adenylate cyclase that acts on nearby cells to cause increased levels of cyclic AMP and disruption of cell permeability (Leppla, 1982). One of the toxin components of *Bordetella pertussis*, pertussis adenylate cyclase, has a similar effect. These toxins may contribute to invasion through their effects on macrophages or lymphocytes in the vicinity which are playing an essential role to contain the infection.

Co-ordination of expression of virulence factors: Quorum sensing (QS)

To establish an infection, bacteria carefully orchestrate a number of bacterial factors and determinants which have a role in determining pathogenicity. Proficient co-ordination of these factors is required for bacterial survival and successful colonisation. Thus, bacteria have developed sophisticated regulatory systems to adapt gene expression to changing environmental conditions. The notion that bacteria can signal each other and co-ordinate their assault patterns against susceptible hosts is now well established (Miller and Bassler, 2001). When invading their host, bacteria do not operate in isolation. Pathogens employ a series of chemical signals and sensing systems that jointly engage bacterial communities to genetically respond in concert to specific conditions in the host and promote an advantageous lifestyle within a given environmental niche. A central component in this process is a sophisticated communication system known as quorum sensing (QS) (Ng and Bassler, 2009). QS systems regulate microbial pathogenesis through the following points: 1) helping pathogens' invasion and colonisation; 2) regulating production of the virulent factor; 3) giving pathogens the ability of immunity or drug resistance (Wu and Xie, 2009).

QS was first observed in the marine halophilic bioluminescent bacterium *Vibrio fischeri* (Nealson et al., 1970), in which the bacterial light-emitting luciferase operon is activated when the population reaches a threshold concentration. It was later realised that QS is achieved through the production, release, and subsequent detection of and response to threshold concentrations of signal molecules called autoinducers, which are synthesised throughout the growth of the bacterium. When a threshold concentration is reached, these signals interact with a transcriptional regulator, allowing the expression of specific genes (Bassler, 2002).

QS systems were shown to regulate a multitude of transcriptional programmes in bacteria *in vitro* and probably *in vivo*, which are relevant for the pathogenic phenotype. These include biofilm formation, growth potential, antibiotic resistance expression and genetic determinants of virulence (Kendall and Sperandio, 2007; Yarwood and Schlievert, 2003; Mack et al., 2007; Kong, Vuong and Otto, 2006; Costerton et al., 2003; Bjarnsholt et al., 2010). That QS has a fundamental role in bacterial pathogenesis was confirmed as researchers began to find that many clinically relevant microbial pathogens displayed auto-inducer systems homologous to the one discovered in *V. fischeri*. Many common bacterial pathogens, including *Escherichia coli*, *Pseudomonas aeruginosa*, *Bacteroides*, *Yersinia*, *Burkholderia* and *Enterococcus* spp., and many clinically important staphylococcal and streptococcal pathogens were shown to contain QS genes, which participate in the regulation of multiple bacterial genes, including virulence genes (Miller and Bassler, 2001; Greenberg, 2003; Cámara, Williams and Hardman, 2002; Shiner, Rumbaugh and Williams, 2005; Qazi et al., 2006; Parsek and Greenberg, 2000; Brady et al., 2008; Williams, 2007).

QS circuits can also regulate human transcriptional programmes to the advantage of the pathogen. Human stress hormones and cytokines can be detected by bacterial quorum sensing systems. By this mechanism, the pathogen can detect the physiologically stressed host, providing an opportunity to invade when the patient is most vulnerable. (Li et al., 2009).

QS systems are broadly grouped into three categories. The quorum sensing systems identified in many gram-negative bacteria mostly resemble the typical quorum sensing circuit of the bioluminescent bacterium *V. fischeri* (Miller and Bassler, 2001; Smith et al., 2006) in which they consist, at a minimum, of homologues of the two *V. fischeri* regulatory proteins called LuxI and LuxR. The LuxI-like proteins (the auto-inducer synthases) are responsible for the biosynthesis of a specific acylated homoserine lactone signaling molecule, termed type 1 autoinducers (AI-1). The autoinducer concentration increases with increasing cell-population density. The LuxR-like proteins (the transcription factors) bind cognate AI-1 autoinducers that have achieved a critical threshold concentration and the LuxR-autoinducer complexes activate target gene transcription, including virulence genes (Wagner et al., 2007). Over 50 species of gram-negative bacteria produce acylated homoserine lactones that differ only in the acyl side chain moiety, and each LuxR-type protein is highly selective for its cognate autoinducer signal molecule (Bassler, 2002).

The autoinducers in the QS system of a gram-positive bacterium are short, usually modified peptides processed from precursors. In contrast with the diffusible behaviour of AI-1 autoinducers, these signals are actively exported out of the cell (through an ATP-binding cassette transporter, ABC-transporter), and interact with the external domains of membrane bound sensor proteins. Signal transduction triggers a phosphorylation cascade that culminates in the activation of a DNA binding protein that controls transcription of target genes. Similar to gram-negative bacteria, gram-positive bacteria can use multiple autoinducers and sensors (Bassler, 2002).

Finally, a third QS pathway, initially discovered in the *V. harveyi* bioluminescence system, is mediated by the *luxS* gene locus (the autoinducer synthase gene) and related homologues. Signaling elements in this system, termed type 2 autoinducers (AI-2), are composed of rather complex, unusual, multiple-ringed, cyclical furanosyl-borate diester molecules. The AI-2 pathway uses a more complex, two-component membrane receptor, LuxPQ, comprised of periplasmic binding protein (LuxP) and histidine sensor kinase (LuxQ) subunits (Neiditch et al., 2006). Components of this system are detectable in almost one-half of all sequenced bacterial genomes, so this system is now recognised as the most ubiquitous signaling system employed by both gram-negative and gram-positive bacteria. It has been proposed that the AI-2 pathway is a more universal, interspecies chemical language (Bassler, 2002).

Quorum sensing molecules and systems show a remarkable array of very complex properties. These systems are also capable of influencing environmental processes. Geochemical and biological modifications of signals probably occur in extracellular environments, and these could disrupt or interfere with intended communication signals. It has been postulated that quorum sensing occurs within cell clusters, where signal dispersion might be significantly influenced by existing extracellular polymers (Decho, Norman and Visscher, 2010).

Molecular aspects of pathogenicity

Molecular genetic definition of bacterial virulence

The application of molecular biology to microbial pathogenesis was described by Falkow (1988) in a molecular form of Koch's postulates: 1) the phenotype or property under investigation should be associated with pathogenic members of a genus or pathogenic strains of a species; 2) specific inactivation of the gene(s) associated with the suspected virulence trait should lead to a measurable loss in pathogenicity or virulence; and 3) reversion or allelic replacement of the mutated gene should lead to restoration of pathogenicity. Meeting these postulates requires the technical possibility to directly affect the genes in question, and, even more important, the availability of models to measure virulence. As this is not always feasible, an alternative approach was added: 4) the induction of specific antibodies to a defined gene product should neutralise pathogenicity. This fourth postulate is sometimes taken alone: when antibodies against a certain factor protect an animal from disease, this is sufficient to call this factor a virulence factor.

Like Koch's postulates, the "molecular Koch's postulates" cannot always be applied rigidly. If the virulence phenotype is multifactorial, as will usually be the case, the gene products identified as virulence factors may either be a "classical" virulence factor or an accessory factor that is essential for expression of the phenotype, but not directly involved in it. As an example: the fimbriae, hairlike surface structures, that are virulence factors of uropathogenic *Escherichia coli* strains carry an adhesin molecule at their tip that performs the directly virulence related task of adherence to epithelial cells of the host. They can, however, only efficiently perform this task when carried at the tip of the fimbriae that are composed of other protein molecules that lack the adhesive property. The gene identified as a virulence factor may not even be a structural gene, coding for a gene product, but may have a regulatory function in the expression of the structural gene. In the literature there is a tendency to describe all genes that pass the tests described in the molecular Koch's postulates as virulence genes. This approach has resulted in the identification as "virulence genes" genes that are not directly involved in virulence as such, but are indispensable for the expression of the virulent phenotype because they are required in some way for correct expression of virulence genes. In fact, the molecular approach may detect a whole spectrum of "virulence genes" ranging from "true" virulence genes to genes encoding "house-keeping enzymes" that through some remote mechanism influence the virulence phenotype. This may indicate a need for a more restrictive definition of virulence genes than simply genes that are detected in virulence assays.

A definition of bacterial virulence should enable the discrimination between "true" virulence genes that are directly associated with the virulent phenotype, and accessory genes, that will also be identified as virulence genes by screening methodologies that rely on gene inactivation resulting in attenuation of virulence. A well-known example of a housekeeping gene identified as a virulence factor is the gene *aroA* (as well as other "housekeeping" genes; see Uzzau et al., 2005), inactivation of which results in attenuation of pathogenicity. The *aroA* gene, however, is involved in aromatic amino acid biosynthesis, and as such is present in both pathogens and non-pathogens and is not considered a virulence gene. This is easily understood in the case of *aroA*, but when the gene product has no known housekeeping function, such genes would be described in the literature as virulence genes. The problem is where to draw the line in the continuum between "virulence-associated genes" and "housekeeping genes". In order to exclude housekeeping genes from the set of "virulence" genes, the requisite is often added to Falkow's molecular postulates that virulence genes should not be expressed outside the host. However, this would exclude certain well-characterised virulence genes, for instance lipopolysaccharide (LPS)-producing enzymes are expressed under all circumstances, and yet LPS is a generally accepted virulence factor. Moreover, lack of expression outside the host may be a reflection of the applied culture conditions. In conclusion, the border between virulence-associated genes and housekeeping genes cannot be sharply defined.

Molecular approaches to identify virulence genes

Three basic approaches are used to identify virulence genes: genetic methods to obtain phenotypic evidence for virulence, methods based on the proposed immunogenicity of virulence factors for immunological evidence and comparative genetic methods that yield additional indirect evidence.

Phenotypic evidence: Within the genetic methods, two approaches are used: 1) inactivation of a virulence gene must result in loss of virulence; or 2) introduction of a virulence gene into a non-virulent strain must add virulent properties. It should be noted that both principles are heavily dependent on models to determine the virulent phenotype. Models to determine virulence are ideally animal models that minutely mimic natural disease, but these are not always available. More often, animal models have to be used that display only some of the naturally occurring characteristics, or *in vitro* models that only poorly resemble disease characteristics. Most processes leading to virulence are multi-factorial. The complicated interaction of host and bacteria is often ignored when *in vitro* models are applied. Even under simplified conditions of *in vitro* models, a presumably straightforward process such as bacterial invasion can be driven and regulated by multiple genes and gene loci, which work in concert or complementarity. Inactivation of one link of the chain may eliminate invasiveness, but complementation in a different setting may require several genetic loci. Alternatively, inactivation of a factor may be overcome by alternative factors so that loss of virulence is not observed, but complementation in a different environment may have strong phenotypic effects. The relevance of the applied models to extrapolate their outcome as phenotypic evidence of virulence is a point of debate, which is pragmatically ignored by lack of alternatives.

Immunological evidence: A second approach for identifying virulence genes is based on the proposed immunogenicity of virulence factors. Knowing that acquired immunity can protect against disease, it is assumed that protective antibodies are directed against virulence-associated genes. When an infection results in an antigenic response directed against one or more specific antigens, this is taken as a strong indication that these antigens represent virulence-associated factors.

Comparative genetical evidence: Examples of the genetic approach to identification of potential virulence-associated genes are the identification of: 1) genes with a degree of homology to known virulence-associated genes that is considered significant in bioinformatics surveys; 2) related genes that show variation that can be interpreted as antigenic variation; 3) genes that are shown to be present in (more) virulent strains, while absent in avirulent strains. Comparative genetic approaches are further discussed in the section on trends in virulence gene identification.

In addition to these approaches, several techniques have been developed to identify and characterise bacterial genes that are induced during the intracellular infection and therefore, potentially, may play a role in pathogenesis. Examples are the search for genes that are specifically induced in the host, and "signature-tagged mutagenesis" (STM), involving comparative hybridisation to isolate mutants unable to survive the environmental conditions in the host (Mahan, Slauch and Mekalanos, 1993; Chiang, Mekalanos and Holden, 1999; Harb and Abu Kwaik, 1999). A very powerful approach to isolate mutants that may be affected in a virulence gene is STM as discussed by Autret and Charbit (2005). The general technique of STM can be applied to find specific genes involved in survival persistence of a bacterium in a host; virulence genes would fall into this class of genes (Wassenaar and Gaastra, 2001). The only prerequisite for a gene to be found by STM is that its loss of function should not result in a lethal phenotype under conditions of growth *in vitro*, in broth. This is probably not an impediment for most virulence genes to be identified by this technique. The STM approach involves transposon (usually) mutagenesis of a bacterial strain, followed by pooling of a number of mutants that can be individually recognised by a polymerase chain reaction amplifiable tag. The pooled mutants are inoculated in an animal model, and bacteria retrieved from the animal are analysed for mutants that are present, as shown by the presence of their tag. Mutants that are lost have been mutated in genes that have a function in the pathogenic process, or which at least have a function that is needed to survive and be retrieved in the experiment.

Ideally, for the identification of virulence, several approaches should lead to the same gene or set of genes, and the characterisation of a gene as virulence-associated should be based on evidence from more than one approach. Even then, the controversy between housekeeping genes and virulence genes is not always solved. For example, the housekeeping magnesium transport system of *Salmonella typhimurium, mgtA/B*, is under *PhoP/PhoQ* regulation, and is activated during invasion *in vitro* (Smith and Maguire, 1998). One example is glutamine synthetase of *Salmonella typhimurium*, which is under the regulation of *ntrC* (an alternative sigma factor that can be indicative for *in vivo* regulation of expression) and which was identified as a virulence gene based on phenotypic evidence, since inactivation resulted in attenuation (Klose and Mekalanos, 1997). The enzyme presumably provides glutamine to the organism while surviving in the host, and could for that reason be considered a virulence-associated gene that enables colonisation. Since glutamine synthetase is also present in non-pathogenic bacteria, it is not considered a virulence gene in the comparative genetic approach. As the absence of virulence genes in non-pathogenic bacteria receives a lot of weight in this approach, two points need to be considered: 1) the outcome of such comparative genetics is heavily dependent on the content of the databases used; and 2) gene function is not always correctly predicted by comparative genetics alone. Putative virulence gene candidates identified in this way should therefore at least be confirmed by phenotypic evidence, despite the mentioned shortcomings of such evidence.

Trends in virulence gene identification

Due to explosive developments in genomics it is now feasible to analyse the complete genome of bacterial pathogens by *in silicio* subtractive hybridisation. With the expanding annotation of genes from genome sequences, this can lead to the identification of new virulence genes (Field, Hood and Moxon, 1999; Frosch, Reidl and Vogel, 1998). The annotation of these newly sequenced genes is based on sequence identity. This identification of virulence genes by comparative genomics, based on genetic similarity is, however, risky for several reasons.

An acceptable level of sequence conservation is seen as (indirect) evidence of conserved function, so that the gene function of a newly sequenced gene is extrapolated from a well-characterised analogue in another species. However, genes may have a niche-adapted function in a particular organism, and this may have its effect on the role of the gene product in virulence. Functional domains may not be conserved (Lee and Klevit, 2000). Therefore, sequence homology does not always predict function, and even when there is a high degree of genetic conservation between a non-characterised gene and a known virulence gene, the function of the gene product of the non-characterised gene as a virulence factor should first be experimentally tested before functional homology is assigned. Until then, the newly identified virulence gene should be annotated as "putative". Misannotation based on "putativism" is quite common, since it is now easier to generate sequencing data than to experimentally prove a function of the given gene product.

Another, diametrically opposed, pitfall of comparative genetics is that genes that share no sequence homology can have identical functions, as is demonstrated for *actA* of *Listeria monocytogenes* and *IcsA* in *Shigella flexneri*, whose gene products recruit host cell actin (Strauss and Falkow, 1997). This type of functional similarity will go unnoticed by genome comparison.

Many virulence genes display antigenic polymorphisms, presumably to evade the selection pressure of the host immune system (Deitsch, Moxon and Wellems, 1997). The correlation between polymorphism and virulence is so strong that polymorphisms observed *in silicio* are taken as indirect evidence for a role in virulence. It should be noted that the term polymorphism is used for different phenomena. The term is used when one isolate of a bacterial species can produce antigenic variants of a gene product by means of gene multiplication, alternative expression or post-translational modification. "Polymorphism" is also used for antigenic or genetic differences observed between isolates of the same species, for which the term "allelic polymorphism" is more exact. In addition, slippage during replication or translation can cause variation in the number of DNA repeats (with units of one to seven nucleotides) present within a gene, leading to polymorphic offspring (either represented in DNA or in protein) of a given cell (Van Belkum et al., 1998). All of these polymorphic mechanisms serve the general goal of adaptation to varying conditions. In the case of pathogens this is often, though not exclusively, a mechanism to avoid host defense responses. With the high throughput of sequencing data, it becomes possible to identify putative virulence properties for genes based on the polymorphic nature of their predicted translation products.

In conclusion, different paths lead to the identification of virulence genes. A "top-down" approach, starting from a single pathogen, will start by defining the pathogenic phenotype of the organism ("adhering", "invasive", "toxin producing", "phagocytic survival"), and subsequently zoom in on the virulence genes responsible for this phenotype. A "bottom-up" approach will start from an annotated genome sequence, from which putative virulence genes can be identified by comparative genetics. The relevance of such identified putative virulence genes for the pathogenic phenotype then remains to be proven. For this, a "lateral" approach can be useful, as pathogens often employ similar pathogenic mechanisms, and analogies between virulence factors can be used for identification strategies. In parallel, genetically related organisms that have a different pathogenic repertoire can be compared to identify the genes responsible for the differences in virulence. The second section of this chapter presented an overview of genes that are involved in different stages of pathogenicity: host recognition and adherence, host invasion, multiplication in the host, the ability to overcome the host immune response and host defense systems, and the ability to damage or kill the host.

The perspective of virulence genes

Understanding of bacterial virulence factors can be biased because of the experimental setup applied to identify or study the factor (Quinn, Newman and King, 1997). For instance, many bacterial toxins are described as "haemolysin", because they have been originally recognised as cytolytic to erythrocytes. However, in real life these toxins may not be targeted at erythrocytes, but at leukocytes or other host cells instead. This is just one example of how the perception of bacterial virulence factors is influenced by experimental design.

Pathogenicity and virulence are often addressed in an anthropomorphic manner, i.e. the incorrect concept that it is the "aim" of pathogenic bacteria to cause disease in their host. Like every organism, pathogens have adapted to occupy their ecological niche. Their close association with a host causes damage to their host. Often this damage is "coincidental", but it may even be beneficial to the survival or spreading of the pathogen.

Examples are the release of nutrients by cell damage, or enabling contagion of the next host by inducing coughing or diarrhea. The degree of damage is dependent on the equilibrium that results from the interplay of pathogen and host, and may, for instance, be dependent on the immune response of the individual (Casadevall and Pirofski, 1999). Conditions that result in disease can vary among host individuals, and from host species to host species. This adds to the difficulties to identify the bacterial genes that are directly responsible for the disease. Ideally, experimental shortcomings, subjective observations and the anthropomorphic view on pathogenicity should all be considered when establishing the relevance of a certain virulence gene to the pathogenicity of a micro-organism.

Classification of virulence genes

From the previous sections it is clear that there are many ways of defining, identifying and testing virulence genes. But, since each pathogen has evolved to fit its own niche, different pathogens do not necessarily share common pathogenic characteristics. Despite the recognition of common themes in bacterial virulence (Finlay and Falkow, 1989; 1997), a larger part of all virulence genes described in the literature that resulted from over 30 years of research have little in common, other than having some function in virulence. In order to interpret the vast amount of data on this subject these genes need to be classified.

As already stated in the introduction to this chapter, regulators dealing with risk/safety assessment of genetically engineered bacteria need a good understanding of the significance of a given virulence gene in its physiological background; only if the newly acquired gene can have a role in the pathogenic lifestyle of the recipient micro-organism can an interaction be expected between the newly acquired gene and the resident genes contributing to the pathogenic lifestyle. Wassenaar and Gaastra (2001) have proposed a classification of virulence genes according to their potential role in pathogenic lifestyles that should be helpful to evaluate the potential influence of newly acquired genes on virulence.

Wassenaar and Gaastra (2001) discriminate among four lifestyles: exclusive pathogens, host-dependent pathogens, opportunistic pathogens and fully non-pathogenic organisms. Seven types of virulence gene classes are distinguished: true virulence genes, directly involved in interactions with the host and responsible for pathological damage (e.g. toxins); colonisation genes, determining the localisation of the infection; host defense system evasion genes (e.g. immunoglobulin specific proteases); processing genes involved in the biosynthesis of virulence lifestyle factors (e.g. chaperonins; gene products with a virulence lifestyle substrate), secretory genes, virulence housekeeping genes (e.g. urease, catalase) and regulatory genes, involved in the regulation of virulence lifestyle genes. Further subclasses may be identified for these classes.

Evolution and spread of virulence genes: Pathogenicity islands

In general, three mechanisms can be proposed for the evolution of pathogens: acquisition of virulence genes from existing pathogens by horizontal gene transfer; a change in host specificity (host jump) of an existing pathogen, possibly together with, or as a result of, the acquirement of genes to adapt to a new ecological niche; and evolution of new virulence genes from the existing gene pool of a bacterial species, resulting in (an increase of) virulence.

Over the past few years it has become apparent that of these, the evolutionary consequences of horizontal gene transfer are probably the most drastic. There is ample evidence that virulence genes have spread by horizontal gene tranfer, by all processes known to contribute to the process (see OECD, 2010). Of special importance are bacteriophages, that confer virulence factors to bacteria (Boyd and Brüssow, 2002; Wagner and Waldor, 2002).

In the late 1980s, Hacker and colleagues (Dobrindt et al., 2004; see also Schmidt and Hensel, 2004) were the first to notice that pathogenicity related genes are often located on mobile genetic elements, called "pathogenicity islands" (PAIs). PAIs may be identified as strain specific sequences by subtractive hybridisation between virulent and avirulent strains of the same species. They are frequently found integrated in or near to tRNA genes, which have perfect properties for docking sites because they are highly conserved and often present in multiple copies. They are characterised by (the remains of) insertion sequences at their borders that, if still functional, may lead to genetic instability and to the spread of the PAI to other strains by horizontal gene transfer. The guanine-cytosine (GC)-content and codon usage of PAIs is often different from the GC-content and codon usage of the rest of the genome, which is taken as an indication of their recent acquisition in the genome.

PAIs typically contain sequences that code for gene products that have a (putative) role in virulence. The uropathogenic strain 536 of *E. coli* that has been extensively studied by the group of Hacker (Brzuszkiewicz et al., 2006), provides a good example of what might be found on PAIs. *E. coli* 536 contains seven PAIs coding for different types of fimbriae, haemolysins, a capsule, a siderophore system, a Yersiniabactin, proteins involved in intracellular multiplication, and for a hybrid peptide-polyketide genotoxin that causes cell cycle arrest and eventually cell death of eukaryotic cells that are in contact with this *E. coli* strain (Nougayrède et al., 2006).

The ongoing elucidation and analysis of prokaryotic genomes has shown that not only pathogenicity related traits are located on "islands". PAIs are a specific example of a "genomic island" (GEI), the term that has been coined for the phenomenon that bacteria carry in their genome a flexible gene pool that encodes additional traits that can be beneficial under certain circumstances, and that allows them to occupy a specific niche, while the more constant part of the genome takes care of "household" functions. GEIs are commonly found in the "metagenome", i.e. the combined genomes, of bacteria that share a niche. They would appear to facilitate exchange of useful genes between these bacteria that are mutually supportive in occupying the same environment. The traditional view of bacterial evolution occurring through clonal divergence and selection must be broadened to include gene exchange as a major driving force for adaptation to specific niches. PAIs would be an example of this phenomenon of gene transfer, in facilitating bacteria to function as pathogenic organisms.

The genome flexibility that leads to enhanced virulence is not restricted to acquisition of virulence factors; it may also include loss of genomic sequences, as has been shown for *Shigella flexneri* and enteroinvasive *E. coli* (Maurelli et al., 1998). In general, it appears to be evolutionarily profitable to counteract the acquisition of genes that provide selective advantage with loss of genetic information that can be dispensed with in the new niche, as is the case for instance for intracellular symbionts.

In conclusion, pathogenicity is not a singular trait of a singular type of organism, "the pathogens". Rather, pathogenic traits are adaptive traits that equip a bacterium for a specific lifestyle in a specific niche that happens to be the surface or interior of the host's body.

Assessing potential for bacteria-mediated adverse human health effects

In the previous sections it has been argued that deleterious effects that are caused by pathogenic organisms can be understood as effects of a "lifestyle", or constellation of traits, that enables these organisms to colonise and use specific environments in or on the human or, in general, animal body, as a niche. This line of thinking has been broadened by Casadevall (2006), who has pointed out that many micro-organisms in the environment have developed lifestyles that allow them to interact with the other organisms that they encounter in the environment. The same or similar gene products may have a role in different lifestyles, e.g. the interaction with fungi, protozoa as well as vertebrates. Bacterial strains that have no directly apparent role in human pathogenicity, for instance because they do not survive or replicate at 37°C, may still carry genes that code for gene products with a potential role as virulence factors in bacteria that are more compatible with a lifestyle as a human pathogen. DNA exchange between microbial strains may in the end provide bacteria that thrive in the human environment with new virulence factors derived from such bacteria. This complicates the risk/safety assessment of deliberate release of environmental strains, particularly if these strains have been subject to genetic engineering. The following is intended to help the assessor negotiate these complications.

Risk/safety assessment of environmental release of bacterial strains to determine whether these may cause adverse human health effects

Environmental release of bacteria should be preceded by a risk/safety assessment. Risk assessment usually comprises four steps: 1) hazard identification; 2) hazard characterisation (e.g. dose-response assessment); 3) exposure assessment (dose, concentration, survival); and 4) risk characterisation. In the risk assessment of environmental releases of bacterial strains, one aspect that has to be taken into account in hazard identification and hazard characterisation is the pathogenic potential of the bacteria to cause adverse human health effects.[4] The WHO *Laboratory Biosafety Manual* (WHO, 2004, Chapters 2 and 16) provides helpful considerations on the risk/safety assessment of (potentially) pathogenic organisms. These considerations apply primarily to laboratory settings, but they can be easily adapted and applied to environmental settings.

For genetically engineered bacteria, the risk group of the species is a first approximation of the degree of bacterial pathogenicity in humans. But assessing the degree of pathogenicity of a bacterial strain calls for an unequivocal identification of the location of the strain in the spectrum from clear non-pathogen to clear pathogen. This should be done with caution. Truly non-pathogenic bacteria will lack the ability to survive in a human host (with the exception of commensal bacteria), or cause any adverse effects. Bacteria that are incompatible with the human environment e.g. bacteria that cannot survive at temperatures between 30-42°C, or that are exclusively phototrophic or lithotrophic would be expected to be non-pathogenic. Still, one should be careful drawing this conclusion. For instance, lithotrophic bacteria have been found in infections associated with surgical implants in the human body (Dempsey et al., 2007). Indeed, Casadevall (2006) has pointed out that bacterial strains that have lifestyles that do not link them to pathogenicity in humans can carry genes that code for gene products with a potential role as virulence factors in bacteria that are or could develop into human pathogens. Falkow (2008) argues that it is difficult to separate the pathogenic from the commensal lifestyle. What is the difference between a pathogen and a commensal? Pathogens possess the inherent ability to cross anatomic barriers or breach other host defenses that limit the survival or replication of other microbes and commensals. Therefore, most, but certainly not all, pathogens establish themselves in an environment usually devoid of other stable microbial populations. These invasive properties are essential for their survival in nature, and, are often host specific. However, many "commensal" bacteria, that are able to colonise the human host without displaying immediate virulence phenotypes, can cause disease, (e.g. Group A and B streptococci, *S. aureus*, *N. meningitidis*, *S. pneumonia*, *H. influenza*). Many features that are seen as virulence factors of pathogens (pili, antiphagocytic proteins, capsules) may also be found in non-pathogenic bacteria. Virulence factors may simply be examples of a more general class of "adaptive factors" common to all bacteria (Casadevall, 2006).

Bacterial strains that have been derived from wild type isolates from the environment may have changed significantly in their pathogenicity compared to fresh wild type isolates. This may, in particular, be the case if the strains have been mutated, e.g. for attenuation, or if new virulence related genes have been introduced by genetic engineering, have been lost afterwards. But also the fact that strains have been handled in the laboratory during many generations may have led to the occurrence of mutants with changed properties. Losses of properties that have no function in survival under laboratory conditions, like virulence factors, occur quite frequently under these circumstances. As in all cases of attenuation, strains that have been attenuated in this way should be thoroughly tested for stability of the non-pathogenic phenotype.

Predicting the effects of introduced (potential) virulence genes on the pathogenicity of the recipient strain is not straightforward[5] (see below). If "true virulence genes" (as defined above) have been introduced, an effect on pathogenicity is more likely, as these genes are directly responsible for pathological damage, such as toxins. However, the degree of damage that these gene products can cause is highly dependent on the context of the pathogenic "lifestyle" of the bacterium, which depends on the secondary virulence factors available in the bacterial strain. This discussion is especially pertinent for medical and veterinary applications of potentially (non)pathogenic strains. The discussion is also highly relevant for risk/safety assessments of releases of environmental strains that have been engineered in the laboratory. The case of the *Burkholderia cepacia* complex (Bcc) is an example of a group of host species where the effect on pathogenicity of introduced potential virulence genes could be difficult to predict. It has has already been mentioned that these bacteria occur normally in the environment, but are now recognised as an important human health hazard. The species in this complex have for a long time been seen as (potential) plant pathogens, and their potential as human (opportunistic) pathogens has only recently been recognised. It would require more insight into the lifestyle of these bacteria to be able to predict the role of introduced virulence genes in the establishment of pathogenic potential. In cases like this, caution should be used in establishing conclusions on their pathogenicity.

The likelihood of the strain actually causing adverse human health effects will depend on the exposure of humans to the bacterial strain during and after the release. Adverse effects are only to be expected if the exposure is such that it will lead to contact of humans with the bacteria in sufficient numbers in relation to the infectious dose of the specific bacterium, and in such a way that pathogenic effects may ensue, e.g. by ingestion, inhalation, dermal contact. Factors that determine the degree of exposure are:

1) number of bacteria released into the environment; 2) physiological state of the bacteria, e.g. due to fermentations conditions prior to the administration; 3) spread during release, dependent on the method used, e.g. by aerosols, injection or mixing in the soil, seed coating; 4) survival; and 5) dissemination after release, e.g. through surface and subsurface water movement, by soil fauna or by disturbance of the site of application. Spreading of aerosols is dependent on conditions of wind at the time of the application; survival of bacteria in aerosol droplets is dependent on environmental factors, e.g. temperature, humidity and UV radiation. Survival of bacteria in soil is variable for different strains of the same species. In many cases where bacteria have been introduced into the environment, a rapid decrease has been observed, i.e. the number of detectable bacteria drops below the detection limit[6] of a direct viable count within months or even weeks. This is even the case for many strains that are well-adapted to a soil lifestyle, e.g. the root colonising *Pseudomonas* spp. (Glandorf et al., 2001; Weller, 2007). The bacteria are, however, not "lost" from the environment, and may appear again readily if environmental circumstances are favourable, for instance if the plant that they are prone to colonise is again present in the environment; also, in some cases long-term survival and persistence of introduced micro-organisms has been demonstrated (Hirsch, 1996).

If a human health hazard is expected, risk estimations should be made based on worst case assumptions on survival and spreading. Risk estimates may be refined if the results of further research show that the worst case assumptions are not realistic.

General considerations for assessing altered pathogenicity of micro-organisms as a result of genetic engineering

The risk/safety assessment of genetically engineered micro-organisms requires careful consideration of numerous factors, not the least of which is the genetic composition of both the recipient and the donor organisms, and their respective lifestyles and phenotypic expression. While the intended use of the organism is factored into the initial assessment, some foresight should be given to potential unintended uses, in particular if the genetically engineered strains are meant to be commercially available.

Genetic engineering may cause, advertently or inadvertently, changes in the various factors that determine the niche of a bacterium, and may broaden its niche, that then needs to be redefined. As described in the previous sections, pathogenicity is the capacity to cause disease, and is related to the ability of a micro-organism to reach and occupy a particular habitat on or in the host and to subsequently cause harm to the host. Thus, when performing an assessment of pathogenic potential to humans, one should consider how the engineering may change a bacterium's capacity to cause disease.

There are several determinants that should be considered when assessing the potential for bacterial pathogenicity as a result of genetic engineering. Consideration should be given to the biological and ecological characteristics of the non-modified strain, i.e. its "lifestyle", inasfar as it is compatible with causing pathogenicity in humans. Due to the lifestyle of the vast majority of bacterial strains in the environment, e.g. psychrophilic or thermophillic, lithotrophic or phototrophic bacteria, it is not likely that they will turn into potential human pathogens just by the introduction of one virulence factor derived from a human pathogen. On the other hand, genes derived from bacteria that are not suspected human pathogens on the basis of their lifestyle may still code for gene products that can contribute to future virulence (Casadevall, 2006).

Genetic engineering may involve genes whose products are not inherently harmful but adverse effects may still arise from the modification or exacerbation of an existing constellation of traits in the recipient micro-organism. This may arise as the result, for example, of the product of an inserted gene acting alongside existing pathogenic determinants or the addition of a trait completing the suite of traits necessary for a pathogenic lifestyle. Alternatively it is possible that modification of normal genes may also alter pathogenicity. In an assessment for environmental risk/safety of an application intended for environmental release, the following points should be considered:

- characteristics of the recipient, for instance whether the recipient possesses a sufficient number of the constellation of traits that it could be a potential human or mammalian pathogen
- existing traits in the recipient organism that might lead to an increase in pathogenicity or infectivity when altered (e.g. alteration of host range or tissue tropism)
- the likelihood that any disabling mutation within the recipient might be overcome, for example through complementation or reversion, due to the insertion of the foreign DNA, or through the inserted gene encoding an enzyme that would complete an anabolic pathway for a pathogenicity determinant
- the transmissibility of the vector used to introduce relevant genes
- whether the foreign DNA carries a pathogenicity determinant from a related organism (toxin, invasin, integrin, and surface structures such as fimbriae, LPS and capsule)
- when the foreign DNA does carry a virulence factor, the feasibility that this gene could contribute to the pathogenicity of the genetically engineered micro-organism, or whether the virulence factor provides resistance to host defense mechanisms
- whether the foreign DNA carries a gene that renders the recipient resistant to an antibiotic, especially if the specific resistance has not yet spread by natural processes to the genus to which the recipient belongs (see Appelbaum, 2006; Noble, Virani and Cree, 1992), or if the mechanism of resistance has emerged newly, like Qnr determinants (Nordmann and Poirel, 2005)
- whether susceptibility to antibiotics or other forms of therapy may be affected as a consequence of the genetic engineering
- whether attenuated or inactivated strains remain stably attenuated or inactivated
- whether a surface component that might bind to a different receptor than that used by recipient micro-organism could increase virulence
- whether the foreign DNA encodes gene products, e.g. toxins, that even in the absence of live organisms, may cause pathogenic effects
- whether the derived from unrelated bacteria foreign DNA encodes a protein that does not interact with the pathogenic properties of the parental strain but may cause pathogenic reaction, e.g. a modulator of growth (hormone, cytokine), or other protein with a potentially harmful biological activity (see also Bergmans et al., 2008)
- when mutations are introduced that inactivate specific virulence factors of potential pathogens, whether the stability of the mutation has been demonstrated,

and whether attenuation has been shown to be sufficient so that the resulting strain can be considered non-pathogenic.

The above points are illustrative rather than inclusive. Assessors must use good judgment in utilising this list, recognising that additional examples may be pertinent to the case at hand. Although there are a number of considerations that must be taken in the evaluation of the pathogenic potential of genetically engineered bacteria, it is highly unlikely that a pathogen will be inadvertently created from a non-pathogen lacking most or the entire constellation of traits enabling the pathogenic lifestyle by combining virulence factors.

Notes

1. The terms "genetically engineered micro-organism" and "genetically modified micro-organism" are used in different legislative systems for micro-organisms in which genetic information has been added or removed by techniques of modern biotechnology.
2. Toxins may also be low molecular weight metabolites; this type of toxins is, however, not taken into consideration in this chapter.
3. The following paragraphs describing adhesion of bacteria to various surfaces are restricted to the pili, adhesins and secretion systems of gram-negative bacteria.
4. Other aspects that have to be taken into account in various steps of the risk assessment include (but are not limited to) natural background levels; conditions for survival, persistence, growth and reproduction; mode of dispersal; potential for gene transfers, in particular genes associated with pathogenicity, toxicity or persistence; antibiotic resistance.
5. One factor that complicates this prediction is the influence of the condition of the host, e.g. the immune status of the host, on the effectiveness of a virulence factor. In regulatory discussions, however, this complication is usually evaded, as for the outcome of the discussion the conditions of the host is usually supposed to be "normal".
6. For guidance on the detection of bacteria in soil, see OECD (2004).

References

Aebi, C. et al. (1998), "Phenotypic effect of isogenic uspA1 and uspA2 mutations on *Moraxella catarrhalis* 035E", *Infection and Immunity*, Vol. 66, No. 7, pp. 3 113-3 119, July.

Al-Shangiti, A.M. et al. (2004), "Structural relationships and cellular tropism of staphylococcal superantigen-like proteins", *Infection and Immunity*, Vol. 72, No. 7, pp. 4 261-4 270.

Alouf, J.E. (2000), "Bacterial protein toxins: An overview", *Methods in Molecular Biology*, Vol. 145, pp. 1-26, *http://dx.doi.org/10.1385/1-59259-052-7:1*.

An, Y.H., R.B. Dickinson and R.J. Doyle (2000), "Mechanisms of bacterial adhesion and pathogenesis of implant and tissue infections", in: An, Y.H. and R.J. Friedman (eds.), *Handbook of Bacterial Adhesion: Principles, Methods and Applications*, Humana Press Inc., Totowa, New Jersey.

Appelbaum, P.C. (2006), "MRSA – The tip of the iceberg", *Clinical Microbiology and Infection*, Vol. 12, Supplement 2-3, pp. 3-10, April.

Appelberg, R. and I.S. Leal (2000), "Mutants of *Listeria monocytogenes* defective in *in vitro* invasion and cell-to-cell spreading still invade and proliferate in hepatocytes of neutropenic mice", *Infection and Immunity*, Vol. 68, No. 2, pp. 912-914, February.

Armant, M.A. and M.J. Fenton (2002), "Toll-like receptors: A family of pattern-recognition receptors in mammals", *Genome Biology*, Vol. 3, No. 8, pp. 3011.1-3011.6.

Atapattu, D.N. and C.J. Czuprynski (2005), "*Mannheimia haemolytica* leukotoxin induces apoptosis of bovine lymphoblastoid cells (BL-3) via a caspase-9-dependent mitochondrial pathway", *Infection and Immunity*, Vol. 73, No. 9, pp. 5 504-5 513, September.

Athamna, A, M.R. Kramer and I. Kahane (1996), "Adherence of *Mycoplasma pneumoniae* to human alveolar macrophages", *FEMS Immunology and Medical Microbiology*, Vol. 15, Nos. 2-3, pp. 135-41, September.

Athanasopoulos, A.N. et al. (2006), "The extracellular adherence protein (Eap) of *Staphylococcus aureus* inhibits wound healing by interfering with host defense and repair mechanisms", *Blood*, Vol. 107, No. 7, pp. 2 720-2 727, April.

Autret, A. and A. Charbit (2005), "Lessons from signature-tagged mutagenesis on the infectious mechanisms of pathogenic bacteria", *FEMS Microbiology Reviews*, Vol. 29, No. 4, pp. 703-717, September.

Awad, M.M. et al. (1995), "Virulence studies on chromosomal alpha-toxin and theta-toxin mutants constructed by allelic exchange provide genetic evidence for the essential role of alpha-toxin in *Clostridium perfringens*-mediated gas gangrene", *Molecular Microbiology*, Vol. 15, No. 2, pp. 191-202, January.

Azghani, A.O. et al. (2002), "*Pseudomonas aeruginosa* outer membrane protein F is an adhesin in bacterial binding to lung epithelial cells in culture", *Microbial Pathogenesis*, Vol. 33, No. 3, pp. 109-114, September.

Balish, M.F. et al. (2003), "Deletion analysis identifies key functional domains of the cytadherence-associated protein HMW2 of *Mycoplasma pneumonia*", *Molecular Microbiology*, Vol. 50, No. 5, pp. 1 507-1 516, December.

Barenkamp, S.J. and J.W. St. Geme III (1996), "Identification of a second family of high-molecular-weight adhesion proteins expressed by non-typable *Haemophilus influenza*", *Molecular Microbiology*, Vol. 19, No. 6, pp. 1 215-1 223, March.

Bassinet, L. et al. (2000), "Role of adhesins and toxins in invasion of human tracheal epithelial cells by *Bordetella pertussis*", *Infection and Immunity*, Vol. 68, No. 4,, pp. 1 934-1 941, March.

Bassler, B.L. (2002), "Small talk: Cell-to-cell communication in bacteria", *Cell*, Vol. 109, No. 4, pp. 421-424, May.

Bäumler, A.J., R.M. Tsolis and F. Heffron (1997), "Fimbrial adhesins of *Salmonella typhimurium*. Role in bacterial interactions with epithelial cells", *Advances in Experimental Medecine and Biology*, Vol. 412, pp. 149-58.

Benz, I. and M.A. Schmidt (1989), "Cloning and expression of an adhesin (AIDA-I) involved in diffuse adherence of enteropathogenic *Escherichia coli*", *Infection and Immunity*, Vol. 57, No. 5, pp. 1 506-1 511, May.

Bergan, T. (1984), "Pathogenicity of anaerobic bacteria", *Scandinavian Journal of Gastroenterology*, Suppl. 91, pp. 1-11.

Bergmans, H. et al. (2008), "Identification of potentially hazardous human gene products in GMO risk assessment", *Environmental Biosafety Research*, Vol. 7, No. 1, pp. 1-9, January-March, *http://dx.doi.org/10.1051/ebr:2008001*.

Bjarnsholt, T. et al. (2010), "Quorum sensing and virulence of *Pseudomonas aeruginosa* during lung infection of cystic fibrosis patients", *PLoS ONE*, Vol. 5, No. 4, *http://dx.doi.org/10.1371/journal.pone.0010115*.

Blancou, J. et al. (2005), "Emerging or re-emerging bacterial zoonoses: Factors of emergence, surveillance and control", *Veterinary Research*, Vol. 36, No. 3, pp. 507-522, May-June.

Bonacorsi, S.P. et al. (2000), "Identification of regions of the *Escherichia coli* chromosome specific for neonatal meningitis-associated strains", *Infection and Immunity*, Vol. 68, No. 4, pp. 2 096-2 101, April.

Bonci, A. et al. (1997), "Relatedness and phylogeny within the family of periplasmic chaperones involved in the assembly of pili or capsule-like structures of gram-negative bacteria", *Journal of Molecular Evolution*, Vol. 44, No. 3, pp. 299-309, March.

Borriello, S.P. (1998), "Pathogenesis of *Clostridium difficile* infection", *Journal of Antimicrobial Chemotherapy*, Vol. 41, Supplement 3, pp. 13-19, *http://dx.doi.org/10.1093/jac/41.suppl_3.13*.

Boyd, E.F. and H. Brüssow (2002), "Common themes among bacteriophage-encoded virulence factors and diversity among the bacteriophages involved", *Trends in Microbiology*, Vol. 10, No. 11, pp. 521-529, November.

Boyle, E.C. and B.B. Finlay (2003), "Bacterial pathogenesis: Exploiting cellular adherence", *Current Opinion in Cell Biology*, Vol. 15, No. 5, pp. 633-639, October.

Brady, R.A. et al. (2008), "Osteomyelitis and the role of biofilms in chronic infection", *FEMS Immunology and Medical Microbiology*, Vol. 52, No. 1, pp. 13-22, January.

Braun, L., B. Ghebrehiwet and P. Cossart (2000), "gC1q-R/p32, a C1q-binding protein, is a receptor for the InlB invasion protein of *Listeria monocytogenes*", *EMBO Journal*, Vol. 19, No. 7, pp. 1 458-1 466, *http://dx.doi.org/10.1093/emboj/19.7.1458*.

Brophy, P.J. (2002), "Subversion of Schwann cells and the Leper's bell", *Science*, Vol. 296, No. 5 569, pp. 862-863, May.

Brzuszkiewicz, E. et al. (2006), "How to become a uropathogen: Comparative genomic analysis of extraintestinal pathogenic *Escherichia coli* strains", *Proceedings of the National Academy of Sciences USA*, Vol. 103, No. 34, pp. 12 879-12 884, August.

Burgos, R. et al. (2006), "*Mycoplasma genitalium* P140 and P110 cytadhesins are reciprocally stabilized and required for cell adhesion and terminal-organelle development", *Journal of Bacteriology*, Vol. 188, No. 24, pp. 8 627-8 637, December.

Cámara, M., P. Williams and A. Hardman (2002), "Controlling infection by tuning in and turning down volume of bacterial small-talk", *Lancet Infectious Diseases*, Vol. 2, No. 11, pp. 667-676, November.

Cameron, C.E. (2003), "Identification of a *Treponema pallidum* laminin-binding protein", *Infection and Immunity*, Vol. 71, No. 5, pp. 2 525-2 533, May, *http://dx.doi.org/10.1128/IAI.71.5.2525-2533.2003*.

Casadevall, A. (2006), "Cards of virulence and the global virulome for humans", *Microbe Magazine*, Vol. 1, No. 8, pp. 359-364, *www.microbemagazine.org/images/stories/arch2006/aug06/znw00806000359.pdf*.

Casadevall, A. and L.-A. Pirofski (1999), "Host-pathogen interactions: Redefining the basic concepts of virulence and pathogenicity", *Infection and Immunity*, Vol. 67, No. 8, pp. 3 703-3 713.

Cascales, E. (2008), "The type VI secretion toolkit", *EMBO Reports*, Vol. 9, No. 8, pp. 735-741, August, *http://dx.doi.org/10.1038/embor.2008.131*.

Cheng, Q. et al. (2002), "The group B streptococcal C5a peptidase is both a specific protease and an invasion", *Infection and. Immunity*, Vol. 70, No. 5, pp. 2 408-2 413, May.

Chiang, S.L., J.J. Mekalanos and D.W. Holden (1999), "In vivo genetic analysis of bacterial virulence", *Annual Review of Microbiology*, Vol. 53, pp. 129-154.

China, B. et al. (1993), "Role of the YadA protein in prevention of opsonization of Yersinia Enterocolitica by C3b molecules", *Infection and Immunity*, Vol. 61, No. 8, pp. 3 129-3 136, August.

Cirillo, S.L. et al. (2001), "*Legionella pneumophila* entry gene *rtxA* is involved in virulence", *Infection and Immunity*, Vol. 69, No. 1, pp. 508-517, January, *http://dx.doi.org/10.1128/IAI.69.1.508-517.2001*.

Cockeran, R., R. Anderson and C. Feldman (2002), "The role of pneumolysin in the pathogenesis of *Streptococcus pneumoniae* infection", *Current Opinion in Infectious Disease*, Vol. 15, No. 3, pp. 235-239, June.

Connell, H. et al. (1997), "Bacterial attachment to uro-epithelial cells: Mechanisms and consequences", *Advances in Dental Research*, Vol. 11, No. 1, pp. 50-58, April.

Cossart, P. and P.J. Sansonetti (2004), "Bacterial invasion: The paradigms of enteroinvasive pathogens", *Science*, Vol. 304, No. 5 668, pp. 242-248, April.

Costerton, W. et al. (2003), "The application of biofilm science to the study and control of chronic bacterial infections", *Journal of Clinical Investigations*, Vol. 112, No. 10, pp. 1 466-1 477, November.

Craig, L., M.E. Pique and J.A. Tainer (2004), "Type IV pilus structure and bacterial pathogenicity", *Nature Reviews Microbiology*, Vol. 2, pp. 363-378, May, *http://dx.doi.org/10.1038/nrmicro885*.

Crocquet-Valdes, P.A., K. Weiss and D.H. Walker (1994), "Sequence analysis of the 190-kDa antigen-encoding gene of *Rickettsia conorii* (Malish 7 strain)", *Gene*, Vol. 140, No. 1, pp. 115-119, March.

Cue, D. et al. (2000), "A nonpeptide integrin antagonist can inhibit epithelial cell ingestion of *Streptococcus pyogenes* by blocking formation of integrin α5β1-fibronectin-M1 protein complexes", *Proceedings of the National Academy of Sciences USA*, Vol. 97, No. 6, pp. 2 858-2 863, March, *http://dx.doi.org/10.1073/pnas.050587897*.

Cummings, C.A. et al. (2004), "*Bordetella* species are distinguished by patterns of substantial gene loss and host adaptation", *Journal of Bacteriology*, Vol. 186, No. 5, pp. 1 484-1 492, March.

Davies, R.L., S. Campbell and T.S. Whittam (2002), "Mosaic structure and molecular evolution of the Leukotoxin Operon (lktCABD) in *Mannheimia (Pasteurella) haemolytica*, *Mannheimia glucosida*, and *Pasteurella trehalosi*", *Journal of Bacteriology*, Vol. 184, No. 1, pp. 266-267, January.

de Repentigny, L. et al. (2000), "Characterization of binding of *Candida albicans* to small intestinal mucin and its role in adherence to mucosal epithelial cells", *Infection and Immunity*, Vol. 68, No. 6, pp. 3 172-3 179, June.

Decho, A.W., R.S. Norman and P.T. Visscher (2010), "Quorum sensing in natural environments: Emerging views from microbial mats", *Trends in Microbiology*, Vol. 18, No. 2, pp. 73-80, February.

Dehio, C., S.D. Gray-Owen and T.F. Meyer (2000), "Host cell invasion by pathogenic *Neisseriae*", *Subcell Biochemistry*, Vol. 33, pp. 61-96.

Deitsch, K.W., E.R. Moxon and T.E. Wellems (1997), "Shared themes of antigenic variation and virulence in bacterial, protozoal, and fungal infections", *Microbiology and Molecular Biology Reviews*, Vol. 61, No. 3, pp. 281-293, September.

Dempsey, K.E. et al. (2007), "Identification of bacteria on the surface of clinically infected and non-infected prosthetic hip joints removed during revision arthroplasties by 16S rRNA gene sequencing and by microbiological culture", *Arthritis Research & Therapy*, Vol. 9, No. 3, R46.

Desvaux, M., N.J. Parham and I.R. Henderson (2004), "Type V protein secretion: Simplicity gone awry?", *Current Issues in Molecular Biology*, Vol. 6, No. 2, pp. 111-124, July.

Dinges, M.M., P.M. Orwin and P.M. Schlievert (2000), "Exotoxins of *Staphylococcus aureus*", *Clinical Microbiology Review*, Vol. 13, No. 1, pp. 16-34, January.

Dobrindt, U. et al. (2004), "Genomic islands in pathogenic and environmental microorganisms", *Nature Review of Microbiology*, Vol. 2, No. 5, pp. 414-424, May.

Dodson, K.W. et al. (2001), "Structural basis of the interaction of the pyelonephritic *E. coli* adhesin to its human kidney receptor", *Cell*, Vol. 105, No. 6, pp. 733-743, June.

Dombek, P.E. et al. (1999), "High-frequency intracellular invasion of epithelial cells by serotype M1 group A streptococci: M1 protein-mediated invasion and cytoskeletal rearrangements", *Molecular Microbiology*, Vol. 31, No. 3, pp. 859-870, February.

Doran, K.S. et al. (2002), "Group B streptococcal Beta-hemolysin/cytolysin promotes invasion of human lung epithelial cells and the release of Interleukin-8", *Journal of Infectious Diseases*, Vol. 185, No. 2, pp. 196-203, January.

Duclos, S. and M. Desjardins (2000), "Subversion of a young phagosome: The survival strategies of intracellular pathogens", *Cellular Microbiology*, Vol. 2, No. 5, pp. 365-377, October.

Edwards, J.L., D.D. Entz and M.A. Apicella (2003), "Gonococcal phospholipase D modulates the expression and function of complement receptor 3 in primary cervical epithelial cells", *Infection and Immunity*, Vol. 71, No. 11, pp. 6 381-6 391, November, *http://dx.doi.org/10.1128/IAI.71.11.6381-6391.2003*.

El Tahir, Y. and M. Skurnik (2001), "YadA, the multifaceted *Yersinia* adhesin", *International Journal of Medical Microbiology*, Vol. 291, No. 3, pp. 209-218, August.

Ernst, J.D. (2000), "Bacterial inhibition of phagocytosis", *Cell Microbiology*, Vol. 2, No. 5, pp. 379-386, October.

Falkow, S. (2008), "Bacterial pathogenicity in a historical and experimental perspective", presentation at Microbial Evolution and Co-Adaptation: A Workshop in Honor of Joshua Lederberg, 20 May 2008", Institute of Medicine, Washington, DC, *http://iom.edu/~/media/Files/Activity%20Files/PublicHealth/MicrobialThreats/Falkow.pdf*.

Falkow, S. (1988), "Molecular Koch's postulates applied to microbial pathogenicity", *Reviews of Infectious Diseases*, Vol. 10, Supplement 2, Microbial Surfaces: Determinants of Virulence and Host Responsiveness, pp. S274-276, *www.jstor.org/stable/4454582*.

Fällman, M. and A. Gustavsson (2005), "Cellular mechanisms of bacterial internalization counteracted by *Yersinia*", *International Review of Cytology*, Vol. 246, pp. 135-188.

Field, D., D. Hood and R. Moxon (1999), "Contribution of genomics to bacterial pathogenesis", *Current Opinion in Genetics & Development*, Vol. 9, No. 6, pp. 700-703, December.

Finlay, B.B. and P. Cossart (1997), "Exploitation of mammalian host cell functions by bacterial pathogens", *Science*, Vol. 276, No. 5 313, pp. 718-25, Erratum in: *Science*, Vol. 278, No. 5 337, pp. 373.

Finlay, B.B. and S. Falkow (1997), "Common themes in microbial pathogenicity revisited", *Microbiology and Molecular Biology Reviews*, Vol. 61, No. 2, pp. 136-169, June.

Finlay, B.B. and S. Falkow (1989), "Common themes in microbial pathogenicity", *Microbiological Reviews*, Vol. 53, No. 2, pp. 210-230, June.

Finn T.M. and D.F. Amsbaugh (1998), "Vag8, a *Bordetella pertussis* Bvg-regulated protein", *Infection and Immunity*, Vol. 66, No. 8, pp. 3 985-3 989, August.

Finn, T.M. and L.A. Stevens (1995), "Tracheal colonization factor: A *Bordetella pertussis* secreted virulence determinant", *Molecular Microbiology*, Vol. 16, No. 4, pp. 625-634, May, *http://onlinelibrary.wiley.com/doi/10.1111/j.1365-2958.1995.tb02425.x/epdf*.

Foong, S.C. and J.S. Dickson (2004), "Attachment of *Listeria monocytogenes* on ready-to-eat meats", *Journal of Food Protection*, Vol. 67, No. 3, pp. 456-462, March.

Fouet, A. and S. Mesnage (2002), "*Bacillus anthracis* cell envelope components", *Current Topics in Microbiology and Immunology*, Vol. 271, pp. 87-113.

Fournier B. and D.J. Philpott (2005), "Recognition of *Staphylococcus aureus* by the innate immune system", *Clinical Microbiology Reviews*, Vol. 18, No. 3, pp. 521-540, July, *http://dx.doi.org/10.1128/CMR.18.3.521-540.2005*.

Franzon, V.L., A. Barker and P.A. Manning (1993), "Nucleotide sequence encoding the mannose-fucose-resistant hemagglutinin of *Vibrio cholerae* O1 and construction of a mutant", *Infection and Immunity*, Vol. 61, No. 7, pp. 3 032-3 037, July.

Fredricks, D.N. and D.A. Relman (1996), "Sequence-based identification of microbial pathogens: A reconsideration of Koch's postulates", *Clinical Microbiology Reviews*, Vol. 9, No. 1, pp. 18-33, January.

Freemont, A.J. (1998), "Demystified ... adhesion molecules", *Journal of Clinical Pathology Molecular Pathology*, Vol. 51, No. 4, pp. 175-184, August.

Frosch, M., J. Reidl and U. Vogel (1998), "Genomics in infectious diseases: Approaching the pathogens", *Trends in Microbiology*, Vol. 6, No. 9, pp. 346-349, September.

Futagawa-Saito K. et al. (2004), "Prevalence and characterization of leukotoxin-producing *Staphylococcus intermedius* in isolates from dogs and pigeons", *Journal of Clinical Microbiology*, Vol. 42, No. 11, pp. 5 324-5 326, November, *http://dx.doi.org/10.1128/JCM.42.11.5324-5326.2004*.

Gao, L.Y. et al. (2003), "Requirement for kasB in mycobacterium mycolic acid biosynthesis, cell wall impermeability and intracellular survival: Implications for therapy", *Molecular Microbio*logy, Vol. 49, No. 6, pp. 1 547-1 563, September.

Garton N.J. et al. (2002), "A novel lipoarabinomannan from the equine pathogen *Rhodococcus equi*. Structure and effect on macrophage cytokine production", *Journal of Biological Chemistry*, Vol. 277, No. 35, pp. 31 722-31 733, August.

Ghigo, E. et al. (2002), "*Coxiella burnetti* survival in THP-1 monocytes involves the impairment of phagosome maturation: IFN-γ mediates its restoration and bacterial killing", *Journal of Immuno*logy, Vol. 169, No. 8, pp. 4 488-4 495, October.

Gilbert, P., J. Das and I. Foley (1997), "Biofilm susceptibility to antimicrobials", *Advances in Dental Research*, Vol. 11, No. 1, pp. 160-167, April.

Girish, K.S. et al. (2004), "Isolation and characterization of hyaluronidase a 'spreading factor' from Indian cobra (*Naja naja)* venom", *Biochimie*, Vol. 86, No. 3, pp. 193-202, March.

Gladysheva, I.P. et al. (2003), "Coevolutionary patterns in plasminogen activation", *Proceedings of the National Academy of Sciences USA*, Vol. 100, No. 16, pp. 9 168-9 172, *www.pnas.org/content/100/16/9168.full.pdf*.

Glandorf, D.C.M. et al. (2001), "Effect of genetically modified *Pseudomonas putida* WCS358r on the fungal rhizosphere micro of field-grown wheat", *Applied and Environmental Microbiology*, Vol. 67, No. 8, pp. 3 371-3 378, *http://dx.doi.org/10.1128/AEM.67.8.3371-3378.2001*.

Glomski, I.J. et al. (2002), "The *Listeria monocytogenes* hemolysin has an acidic pH optimum to compartmentalize activity and prevent damage to infected host cells", *Journal of Cell Biology*, Vol. 156, No. 6, pp. 1 029-1 038, March.

Goodwin, A.C. et al. (2008), "Expression of the *Helicobacter pylori* adhesin SabA is controlled via phase variation and the ArsRS signal transduction system", *Microbiology*, Vol. 154, Pt. 8, pp. 2 231-2 240, *http://dx.doi.org/10.1099/mic.0.2007/016055-0*.

Goosney, D.L., R. DeVinney and B.B. Finlay (2001), "Recruitment of cytoskeletal and signalling proteins to enteropathogenic and enterohemorrhagic *Escherichia coli* pedestals", *Infection and Immunity*, Vol. 69, No. 5, pp. 3 315-3 322, *http://dx.doi.org/10.1128/IAI.69.5.3315-3322.2001*.

Gouin, E. et al. (1999), "A comparative study of the actin-based motilities of the pathogenic bacteria *Listeria monocytogenes*, *Shigella flexneri* and *Rickettsia conorii*", *Journal of Cell Science*, Vol. 112, Pt. 11, pp. 1 697-1 708, June.

Gradon, J.D., J.G. Timpone and S.M. Schnittman (1992), "Emergence of unusual opportunistic pathogens in AIDS: A review", *Clinical Infectious Diseases*, Vol. 15, No. 1, pp. 134-157, July.

Grange, P.A. et al. (2002), "Evaluation of receptor binding specificity of *Escherichia coli* K88 (F4) fimbrial adhesin variants using porcine serum transferrin and glycosphingolipids as model receptors", *Infecton and Immunity*, Vol. 70, No. 5, pp. 2 336-2 343, May.

Greenberg, E.P. (2003), "Bacterial communication and group behavior", *Journal of Clinical Investigation*, Vol. 112, No. 9, pp. 1 288-1 290, *http://dx.doi.org/10.1172/JCI200320099*.

Guhathakurta, B. et al. (1999), "Adhesion and invasion of a mutant *Shigella flexneri* to an eukaryotic cell line in absence of the 220-kb virulence plasmid", *FEMS Microbiology Letters*, Vol. 181, No. 2, pp. 267-275, December.

Hackstadt, T. and J.C. Williams (1981), "Biochemical strategem for obligate parasitism of eukaryotic cells by *Coxiella burnetti*", *Proceedings of the National Academy of Sciences USA*, Vol. 78, No. 5, pp. 3 240-3 244, May.

Hancock, L.E. and M.S. Gilmore (2002), "The capsular polysaccharide of *Enterococcus faecalis* and its relationship to other polysaccharides in the cell wall", *Proceedings of the National Academy of Sciences USA*, Vol. 99, No. 3, pp. 1 574-1 579, February.

Harb, O.S. and Y. Abu Kwaik (1999), "Probing the microenvironment of intracellular bacterial pathogens", *Microbes and Infection*, Vol. 1, No. 6, pp. 445-453, May.

Hardy, G.G., S.M. Tudor and J.W. St. Geme III (2003), "The pathogenesis of disease due to nontypeable *Haemophilus influenzae*", *Methods of Molecular Medicine*, Vol. 71, pp. 1-28.

Harris, T.O. et al. (2003), "A novel streptococcal surface protease promotes virulence, resistance to opsonophagocytosis, and cleavage of human fibrinogen", *Journal of Clinical Investigation*, Vol. 111, No. 1, pp. 61-70, January, *http://dx.doi.org/10.1172/JCI200316270*.

Hase, C.C. and J.J. Mekalanos (1998), "TcpP protein is a positive regulator of virulence gene expression in *Vibrio cholerae*", *Proceedings of the National Academy of Sciences USA*, Vol. 95, No. 2, pp. 730-734, January.

Hauck, C.R. (2002), "Cell adhesion receptors – Signalling capacity and exploitation by bacterial pathogens", *Medical Microbiology and Immunology*, Vol. 191, No. 2, pp. 55-62, October.

Hauck, C.R. and T.F. Meyer (2003), "'Small' talk: Opa proteins as mediators of *Neisseria*-host-cell communication", *Current Opinions in Microbiology*, Vol. 6, No. 1, pp. 43-49, February.

Hayward, R.D. et al. (2002), "A *Salmonella* SipB-derived polypeptide blocks the 'trigger' mechanism of bacterial entry into eukaryotic cells", *Molecular Microbiology*, Vol. 45, No. 6, pp. 1 715-1 727, September.

Hazenbos, W.L. et al. (1995), "Binding of FimD on *Bordetella pertussis* to very late antigen-5 on monocytes activates complement receptor type 3 via protein tyrosine kinases", *Journal of Immunology*, Vol. 155, No. 8, pp. 3 972-3 978.

Henderson, I.R. and P. Owen (1999), "The major phase-variable outer membrane protein of *Escherichia coli* structurally resembles the immunoglobulin A1 protease class of exported protein and is regulated by a novel mechanism involving dam and OxyR", *Journal of Bacteriology*, Vol. 181, No. 7, pp. 2 132-2 141, April.

Henderson, I.R., R. Cappello and J.P. Nataro (2000), "Autotransporter proteins, evolution and redefining protein secretion", *Trends in Microbiology*, Vol. 8, No. 12, pp. 529-532, December.

Henderson, I.R., F. Navarro-Garcia and J.P. Nataro (1998), "The great escape: Structure and function of the autotransporter proteins", *Trends in Microbiology*, Vol. 6, No. 9, pp. 370-378, September.

Henderson, B., S. Poole and M. Wilson (1996), "Bacterial modulins: A novel class of virulence factors which cause host tissue pathology by inducing cytokine synthesis", *Microbiology Reviews*, Vol. 60, No. 2, pp. 316-341, June.

Hicks, S. et al. (1998), "Role of intimin and bundle-forming pili in enteropathogenic *Escherichia coli* adhesion to pediatric intestinal tissue in vitro", *Infection and Immunity*, Vol. 66, No. 4, pp. 1 570-1 578, April.

Hingley-Wilson, S.M., V.K. Sambandamurthy and W.R. Jacobs Jr. (2003), "Survival perspectives from the world's most successful pathogen, *Mycobacterium tuberculosis*", *Nature Immunology*, Vol. 4, No. 10, pp. 949-955, October.

Hirsch, P.R. (1996), "Population dynamics of indigenous and genetically modified rhizobia in the field", *New Phytologist*, Vol. 133, Issue 1, pp. 159-171, May, *http://dx.doi.org/10.1111/j.1469-8137.1996.tb04351.x*.

Holmes, A.R. et al. (2001), "The *pavA* gene of *Streptococcus pneumoniae* encodes a fibronectin-binding protein that is essential for virulence", *Molecular Microbiology*, Vol. 41, No. 6, pp. 1 395-1 408, September.

Hughey, V.L. and E.A. Johnson (1987), "Antimicrobial activity of lysozyme against bacteria involved in food spoilage and food-borne disease", *Applied Environmental Microbiology*, Vol. 53, No. 9, pp. 2 165-2 170, September.

Hultgren, S.J. et al. (1993), "Pilus and nonpilus bacterial adhesins: Assembly and function in cell recognition", *Cell*, Vol. 73, No. 5, pp. 887-901, June.

Hynes, W.L. et al. (2000), "The extracellular hyaluronidase gene (hylA) of *Streptococcus pyogenes*", *FEMS Microbiology Letters*, Vol. 184, No. 1, pp. 109-112, March.

Ilver, D. et al. (1998), "*Helicobacter pylori* adhesin binding fucosylated histo-blood group antigens revealed by retagging", *Science*, Vol. 279, No. 5 349, pp. 373-377, January.

Imberty, A. et al. (2004), "Structures of the lectins from *Pseudomonas aeruginosa*: Insight into the molecular basis for host glycan recognition", *Microbes and Infection*, Vol. 6, Issue 2, pp. 221-228, February, *http://dx.doi.org/10.1016/j.micinf.2003.10.016*.

Isberg, R.R. and P. Barnes (2001), "Subversion of integrins by enteropathogenic *Yersinia*", *Journal of Cell Science*, Vol. 114, Pt. 1, pp. 21-28, January.

Isberg, R.R. and J.M. Leong (1990), "Multiple Beta 1 chain integrins are receptors for invasin, a protein that promotes bacterial penetration into mammalian cells", *Cell*, Vol. 60, No. 5, pp. 861-871, March.

Ishibashi, Y. et al. (2002), "Role of phosphatidylinositol 3-kinase in the binding of *Bordetella pertussis* to human monocytes", *Cell Microbiology*, Vol. 4, No. 12, pp. 825-833, December.

Ishikawa, K. et al. (2004), "Receptors for *Escherichia coli* adhesins in the genitourinary tract in a non-human primate", *Scandinavian Journal of Urology and Nephrology*, Vol. 38, No. 1, pp. 3-14.

Jarva, H. et al. (2003), "Complement resistance mechanisms of streptococci", *Molecular Immunology*, Vol. 40, Nos. 2-4, pp. 95-107, September.

Jett, B.D., M.M. Huycke and M.S. Gilmore (1994), "Virulence of enterococci", *Clinical Microbiology Reviews*, Vol. 7, No. 4, pp. 462-478, October.

Johnson, J.R. (1991), "Virulence factors in *Escherichia coli* urinary tract infection", *Clinical Microbiology Reviews*, Vol. 4, No. 1, pp. 80-128, January.

Johnson, J.R. et al. (2001), "Novel molecular variants of allele I of the *Escherichia coli* P fimbrial adhesin gene *papG*", *Infection and Immunity*, Vol. 69, No. 4, pp. 2 318-2 327, April.

Kachlany, S.C. et al. (2000), "Nonspecific adherence by *Actinobacillus actinomycetemcomitans* requires genes widespread in bacteria and archaea", *Journal of Bacteriology*, Vol. 182, No. 21, pp. 6 169-6 176, November.

Kathariou, S.A. (2002), "*Listeria monocytogenes* virulence and pathogenicity, a food safety perspective", *Journal of Food Protection*, Vol. 65, No. 11, pp. 1 811-1 829, November.

Kendall, M.M. and V. Sperandio (2007), "Quorum sensing by enteric pathogens", *Current Opinion in Gastroenterology*, Vol. 23, No. 1, pp. 10-15, January.

Kerr, J.R. (1999), "Cell adhesion molecules in the pathogenesis of and host defense against microbial infection", *Journal of Clinical Pathology: Molecular Pathology*, Vol. 52, No. 4, pp. 220-230, August.

Kingsley, R.A. et al. (2004), "The ShdA adhesin binds to the cationic cradle of the fibronectin 13FnIII repeat module: Evidence for molecular mimicry of heparin binding", *Molecular Microbiology*, Vol. 52, No. 2, pp. 345-355, April.

Kingsley, R.A. et al. (2002), "*Salmonella enterica* serotype typhimurium ShdA is an outer membrane fibronectin-binding protein that is expressed in the intestine", *Molecular Microbiology*, Vol. 43, No. 4, pp. 895-905, February.

Kjaergaard, K. et al. (2000), "Antigen 43 from *Escherichia coli* induces inter- and intraspecies cell aggregation and changes in colony morphology of *Pseudomonas fluorescens*", *Journal of Bacteriology*, Vol. 182, No. 17, pp. 4 789-4 796, September.

Klein, M.B. et al. (2000), "The agent of human granulocytic ehrlichiosis induces the production of myelosuppressing chemokines without induction of proinflammatory cytokines", *Journal of Infectious Diseases*, Vol. 182, No. 1, pp. 200-205, July.

Klemm, P. and M.A. Schembri (2000), "Bacterial adhesins: Function and structure", *International Journal of Medical Microbiology*, Vol. 290, No. 1, pp. 27-35, March.

Klose, K.E. and J.J. Mekalanos (1997), "Simultaneous prevention of glutamine synthesis and high-affinity transport attenuates *Salmonella typhimurium* virulence", *Infection and Immunity*, Vol. 65, No. 2, pp. 587-596, February.

Kohler, H., S.P. Rodrigues and B.A. McCormick (2002), "*Shigella flexneri* interactions with the basolateral membrane domain of polarized model intestinal epithelium: Role of lipopolysaccharide in cell invasion and in activation of the mitogen-activated protein kinase ERK", *Infection and Immunity*, Vol. 70, No. 3, pp. 1 150-1 158, March, *http://dx.doi.org/10.1128/IAI.70.3.1150-1158.2002*.

Kong, K.-F., C. Vuong and M. Otto (2006), "*Staphylococcus* quorum sensing in biofilm formation and infection", *International Journal of Medical Microbiology*, Vol. 296, Nos. 2-3, pp. 133-139, April.

Kreikemeyer, B. et al. (2004), "*Streptococcus pyogenes* fibronectin-binding protein F2: Expression profile, binding characteristics, and impact on eukaryotic cell interactions", *Journal of Biological Chemistry*, Vol. 279, No. 16, pp. 15 850-15 859, April.

Kuehn, M.J. and N.C. Kesty (2005), "Bacterial outer membrane vesicles and the host-pathogen interaction", *Genes Development*, Vol. 19, No. 22, pp. 2 645-2 655, November.

Kuhn-Nentwig, L., J. Schaller and W. Nentwig (2004), "Biochemistry, toxicology and ecology of the venom of the spider *Cupiennius salei* (*Ctenidae*)", *Toxicon*, Vol. 43, No. 5, pp. 543-553, April.

Kwok, T. et al. (2002), "Specific entry of *Helicobacter pylori* into cultured gastric epithelial cells via a zipper-like mechanism", *Infection and Immunity*, Vol. 70, No. 4, pp. 2 108-2 120, April, *http://dx.doi.org/10.1128/IAI.70.4.2108-2120.2002*.

Laarmann, S. et al. (2002), "The *Haemophilus influenzae* hia autotransporter harbours two adhesive pockets that reside in the passenger domain and recognize the same host cell receptor", *Molecular Microbiology*, Vol. 46, No. 3, pp. 731-743, November.

Lafontaine, E.R. et al. (2000), "The UspA1 protein and a second type of UspA2 protein mediate adherence of *Moraxella catarrhalis* to human epithelial cells in vitro", *Journal of Bacteriology*, Vol. 182, No. 5, pp. 1 364-1 373, March.

Lally, E.T. et al. (1999), "The interaction between RTX toxins and target cells", *Trends in Microbiology*, Vol. 7, No. 9, pp. 356-361, September.

Lecuit, M. et al. (1997), "Internalin of *Listeria monocytogenes* with an intact leucine-rich repeat region is sufficient to promote internalization", *Infection and Immunity*, Vol. 65, No. 12, pp. 5 309-5 319, December.

Lee, S.Y. and R.E. Klevit (2000), "The whole is not the simple sum of its parts in calmodulin from *S. cerevisiae*", *Biochemistry*, Vol. 39, No. 15, pp. 4 225-4 230, April.

Lee, J.H. et al. (2003), "Receptors for *Treponema pallidum* attachment to the surface and matrix proteins of cultured human dermal microvascular endothelial cells", *Yonsei Medical Journal*, Vol. 44, No. 3, pp. 371-378, June.

Legat, F.J., T. Griesbacher and F. Lembeck (1994), "Mediation by bradykinin of rat paw oedema induced by collagenase from *Clostridium histolyticum*", *British Journal of Pharmacology*, Vol. 112, No. 2, pp. 453-460, June.

Leppla, S.H. (1982), "Anthrax toxin edema factor: A bacterial adenylate cyclase that increases cyclic AMP concentrations in eukaryotic cells", *Proceedings of the National Academy of Sciences USA*, Vol. 79, No. 10, pp. 3 162-3 166, May.

Li, H. and D.H. Walker (1998), "rOmpA is a critical protein for the adhesion of *Rickettsia rickettsii* to host cells", *Microbial Pathogenesis*, Vol. 24, No. 5, pp. 289-298, May.

Li, H. et al. (2009), "Influence of *Pseudomonas aeruginosa* quorum sensing signal molecule N-(3 Oxododecanoyl) homoserine lactone on mast cells", *Medical Microbiology and Immunology*, Vol. 198, No. 2, pp. 113-121, May, *http://dx.doi.org/10.1007/s00430-009-0111-z*.

Li, S. et al. (2000), "Structural basis of hyaluronan degradation by *Streptococcus pneumoniae* hyaluronate lyase", *EMBO Journal*, Vol. 19, No. 6, pp. 1 228-1 240, March, *http://dx.doi.org/10.1093/emboj/19.6.1228*.

Li, J. et al. (1992), "Cloning, nucleotide sequence and heterologous expression of the protective outer-membrane protein P.68 pertactin from *Bordetella bronchiseptica*", *Journal of General Microbiology*, Vol. 138, Pt. 8, pp. 1 697-1 705, August.

Lindén, S.K. et al. (2008), "Four modes of adhesion are used during *Helicobacter pylori* binding to human mucins in the oral and gastric niches", *Helicobacter*, Vol. 13, Issue 2, *http://dx.doi.org/10.1111/j.1523-5378.2008.00587.x/pdf*.

Lindenthal, C. and E.A. Elsinghorst (1999), "Identification of a glycoprotein produced by enterotoxigenic *Escherichia coli*", *Infection and Immunity*, Vol. 67, No. 8, pp. 4 084-4 091, August.

Liu, H. et al. (2002), "Point mutants of EHEC intimin that diminish tir recognition and actin pedestal formation highlight a putative tir binding pocket", *Molecular Microbiology*, Vol. 45, No. 6, pp. 1 557-1 573, September.

Mack, D. et al. (2007), "Microbial interactions in *Staphylococcus epidermidis* biofilms", *Analytical and Bioanalytical Chemistry*, Vol. 387, No. 2, pp. 399-408, January.

Mahan, M.J., J.M. Slauch and J.J. Mekalanos (1993), "Selection of bacterial virulence genes that are specifically induced in host tissues", *Science*, Vol. 259, No. 5 095, pp. 686-688, January.

Mahenthiralingham, E., T.A. Urban and J.B. Goldberg (2005), "The multifarious, multireplicon *Burkholderia cepacia* complex", *Nature Reviews of Microbiology*, Vol. 3, No. 2, pp. 144-156, February.

Maier, T.M. et al. (2007), "Identification of *Francisella tularensis Himar1*-based transposon mutants defective for replication in macrophages", *Infection and Immunity*, Vol. 75, No. 11, pp. 5 376-5 389, November.

Marques, M.A.M. et al. (2001), "Binding of α2-laminins by pathogenic and non-pathogenic mycobacteria and adherence to Schwann cells", *Journal of Medical Microbiology*, Vol. 50, pp. 23-28, *http://dx.doi.org/10.1099/0022-1317-50-1-23*.

Martinez de Tejada, G. et al. (1995), "The outer membranes of *Brucella* spp. are resistant to bactericidal cationic peptides", *Infection and Immunity*, Vol. 63, No. 8, pp. 3 054-3 061, August.

Mattoo, S. et al. (2001), "Mechanisms of *Bordetella* pathogenesis", *Frontiers in Bioscience*, Vol. 6, pp. 168-186.

Maurelli, A.T. et al. (1998), "Black holes' and bacterial pathogenicity: A large genomic deletion that enhances the virulence of *Shigella* spp. and enteroinvasive *Escherichia coli*", *Proceedings of the National Academy of Sciences USA*, Vol. 95, No. 7, pp. 3 943-3 948.

McCaw, S.E., E.H. Liao and S.D. Gray-Owen (2004), "Engulfment of *Neisseria gonorrhoeae*: Revealing distinct processes of bacterial entry by individual carcinoembryonic antigen-related cellular adhesion molecule family receptors", *Infection and Immunity*, Vol. 72, No. 5, pp. 2 742-2 752, May.

McCrea, K.W. et al. (1997), "Identification of two minor subunits in the pilus of *Haemophilus influenzae*", *Journal of Bacteriology*, Vol. 179, No. 13, pp. 4 227-4 231, July.

McDowell, J.V. et al. (2004), "Demonstration of the involvement of outer surface protein E coiled coil structural domains and higher order structural elements in the binding of infection-induced antibody and the complement-regulatory protein, Factor H", *Journal of Immunology*, Vol. 173, No. 12, pp. 7 471-7 480, December.

McGuirk, P., C. McCann and K.H.G. Mills (2002), "Pathogen-specific T regulatory 1 cells induced in the respiratory tract by a bacterial molecule that stimulates interleukin 10 production by dendritic cells: A novel strategy for evasion of protective T helper type 1 responses by *Bordetella pertussis*", *Journal of Experimental Medecine*, Vol. 195, No. 2, pp. 221-231, January.

Menzies, B.E. and I. Kourteva (2000), "*Staphylococcus aureus* Alpha-toxin induces apoptosis in endothelial cells", *FEMS Immunology and Medical Microbiology*, Vol. 29, No. 1, pp. 39-45, September.

Merrell, D.S. et al. (2003), "Growth phase-dependent response of *Helicobacter pylori* to iron starvation", *Infection and Immunity*, Vol. 71, No. 11, pp. 6 510-6 525, November.

Merz, A.J. and M. So (2000), "Interactions of pathogenic *Neisseriae* with epithelial cell membranes", *Annual Review of Cell and Developmental Biology*, Vol. 16, pp. 423-457.

Meyer, D.H., K.P. Mintz and P.M. Fives-Taylor (1997), "Models of invasion of enteric and periodontal pathogens into epithelial cells: A comparative analysis", *Critical Reviews in Oral Biology & Medicine*, Vol. 8, No. 4, pp. 389-409.

Middleton, A.M. et al. (2000), "The role of *Mycobacterium avium* complex fibronectin attachment protein in adherence to the human respiratory mucosa", *Molecular Microbiology*, Vol. 38, No. 2, pp. 381-391, October.

Miller, M.B. and B.L. Bassler (2001), "Quorum sensing in bacteria", *Annual Reviews of Microbiology*, Vol. 55, pp. 165-199.

Mims, C., A. Nash and J. Stephen (eds.) (2001), "Attachment to and entry of microorganisms into the body", in: *Mims' Pathogenesis of Infectious Disease*, Academic Press, San Diego, California.

Miyamoto, Y. et al. (2006), "Role of Shiga toxin versus H7 flagellin in enterohaemorrhagic *Escherichia coli* signaling of human colon epithelium *in vivo*", *Cellular Microbiology*, Vol. 8, No. 5, pp. 869-879, May.

Molinari, G. et al. (2000), "Two distinct pathways for the invasion of *Streptococcus pyogenes* in non-phagocytic cells", *Cellular Microbiology*, Vol. 2, No. 2, pp. 145-154, April.

Molinari, G. et al. (1997), "The fibronectin-binding protein of *Streptococcus pyogenes*, SfbI, is involved in the internalization of group A Streptococci by epithelial cells", *Infection and Immunity*, Vol. 65, No. 4, pp. 1 357-1 363.

Moncada, D. et al. (2000), "*Entamoeba histolytica* cysteine proteinases degrade human colonic mucin and alter its function", *Archives of Medical Research*, Vol. 31, Issue 4, Supplement 1, pp. S224-225.

Morinaga, N., Y. Kaihou and M. Noda (2003), "Purification, cloning and characterization of variant LukE-LukD with strong leukocidal activity of staphylococcal bi-component leukotoxin family", *Microbiology and Immunology*, Vol. 47, No. 1, pp. 81-90, January, *http://dx.doi.org/10.1111/j.1348-0421.2003.tb02789.x*.

Mueller-Ortiz, S.L., A.R. Wanger and S.J. Norris (2001), "Mycobacterial protein HbhA binds human complement component C3", *Infection and Immunity*, Vol. 69, No. 12, pp. 7 501-7 511, December.

Nagy, B. and P.Z. Fekete (1999), "Enterotoxigenic *Escherichia coli* (ETEC) in farm animals", *Veterinary Research*, Vol. 30, Nos. 2-3, pp. 259-284, March-June.

Nakagawa, I. et al. (2006), "Invasion of epithelial cells and proteolysis of cellular focal adhesion components by distinct types of *Porphyromonas gingivalis* fimbriae", *Infection and Immunity*, Vol. 74, No. 4, pp. 3 773-3 782, July, *www.ncbi.nlm.nih.gov/pmc/articles/PMC1489697/*.

Nakagawa, I. et al. (2002), "Functional differences among FimA variants of *Porphyromonas gingivalis* and their effects on adhesion to and invasion of human epithelial cells", *Infection and Immunity*, Vol. 70, No. 1, pp. 277-285, January.

Nallapareddy, S.R., K.V. Singh and B.E. Murray (2008), "Contribution of the collagen adhesin Acm to pathogenesis of *Enterococcus faecium* in experimental endocarditis", *Infection and Immunity*, Vol. 76, No. 9, pp. 4 120-4 128, September, *http://dx.doi.org/10.1128/IAI.00376-08*.

Narayanan, S.K. et al. (2002), "Leukotoxins of gram-negative bacteria", *Veterinary Microbiology*, Vol. 84, No. 4, pp. 337-356, February.

Nealson, K.H., T. Platt and J.W. Hastings (1970), "Cellular control of the synthesis and activity of the bacterial luminescent system", *Journal of Bacteriology*, Vol. 104, No. 1, pp. 313-322, October, *www.ncbi.nlm.nih.gov/pmc/articles/PMC248216/pdf/jbacter00376-0343.pdf*.

Neiditch, M.B. et al. (2006), "Ligand-induced asymmetry in histidine sensor kinase complex regulates quorum sensing", *Cell*, Vol. 126, No. 6, pp. 1 095-1 108, September.

Ng W.L. and B.L. Bassler (2009), "Bacterial quorum-sensing network architectures", *Annual Review of Genetics*, Vol. 43, pp. 197-222, *http://dx.doi.org/10.1146/annurev-genet-102108-134304*.

Nitsche, D.P. et al. (2006), "Streptococcal protein FOG, a novel matrix adhesin interacting with collagen I *in vivo*", *Journal of Biological Chemistry*, Vol. 281, No. 3, pp. 1 670-1 679, January.

Nizet, V. (2002), "Streptococcal Beta-hemolysins: Genetics and role in disease pathogenesis", *Trends in Microbiology*, Vol. 10, No. 12, pp. 575-580, December.

Noble, W.C., Z. Virani and R.G.A. Cree (1992), "Co-transfer of vancomycin and other resistance genes from *Enterococcus faecalis* NCTC 12201 to *Staphylococcus aureus*", *FEMS Microbiology Letters*, Vol. 72, No. 2, pp. 195-198, June.

Nordmann P. and L. Poirel (2005), "Emergence of plasmid-mediated resistance to quinolones in enterobacteriaceae", *Journal of Antimicrobial Chemotherapy*, Vol. 56, No. 3, pp. 463-699, September.

Nougayrède, J.-P. et al. (2006), "*Escherichia coli* induces DNA double-strand breaks in eukaryotic cells", *Science*, Vol. 313, No. 5 788, pp. 848-851, August.

Nowicki, B., R. Selvarangan and S. Nowicki (2002), "Family of *Escherichia coli* Dr adhesins: Decay-accelerating factor receptor recognition and invasiveness", *Journal of Infectious Diseases*, Vol. 183, Supplement 1, pp. S24-S27, March.

OECD (2010), "Guidance Document on Horizontal Gene Transfer between Bacteria", *Series on Harmonisation of Regulatory Oversight in Biotechnology No. 50*, OECD, Paris, available at: *www.oecd.org/science/biotrack/46815958.pdf*.

OECD (2004), "Guidance document on methods for detection of micro-organisms introduced into the environment: Bacteria", *Series on Harmonisation of Regulatory Oversight in Biotechnology No. 30*, OECD, Paris, available at: *www.oecd.org/env/ehs/biotrack/46815788.pdf*.

OECD (1986), *Recombinant DNA Safety Considerations. Safety Considerations for Industrial, Agricultural and Environmental Applications of Organisms Derived by Recombinant DNA Techniques* ("The Blue Book"), OECD, Paris, at: *www.oecd.org/env/ehs/biotrack/Recombinant-DNA-Safety-Considerations.pdf*.

O'Neill, J.M. et al. (2003), "Invasive disease due to nontypeable *Haemophilius influenzae* among children in Arkansas", *Journal of Clinical Microbiology*, Vol. 41, No. 7, pp. 3 064-3 069, July.

Parkkinen, J. et al. (1988), "Binding sites in the rat brain for *Escherichia coli* S fimbriae associated with neonatal meningitis", *Journal of Clinical Investigations*, Vol. 81, No. 3, pp. 860-865, March.

Parsek, M.R. and E.P. Greenberg (2000), "Acyl-homoserine lactone quorum sensing in gram-negative bacteria: A signaling mechanism involved in associations with higher organisms", *Proceedings of the National Academy of the Sciences USA*, Vol. 97, No. 16, pp. 8 789-8 793, August.

Paton, J.C. and A. Ferrante (1983), "Inhibition of human polymorphonuclear leukocyte respiratory burst, bactericidal activity, and migration by pneumolysin", *Infection and Immunity*, Vol. 41, No. 3, pp. 1 212-1 216, September.

Paton, J.C. et al. (1993), "Molecular analysis of the pathogenicity of *Streptococcus pneumoniae*: The role of pneumococcal proteins", *Annual Review of Microbiology*, Vol. 47, pp. 89-115.

Poilane, I. et al. (1998), "Protease activity of *Clostridium difficile* strains", *Canadian Journal of Microbiology*, Vol. 44, No. 2, pp. 157-161, February.

Ponnuraj, K. et al. (2003), "A 'dock, lock, and latch' structural model for a staphylococcal adhesin binding to fibrinogen", *Cell*, Vol. 115, No. 2, pp. 217-228, October.

Porte, F., J.P. Liautard and S. Kohler (1999), "Early acidification of phagosomes containing *Brucella suis* is essential for intracellular survival in murine macrophages", *Infection and Immunity*, Vol. 67, No. 8, pp. 4 041-4 047, August.

Pukatzki, S., S.B McAuley and S.T. Miyata (2009), "The type VI secretion system: Translocation of effectors and effector-domains", *Current Opinion in Microbiology*, Vol. 12, No. 1, pp. 11-17, February, *http://dx.doi.org/10.1016/j.mib.2008.11.010.*

Qazi, S. et al. (2006), "N-acyl homoserine lactones antagonize virulence gene expression and quorum sensing in *Staphylococcus aureus*", *Infection and Immunity*, Vol. 74, No. 2, pp. 910-919, February.

Quinn, F.D., G.W. Newman and C.H. King (1997), "In search of virulence factors of human bacterial disease", *Trends in Microbiology*, Vol. 5, No. 1, pp. 20-26, January.

Rajkumar, R., H. Devaraj and S. Niranjali (1998), "Binding of *Shigella* to rat and human intestinal mucin", *Molecular and Cellular Biochemistry*, Vol. 178, Nos. 1-2, pp. 261-268, January.

Ram, S. et al. (2001), "Binding of C4b-binding protein to porin: A molecular mechanism of serum resistance of *Neisseria gonorrhoeae*", *Journal of Experimental Medicine*, Vol. 193, No. 3, pp. 281-296, February.

Ram, S. et al. (1998), "Binding of complement factor H to loop 5 of porin protein 1A: A molecular mechanism of serum resistance of nonsialylated *Neisseria gonorrhoeae*", *Journal of Experimental Medicine*, Vol. 188, No. 4, pp. 671-680, August.

Rathman, R., M.D. Sjaastad and S. Falkow (1996), "Acidification of phagosomes containing *Salmonella typhimurium* in murine macrophages", *Infection and Immunity*, Vol. 64, No. 7, pp. 2 765-2 773, July.

Rezcallah, M.S. et al. (2005), "Engagement of CD46 and α5β1 integrin by group A Streptococci is required for efficient invasion of epithelial cells", *Cellular Microbiology*, Vol. 7, No. 5, pp. 645-653, May, *http://dx.doi.org/10.1111/j.1462-5822.2004.00497.x.*

Ringdahl, U. et al. (1998), "Molecular co-operation between protein PAM and streptokinase for plasmin acquisitionby *Streptococcus pyogenes*", *Journal of Biological Chemistry*, Vol. 273, No. 11, pp. 6 424-6 430, March.

Roche, F.M. et al. (2004), "The N-terminal A domain of fibronectin-binding proteins A and B promotes adhesion of *Staphylococcus aureus* to elastin", *Journal of Biological Chemistry*, Vol. 279, No. 37, pp. 38 433-38 440, September.

Rogolsky, M. (1979), "Nonenteric toxins of *Staphylococcus aureus*", *Microbiology Reviews*, Vol. 43, No. 3, pp. 320-60, September.

Rood, J.I. (1998), "Virulence genes of *Clostridium perfringens*", *Annual Reviews of Microbiology*, Vol. 52, pp. 333-360.

Rubens, C.E. et al. (1987), "Transposon mutagenesis of type III group B *Streptococcus*: Correlation of capsule expression with virulence", *Proceedings of the National Academy of Sciences USA*, Vol. 84, No. 20, pp. 7 208-7 212, October.

Sajjan, U.S. et al. (1995), "Cable (cbl) type II pili of cystic fibrosis-associated *Burkholderia (Pseudomonas) cepacia*: Nucleotide sequence of the cblA major subunit pilin gene and novel morphology of the assembled appendage fibers", *Journal of Bacteriology*, Vol. 177, No. 4, pp. 1 030-1 038, February.

Salyers, A.A. and D.D. Whitt (eds.) (2002), *Bacterial Pathogenesis, a Molecular Approach*, 2nd edition, pp. 118-119, ASM Press, Washington, DC.

Samrakandi, M.M. et al. (2002), "Entry into host cells by *Legionella*", *Frontiers in Bioscience*, Vol. 7, pp. d1-11, January.

Santi, I. et al. (2007), "BibA: A novel immunogenic bacterial adhesin contributing to group B *Streptococcus* survival in human blood", *Molecular Microbiology*, Vol. 63, No. 3, pp. 754-767, February.

Saren, A. et al. (1999), "The cellular form of human fibronectin as an adhesion target for the S fimbriae of meningitis-associated *Escherichia coli*", *Infection and Immunity*, Vol. 67, No. 5, pp. 2 671-2 676, May.

Sasmal, D. et al. (2002), "N-acetyl-D-glucosamine specific hemagglutinin receptor of *Vibrio cholerae* O1 in chicken erythrocyte membranes", *FEMS Immunology and Medical Microbiology*, Vol. 32, No. 3, pp. 187-189, February.

Schmidt, H. and M. Hensel (2004), "Pathogenicity islands in bacterial pathogenesis", *Clinical Microbiology Reviews*, Vol. 17, No. 1, pp. 14-56, January.

Schneider, M.C. et al. (2006), "Functional significance of factor H binding to *Neisseria meningitidis*", *Journal of Immunology*, Vol. 176, No. 12, pp. 7 566-7 575, June.

Schnupf, P. and D.A. Portnoy (2007), "Listeriolysin O: A phagosome-specific lysine", *Microbes and Infection*, Vol. 9, No. 10, pp. 1 176-1 187, August.

Schorey, J.S. et al. (1996), "Characterization of the fibronectin-attachment protein of *Mycobacterium avium* reveals a fibronectin-binding motif conserved among mycobacteria", *Molecular Microbiology*, Vol. 21, No. 2, pp. 321-329, July.

Schulte, R. et al. (2000), "Translocation of *Yersinia entercolitica* across reconstituted intestinal epithelial monolayers is triggered by *Yersinia* invasin binding to β_1 integrins apically expressed on M-like cells", *Cellular Microbiology*, Vol. 2, No. 2, pp. 173-185, April, *http://dx.doi.org/10.1046/j.1462-5822.2000.00047.x*.

Schwan, W.R. et al. (2002), "Osmolarity and pH growth conditions regulate Fim gene transcription and type 1 pilus expression in uropathogenic *Escherichia coli*", *Infection and Immunity*, Vol. 70, No. 3, pp. 1 391-1 402, March.

Seto, S. and M. Miyata (2003), "Attachment organelle formation represented by localization of cytadherence proteins and formation of the electron-dense core in wild-type and mutant strains of *Mycoplasma pneumoniae*", *Journal of Bacteriology*, Vol. 185, No. 3, pp. 1 082-1 091, February, *http://dx.doi.org/10.1128/JB.185.3.1082-1091.2003*.

Seto, S. et al. (2001), "Visualization of the attachment organelle and cytadherence proteins of *Mycoplasma pneumoniae* by immunofluorescence microscopy", *Journal of Bacteriology*, Vol. 183, No. 5, pp. 1 621-1 630, March.

Shabtai, Y. and G. Fleminger (1994), "Adsorption of *Rhodococcus* strain GIN-1 (NCIMB 40340) on titanium dioxide and coal fly ash particles", *Applied Environmental Microbiology*, Vol. 60, No. 9, pp. 3 079-3 088, September.

Shiner, E.K., K.P. Rumbaugh and S.C. Williams (2005), "Interkingdom signaling: Deciphering the language of acyl homoserine lactones", *FEMS Microbiology Reviews*, Vol. 29, pp. 935-947.

Shuter, J., V.B. Hatcher and F.D. Lowy (1996), "*Staphylococcus aureus* binding to human nasal mucin", *Infection and Immunity*, Vol. 64, No. 1, pp. 310-318, January.

Sierig, G. et al. (2003), "Cytotoxic effects of Streptolysin O and Streptolysin S enhance the virulence of poorly encapsulated group A Streptococci", *Infection and Immunity*, Vol. 71, No. 1, pp. 446-455, January, *http://dx.doi.org/10.1128/IAI.71.1.446-455.2003.*

Smith, R.L. and M.E. Maguire (1998), "Microbial magnesium transport: Unusual transporters searching for identity", *Molecular Microbiology*, Vol. 28, No. 2, pp. 217-226, April.

Smith, D. et al. (2006), "Variations on a theme: Diverse N-acyl homoserine lactone-mediated quorum sensing mechanisms in gram-negative bacteria", *Science Progress*, Vol. 89, Pt. 3-4, pp. 167-211.

Soto, G.E. and S.J. Hultgren (1999), "Bacterial adhesins: Common themes and variations in architecture and assembly", *Journal of Bacteriology*, Vol. 181, No. 4, pp. 1 059-1 071, February.

St. Geme III, J.W. (2002), "Molecular and cellular determinants of non-typeable *Haemophilus influenzae* adherence and invasion", *Cellular Microbiology*, Vol. 4, No. 4, pp. 191-200, April.

St. Geme III, J.W., D. Cutter and S.J. Barenkamp (1996), "Characterization of the genetic locus encoding *Haemophilus influenzae* type b surface fibrils", *Journal of Bacteriology*, Vol. 178, No. 21, pp. 6 281-6 287, November.

St. Geme III, J.W., M.L. de la Morena and S. Falkow (1994), "A *Haemophilus influenzae* IgA protease-like protein promotes intimate interaction with human epithelial cells", *Molecular Microbiology*, Vol. 14, No. 2, pp. 217-233, October.

Stathopoulos, C. et al. (2000), "Secretion of virulence determinants by the general secretory pathway in gram-negative pathogens: An evolving story", *Microbes and Infection*, Vol. 2, No. 9, pp. 1 061-1 072, July.

Stewart-Tull, D.E., C. Lucas and C.R. Bleakley (2004), "Experimental immunisation and protection of guinea pigs with *Vibrio cholerae* toxoid and mucinases, neuraminidase and proteinase", *Vaccine*, Vol. 22, Nos. 17-18, pp. 2 137-2 145, June.

Stratton, H.M. et al. (2002), "Cell surface hydrophobicity and mycolic acid composition of *Rhodococcus* strains isolated from activated sludge foam", *Journal of Industrial Microbiology & Biotechnology*, Vol. 28, No. 5, pp. 264-267, May.

Strauss, E.J. and S. Falkow (1997), "Microbial pathogenesis: Genomics and beyond", *Science*, Vol. 276, No. 5 313, pp. 707-711, May.

Strauss, E.J., N. Ghori and S. Falkow (1997), "An *Edwardsiella tarda* strain containing a mutation in a gene with homology to shlB and hpmB is defective for entry into epithelial cells in culture", *Infection and Immunity*, Vol. 65, No. 9, pp. 3 924-3 932, September.

Strom, M.S. and S. Lory (1993), "Structure-function and biogenesis of the type IV pili", *Annual Review of Microbiology*, Vol. 47, pp. 565-596.

Tan, T.T., A. Forsgren and K. Riesbeck (2006), "The respiratory pathogen *Moraxella catarrhalis* binds to laminin via ubiquitous surface proteins A1 and A2", *Journal of Infectious Diseases*, Vol. 194, No. 4, pp. 493-497, August.

Thankavel, K. et al. (1999), "Molecular basis for the enterocyte tropism exhibited by *Salmonella typhimurium* type 1 fimbriae", *Journal of Biological Chemistry*, Vol. 274, No. 9, pp. 5 797-5 809, February.

Touze, T. et al. (2004), "Self-association of EPEC intimin mediated by the Beta-barrel-containing anchor domain: A role in clustering of the tir receptor", *Molecular Microbiology*, Vol. 51, No. 1, pp. 73-87, January.

Towers, R.J. et al. (2003), "Evolution of *sfbI* encoding streptococcal fibronectin-binding protein I: Horizontal genetic transfer and gene mosaic structure", *Journal of Clinical Microbiology*, Vol. 41, No. 12, pp. 5 398-5 406, *http://dx.doi.org/10.1128/JCM.41.12.5398-5406.2003.*

Toyooka, K., S. Takai and T. Kirikae (2005), "*Rhodococcus equi* can survive a phagolysosomal environment in macrophages by suppressing acidification of the phagolysosome", *Journal of Medical Microbiology*, Vol. 54, Pt. 11, pp. 1 007-1 015, November.

Tribble, G.D. et al. (2006), "A *Porphyromonas gingivalis* haloacid dehalogenase family phosphatase interacts with human phosphoproteins and is important for invasion", *Proceedings of the National Academy of Sciences USA*, Vol. 103, No. 29, pp. 11 027-11 032, July.

Tsukano, H. et al. (1999), "*Yersinia pseudotuberculosis* blocks the phagosomal acidification of B10.A mouse macrophages through the inhibition of vacuolar H(+)-ATPase activity", *Microbial Pathogenesis*, Vol. 27, No. 4, pp. 253-263, October.

Tu, A.T. et al. (1999), "Pneumococcal surface protein A inhibits complement activation by *Streptococcus pneumonia*", *Infection and Immunity*, Vol. 67, No. 9, pp. 4 720-4 724, September.

Tullus, K. et al. (1992), "Host factors versus virulence-associated bacterial characteristics in neonatal and infantile bacteraemia and meningitis caused by *Escherichia coli*", *Journal of Medical Microbiology*, Vol. 36, No. 3, pp. 203-208, March.

Ukuku, D.O. and W.F. Fett (2002), "Relationship of cell surface charge and hydrophobicity to strength of attachment of bacteria to cantaloupe rind", *Journal of Food Protection*, Vol. 65, No. 7, pp. 1 093-1 099, July.

Uzzau, S. et al. (2005), "Virulence attenuation and live vaccine potential of *aroA*, *crp cdt cya*, and plasmid-cured mutants of *Salmonella enterica* serovar Abortusovis in mice and sheep", *Infection and Immunity*, Vol. 73, No. 7, pp. 4 302-4 308, July.

van Alphen, L. et al. (1991), "Blocking of fimbria-mediated adherence of *Haemophilus influenzae* by sialyl gangliosides", *Infection and Immunity*, Vol. 59, No. 12, pp. 4 473-4 477, December.

Van Belkum, A. et al. (1998), "Short sequence DNA repeats in prokaryotic genomes", *Microbiology and Molecular Biology Reviews*, Vol. 62, No. 2, pp. 275-293, June.

Van den Broeck, W. et al. (2000), "The F4 fimbrial antigen of *Escherichia coli* and its receptors", *Veterinary Microbiology*, Vol. 71, Nos. 3-4, pp. 223-244, February, *http://dx.doi.org/10.1016/S0378-1135(99)00174-1.*

Van der Goot, F.G. et al. (2004), "Rafts can trigger contact-mediated secretion of bacterial effectors via a lipid-based mechanism", *Journal of Biological Chemistry*, Vol. 279, No. 46, pp. 47 792-47 798, November.

Vasil, M.L. (1986), "*Pseudomonas aeruginosa*: Biology, mechanisms of virulence, epidemiology", *Journal of Pediatrics*, Vol. 108, No. 5, Pt. 2, pp. 800-805, May.

Veiga, E., V. de Lorenzo and L.A. Fernandez (2003), "Autotransporters as scaffolds for novel bacterial adhesins: Surface properties of *Escherichia coli* cells displaying Jun/Fos dimerization domains", *Journal of Bacteriology*, Vol. 185, No. 18, pp. 5 585-5 590, September.

Velasco-Velázquez, M.A. et al. (2003), "Macrophage-mycobacterium tuberculosis interactions: Role of complement receptor 3", *Microbial Pathogenesis*, Vol. 35, No. 3, pp. 125-131, September.

Wagner, P.L. and M.K. Waldor (2002), "Bacteriophage control of bacterial virulence", *Infection and Immunity*, Vol. 70, No. 8, pp. 3 985-3 993, August.

Wagner, V.E. et al. (2007), "Analysis of the hierarchy of quorum-sensing regulation in *Pseudomonas aeruginosa*", *Analytical and Bioanalytical Chemistry*, Vol. 387, No. 2, pp. 469-479, January.

Ward, A.C. et al. (2002), "Characterization of *Pasteurella* spp isolated from healthy domestic pack goats and evaluation of the effects of a commercial *Pasteurella* vaccine", *American Journal of Veterinary Research*, Vol. 63, No. 1, pp. 119-123, January.

Wassenaar, T.M. and W. Gaastra (2001), "Bacterial virulence: Can we draw the line", *FEMS Micrbiology Letters*, Vol. 201, No. 1, pp. 1-7, July.

Watnick, P. and R. Kolter (2000), "Biofilm, city of microbes", *Journal of Bacteriology*, Vol. 182, No. 10, pp. 2 675-2 679, May.

Wei, L. et al. (2005), "Impact of the SpeB protease on binding of the complement regulatory proteins factor H and factor H-like protein 1 by *Streptococcus pyogenes*", *Infection and Immunity*, Vol. 73, No. 4, pp. 2 040-2 050, April.

Weller, D.M. (2007), "*Pseudomonas* biocontrol agents of soilborne pathogens: Looking back over 30 years", *Phytopathology*, Vol. 97, No. 2, pp. 250-256, *http://dx.doi.org/10.1094/PHYTO-97-2-0250*.

Wexler, D.E., D.E. Chenoweth and P.P. Cleary (1985), "Mechanism of action of the group A streptococcal C5a inactivator", *Proceedings of the National Academy of Sciences USA*, Vol. 82, No. 23, pp. 8 144-8 148, December.

WHO (2004), *Laboratory Biosafety Manual*, Third edition, World Health Organization, Geneva, available at: *www.who.int/csr/resources/publications/biosafety/Biosafety7.pdf*.

Wiggins, R. et al. (2001), "Mucinases and sialidases: Their role in the pathogenesis of sexually transmitted infections in the female genital tract", *Sexually Transmitted Infections*, Vol. 77, No. 6, pp. 402-408, December.

Williams, P. (2007), "Quorum sensing, communication and cross-kingdom signaling in the bacterial world", *Microbiology*, Vol. 153, Pt. 12, pp. 3 923-3 928, December.

Wilson, J.W. et al. (2002), "Mechanisms of bacterial pathogenicity", *Postgraduate Medical Journal*, Vol. 78, No. 918, pp. 216-224, April, *http://dx.doi.org/10.1136/pmj.78.918.216*.

Winther-Larsen, H.C. et al. (2001), "*Neisseria gonorrhoeae* PilV, a type IV pilus-associated protein essential to human epithelial cell adherence", *Proceedings of the National Academy of Sciences USA*, Vol. 98, No. 26, pp. 15 276-15 281, December.

Winzer, K. et al. (2000), "The *Pseudomonas aeruginosa* lectins PA-IL and PA-IIL are controlled by quorum sensing and by RpoS", *Journal of Bacteriology*, Vol. 182, No. 22, pp. 6 401-6 411, November.

Wolfgang, M. et al. (2000), "Components and dynamics of fiber formation define a ubiquitous biogenesis pathway for bacterial pili", *EMBO Journal*, Vol. 19, No. 23, pp. 6 408-6 418, December, *http://dx.doi.org/10.1093/emboj/19.23.6408*.

Wu, J. and J. Xie (2009), "Magic spot: (p) ppGpp", *Journal of Cellular Physiology*, Vol. 220, No. 2, pp. 297-302, August, *http://dx.doi.org/10.1002/jcp.21797*.

Yarwood, J.M. and P.M. Schlievert (2003), "Quorum sensing in *Staphylococcus* infections", *Journal of Clinical Investigations*, Vol. 112, No. 11, pp. 1 620-1 625, December, *http://dx.doi.org/10.1172/JCI200320442*.

Yeo, H.J. et al. (2004), "Structural basis for host recognition by the *Haemophilus influenzae* hia autotransporter", *EMBO Journal*, Vol. 23, No. 6, pp. 1 245-1 256, March.

Yoshiie, K. et al. (2000), "Intracellular infection by the human granulocytic ehrlichiosis agent inhibits human neutrophil apoptosis", *Infection and Immunity*, Vol. 68, No. 3, pp. 1 125-1 133, March.

Zecchinon, L., T.T. Fett and D. Desmecht (2005), "How *Mannheimia haemolytica* defeats host defense through a kiss of death mechanism", *Veterinary Research*, Vol. 36, No. 2, pp. 133-156, March-April.

Part II:

Biology of crops

Chapter 2.

Squashes, pumpkins, zucchinis and gourds (*Curcurbita* species)

*This chapter deals with the biology of squashes, pumkins, zucchinis and gourds (*Cucurbita *species). It contains information for use during the risk/safety regulatory assessment of genetically engineered varieties intended to be grown in the environment (biosafety). It includes elements of taxonomy, centres of origin and distribution, morphological characters, reproductive biology, genetics, hybridisation and introgression, crop production and cultivation practices, interactions with other organisms, pests and pathogens, and biotechnological developments.*

The chapter was prepared by the OECD Working Group on the Harmonisation of Regulatory Oversight in Biotechnology, with Mexico as the lead country and the collaboration of the United States. It was initially issued in August 2012. Data from FAOSTAT have been updated.

Species or taxonomic group

The genus *Cucurbita* is well known for its cultivated species which in English are called by the general names of squashes, pumpkins, marrows and gourds, and in Spanish are commonly known as *calabazas* and *zapallos* or in some areas of the Americas by indigenous names. Other names have been given to *Cucurbita* species in countries where these species have been adopted into local agriculture, e.g. in the French language, *citrouille*, *courgette* and *potiron*. The cultivated *Cucurbita* species are: *C. argyrosperma*, which includes the cushaw type (long curved neck) of squash; *C. maxima*, which includes pumpkins and hubbard, turban and buttercup squash; *C. moschata*, which includes the winter squashes; and *C. pepo*, which includes both summer squash (e.g. zucchini, scallop, scallopini, crookneck and cocozelle squash) and winter squash (e.g. the common or "true" pumpkin, delicata, acorn and spaghetti squash) as well as ornamental gourds. *C. ficifolia* includes the fig leaf gourd and lacayote. The most important of these species in terms of agricultural production worldwide are *C. maxima*, *C. moschata* and *C. pepo*.

The genus *Cucurbita* is exclusively native to the Americas. Prior to the European contact (i.e. prior to 1492 of the Common Era), various *Cucurbita* were a mainstay of traditional Native American agriculture. In the Americas, *Cucurbita* have traditionally been one of the "three sisters" of native agriculture together with beans (*Phaseolus vulgaris*) and maize (*Zea maize*). The three crops were grown together, with the maize providing support for the climbing beans, and shade for the *Cucurbita*. The *Cucurbita* provide ground cover to limit weeds and keep the soil moist. The beans fix nitrogen for all three crops. Subsequent to the European contact, *Cucurbita* species were disseminated worldwide and cultivated on other continents (Paris, 1989). Today, *Cucurbita* species are cultivated worldwide, and are an important food source for much of the world's population. In 2013, Asia region (primarily the People's Republic of China, and India) was the largest producer of squashes, pumpkins and gourds with more than 65% of the total hectares planted to *Cucurbita* species worldwide, followed by Africa with 15% of the global acreage, then North and Central America and the Caribbean with 8%, Europe with 7.5%, South America with 3% and Oceania with 0.9 % (FAOSTAT, 2013).

Domesticated species are cultivated in large production areas for processing and the fresh market, as well as in home gardens. Cultivated *Cucurbita* species are grown primarily for their seeds, fruits and gourds. *Cucurbita* seeds can be eaten directly, ground into paste, meal, "nut" butter or fine flour. The seeds are also sources of oils and proteins (Robinson and Decker-Waters, 1997). After the hull is removed, the seeds contain about 50% oil and up to 35% protein. Most of the oil is made up of non-saturated fatty acids, and thus is of high nutritional value. Conjugated fatty acids among some *Cucurbita* oils make them highly useful as drying oils, as they combine readily with oxygen to form an elastic, waterproof film. Mature and immature *Cucurbita* fruits are edible and eaten as a vegetable. The flesh of some fruits (e.g. *C. maxima*) can be dried, ground into a powder and used with cereals to make bread, cakes, etc. (Chiej, 1984; Facciola, 1990; See, Wan Nadiah and Noor Aziah, 2007). The fruits of one *Cucurbita* (*C. ficifolia*) are used to make confections and beverages, some alcoholic. In addition, in many places in Latin America and elsewhere in the world, flowers and some vegetative parts (e.g. tender stem tips, leaves, tendrils) are also eaten as vegetables (Nee, 1990; Merrick, 1991; Lira, Andres and Nee, 1995). The vines and fruits can also be used as fodder for domesticated animals (Mariano and Dirzo, 2002). The gourds can be used for ornamental purposes or, as has been the case historically, as containers. Few people grow *Cucurbita* flowers for decorative purposes as a flower lasts only one morning. Only one species (*C. maxima*) is known to have very aromatic flowers (Lira, Andres and Nee, 1995). Several *Cucurbita* species are used in traditional medicine; as an anthelmintic (Chou and Huangfu, 1960; Schabort, 1978; Chiej, 1984; Argueta, 1994); to treat benign prostatic hypertrophy (Duke and Ayensu, 1985; Kreuter, 2000; Gossell-Williams, Davis and O'Connor, 2006); as a diuretic (Chiej, 1984); as a nerve tonic and to sooth burns, inflammations and boils (Chopra, Nayar and Chopra, 1956); and as an antihyperglycemic agent (Andrade-Cetto and Heinrich, 2005; Xia and Wang, 2006). Lira, Andres and Nee (1995) discuss several of these uses.

Cultivated species of *Cucurbita* are very diverse locally, regionally and worldwide, having a wealth of innumerable strains, landraces and varieties (e.g. Esquinas-Alcazar and Gulick, 1983; Jeffrey, 1990; Nee, 1990; Lira, Andres and Nee, 1995; Sanjur et al., 2002). It should be noted that commonly used descriptors such as "pumpkin" or "squash" apply broadly to morphotypes across all of the cultivated *Cucurbita*, as does the term "gourd". Thus, the term "pumpkin" is applied to certain fruits of *C. maxima* as well as certain fruits of *C. moschata*, *C. argyrosperma* and *C. pepo*; and the term "squash" is applied to certain fruits of *C. pepo* as well as certain fruits of *C. maxima*, *C. moschata* and *C. argyrosperma*. Similarly, in English the word "pumpkin" or "squash" as used colloquially could be describing a variety, a subspecies, a species or, at the extreme, a genus. In order to avoid confusion, this chapter will rely on taxonomic identifiers to the extent possible.

Taxonomy

The genus *Cucurbita*, which is native to the Americas, is in the order Cucurbitales, the family Cucurbitaceae, the subfamily Cucurbitoideae and the tribe Cucurbiteae (Jeffrey, 1990). The *Cucurbita* are not closely related to other Cucurbitaceae genera.

Cucurbita traditionally has been considered a distinct genus of 20-27 species (Bailey, 1943, 1948; Cutler and Whitaker, 1961; Esquinas-Alcazar and Gulick, 1983), which, due to the ecological characteristics of their habitats and duration of their life cycles, are divided into two large groups: the xerophytic species, perennials adapted to dry climates with tuberous storage roots; and the species adapted to moister or mesophytic environments, either annuals or short-lived perennials with fibrous roots. Within this second group are the five cultivated species, whose amazing morphological variation – especially of fruits and seeds – has led to multiple names and classification schemes. Crossability experiments have more recently revealed that some of the species considered separate members of the 27-member cohort were merely from different wild populations. Nee (1990) recognises 12 or 13 species. Lira, Andres and Nee (1995) recognise 15 species and 20 taxa.

The *Cucurbita* can be categorised into seven groups (Table 2.1): the *Argyrosperma* group, the *Maxima* group, the *Pepo* group, the *Okeechobeensis* group, the *Digitata* group, the *Foetidissima* group and a seventh group, those species with no defined group (i.e. *C. ecuadorensis*, *C. ficifolia*, *C. lundelliana* and *C. moschata*). The five cultivated *Cucurbita* species are listed amongst these seven groups, and highlighted in Table 2.1 in bold.

The cultivated species *C. argyrosperma*, *C. maxima*, *C. moschata* and *C. pepo* each are comprised of a large series of convarieties, varieties and forms, including numerous local races (landraces) or varieties, and commercial cultivars. At least three of the cultivated species, *C. argyrosperma*, *C. maxima* and *C. pepo* are known to interbreed naturally with wild relatives, and the appropriate taxonomic assignments for these wild relatives *vis-à-vis* the cultivated species is still in flux as new information becomes available. In some instances the wild relative has been classified as a distinct species, in others it has been classified as a subspecies of the cultivated species. Annex 2.A1 provides a listing of the various names associated with 13 of the species. The taxonomic relationships of two *Cucurbita* species, *C. argyrosperma* and *C. pepo*, have been studied more intensively, and these studies have provided a much clearer perspective of the relationships within these cultivated species, and between the cultivated species and their wild relatives. The other three domesticated species have been less studied in terms of relationships within the species and between the cultivated species and their wild relatives.

Morphological, ecogeographical, archaeological and biosystematic evidence, such as those derived from studies on hybridisation and molecular biology (e.g. Decker, 1988, 1986; Andres, 1990, 1987a, 1987b, 1987c; Decker-Walters et al., 2002, 1993; Decker and Wilson, 1987; Kirkpatrick and Wilson, 1988; Wilson, 1990, 1989; Merrick, 1990; Rodríguez and Lira, 1992; Wilson, Doebley and Duvall, 1992; Wilson, Lira and Rodríguez, 1994; Sanjur et al., 2002), as well as field observations, contribute to the development of the currently recognised limits and relationships of the five cultivated species. Table 2.1 and Annex 2.A1 are syntheses of information from various sources describing the most commonly referred to *Cucurbita* taxa, both wild and domesticated.

Table 2.1. **List of taxa in the genus *Cucurbita* and their natural distributions**

Cucurbita group	Taxa	Natural distribution
Argyrosperma group	**C. argyrosperma Huber ssp. argyrosperma**	Southwestern United States, Mexico, Mesoamerica
	C. argyrosperma ssp. *sororia* (L.H. Bailey) Merrick & Bates	Mesoamerica (Pacific coast from Mexico to Nicaragua)
Maxima group	**C. maxima Duchesne ex Lam. ssp. maxima**	South America (Argentina, Bolivia , Chile)
	C. maxima ssp. *andreana* (Naudin) I.A. Filov	South America (Argentina, Bolivia, Uruguay)
Pepo group	**C. pepo L.**	Northern Mexico, southern and central United States
	C. pepo ssp. *fraterna* (L.H. Bailey) Andres	Northeastern Mexico
	C. pepo ssp. *texana* (Scheele) I.A. Filov	Central, southcentral, southeastern United States
Okeechobeensis group	*C. okeechobeensis* (J.K. Small) L.H. Bailey ssp. *okeechobeensis*	Southern United States (Florida)
	C. okeechobeensis ssp. *martinezii* (L.H. Bailey) Walters & Decker-Walters	Mexico
Digitata group	*C. digitata* A. Gray *	Mexico, United States
	C. cordata S. Watson *	Mexico
	C. palmata S. Watson *	Mexico, United States
Foetidissima group	*C. foetidissima* H.B.K. *	Mexico, United States
	C. pedatifolia L.H. Bailey *	Central Mexico
	C. scabridifolia L.H. Bailey *	Northeastern Mexico
	C. radicans Naudin *	Mexico
Species with no defined group	*C. ecuadorensis* Cutler & Whitaker	Ecuador (Pacific coast)
	C. lundelliana L.H. Bailey	Belize, Costa Rica, El Salvador, Guatemala, Honduras, Mexico, Nicaragua
	C. ficifolia Bouché	Mexico to Argentina and Chile
	C. moschata (Duchesne ex Lam.) Duchesne ex Poiret	Mesoamerica and South America

Note: Names in bold correspond to domesticated taxa; those with an asterisk (*) correspond to perennial species. Names used preferentially in this table and in the text are based on the nomenclature used by Lira, Andres and Nee (1995); Lira Saade, Eguiarte Fruns and Montes Hernández(2009); and Gong et al. (2012).

Cucurbita argyrosperma

A number of researchers have investigated *Cucurbita argyrosperma*, which was formerly known as *C. mixta* Pang (e.g. Nabhan, 1984; Merrick and Nabhan, 1985; Merrick and Bates, 1989; Merrick, 1991, 1990). *C. argyrosperma* is a collection of interfertile domesticated, feral and wild plants. Based on morphological variation, geographical distribution and allozyme variation, Merrick and Bates (1989) and Merrick (1990) have divided *C. argyrosperma* into two subspecies: 1) ssp. *argyrosperma*, with four varieties, three of which (var. *argyrosperma*, var. *callicarpa*, var. *stenosperma*) encompass all the cultivated types of the species, and one that encompasses the wild populations from northern Mexico (var. *palmeri*); and 2) ssp. *sororia*, which is considered the evolutionarily ancestral wild stock (based on its ecogeographical distribution, morphological similarity and reproductive compatibility) and is comprised of the wild populations from Mexico through Central America. Hybridisation studies (Merrick, 1990) and field data (Nabhan, 1984; Merrick and Nabhan, 1985; Decker, 1986; Lira, 1991) have revealed that all five of these taxa are completely interfertile and thus belong to the same biological species.

Spontaneous hybrids between the wild and cultivated plants of the group have been documented for some regions of Mexico in populations quite separated from each other, as reported by Decker (1986), between plants from various taxa of ssp. *argyrosperma* in the state of Jalisco. Moreover, the characteristics of cultivated varieties do not differ much from those that can be found in the wild plants of the group, nor from those of spontaneous hybrids between the wild and cultivated plants of the group. It is therefore not easy to readily distinguish between the various subspecies and varieties. For these reasons, this chapter follows the general convention and recognises only the two subspecies, placing all the cultivated types within ssp. *argyrosperma*, and the wild or spontaneous plants within ssp. *sororia* (as in Table 2.1).

Cucurbita pepo

Cucurbita pepo is probably the most well-studied species of the genus. Systematic, ethnobotanical and morphometric research, together with archaeological information, constitute the main sources of information concerning its origin and domestication. Similar to *C. argyrosperma, C. pepo* is a collection of interfertile domesticated, feral and wild plants.

Morphometric and molecular research, as well as studies on artificial and spontaneous hybridisation, indicate that the wild taxa most closely related to *C. pepo* L. are: 1) *C. pepo* ssp. *fraterna* (also known as *C. fraterna*), only found currently in a few localities in northeastern Mexico (in the states of Tamaulipas and Nuevo León; Bailey, 1943; Andres, 1987a; Nee, 1990; Rodríguez and Lira, 1992; Wilson, Lira and Rodríguez, 1994); and 2) two wild *Cucurbita* originally identified as *C. texana* (identified now as *C. pepo* ssp. *ovifera* var. *texana* and *C. pepo* ssp. *ovifera* var. *ozarkana*), which are endemic to parts of the United States (Gray, 1850; Bailey, 1943; Fursa and Filov, 1982; Decker, 1988, 1986; Andres, 1987a; Decker and Wilson, 1987; Kirpatrick and Wilson, 1988; Wilson, 1990, 1989; Decker-Walters et al., 1990; Wilson, Doebley and Duvall, 1992; Sanjur et al., 2002).

There is still some disagreement concerning the appropriate taxonomic relationship of the members of the *Pepo* group, with no fewer than three different reclassifications suggested, including those from Decker (1988, 1986); Andres (1987a); Decker-Walters et al. (1993, 1990). The first proposal by Decker (1988, 1986) recognised two subspecies: 1) ssp. *pepo*, composed of two varieties: var. *pepo*, in which all edible cultivars are included, and var. *fraterna*, corresponding to the wild ancestor of this part of the group; and 2) ssp. *ovifera*, also with two varieties: var. *ovifera*, which includes cultivars used as decoration, and var. *texana*. The second classification (Andres, 1987a) is simpler and proposes that *C. pepo* is only constituted by three subspecies: ssp. *pepo*, which includes all edible and ornamental cultivated types, and the subspecies *texana* and *fraterna*, in which the wild ancestors of the group are located. The third proposal (Decker-Walters et al., 1993) is a modification of the first, as it has three subspecies: ssp. *pepo*, which includes local races and commercial cultivars, ssp. *fraterna* and ssp. *ovifera*. Under this proposal, subspecies *ovifera* is composed of three varieties: 1) var. *ozarkana* (wild plants in the states of Arkansas, Illinois, Louisiana, Missouri and Oklahoma in the United States), 2) var. *texana* (wild plants in the states of Louisiana, Mississippi, New Mexico and Texas in the United States), and 3) var. *ovifera* (mainly ornamental cultivars). For simplicity, this chapter, in developing the *C. pepo* grouping in Table 2.1, follows the classification proposed by Andres (1987a), but recognises that the classification is still in flux as new molecular information is developed, as discussed below.

Work by Wilson, Doebley and Duvall in 1992 using restriction fragment length polymorphism analysis on 15 species in the genus supports the separation of cultivars of *C. pepo* into two distinct lineages, *C. pepo* ssp. *pepo* and *C. pepo* ssp. *ovifera*, as do isozyme studies of these taxa (Decker-Walters et al., 1993; Jobst, King and Hemleben, 1998). Analysis of the sequence of an intron of the mitochondrial *nad1* gene has also been used to elucidate the relationship between the various members of the *Pepo* group (Sanjur et al., 2002), and this analysis also suggests that *C. pepo* can be subdivided into two subspecies: *C. pepo* ssp. *pepo* and *C. pepo* ssp. *ovifera*.

Smith (2006) agrees that *C. pepo* is comprised of two subspecies: *C. pepo* ssp. *pepo* and *C. pepo* ssp. *ovifera*. *C. pepo* ssp. *pepo* includes pumpkin, zucchini and other marrow squashes, Mexican landraces and a few ornamental gourds. *C. pepo* ssp. *ovifera* comprises both domesticated and free-living populations, and is further divided into three taxa: *C. pepo* ssp. *ovifera* var. *ovifera*, which includes some cultivars (e.g. acorn, crookneck and scallop squash and some pumpkin) and most ornamental gourds; and the free-living populations in the United States, which represent two molecularly distinct populations; *C. pepo* ssp. *ovifera* var. *texana* and *C. pepo* ssp. *ovifera* var. *ozarkana* (Decker-Walters et al., 1993, 2002).

The analysis by Sanjur et al. (2002) of an intron of the mitochondrial *nad1* gene shows that *C. fraterna*, *C. pepo* var. *texana*, *C. pepo* var. *ozarkana* and cultivated *C. pepo* ssp. *ovifera* form a closely related clade, with any of the three wild species a potential progenitor of the domesticated species. The *C. pepo* ssp. *pepo* lineage is separated from the *ovifera* clade on the basis of a three base pair difference in an intron of the gene *nad1* of the mitochondrial DNA, a finding which supports the hypothesis that *C. pepo* ssp. *pepo* and *C. pepo* ssp. *ovifera* arose from two separate domestication events.

All of the *C. pepo* subspecies and variants can successfully hybridise with each other, suggesting that the *C. pepo* progenitors for both subspecies *pepo* and *ovifera* were once part of an extended contiguous population reaching from Mexico through the eastern United States. Whether this extended range occurred naturally or was influenced by humans is still uncertain (Newsom, Webb and Dunbar, 1993; Smith, 2006). Upon reviewing archaeological evidence found in the state of Florida (United States), Hart, Daniels and Sheviak (2004) suggest that the pepo gourd may have first been employed in North America as a float for fish nets, and this use may have facilitated spread of the species in North America.

Cucurbita maxima

Taxonomically, the species *C. maxima* is composed of two subspecies, *maxima* and *andreana*. *C. maxima* ssp. *maxima* contains ornamental and cultivated forms, while *C. maxima* ssp. *andreana* contains only the wild forms. Gene sequence analysis work by Sanjur et al. (2002) on the mitochondrial *nad1* gene detects no base pair differences in an intron of the mitochondrial *nad1* gene between *C. maxima* and *C. andreana*, supporting the assignment based on ecological and morphological evidence that these two species form a wild/domesticated species pair (Nee, 1990). The work of Sanjur et al. (2002) also shows that *C. maxima* is closely related to the free-living South American species, *C. ecuadorensis*. This supports research using chloroplast DNA analysis which shows that *Cucurbita ecuadorensis* groups with *C. maxima* and *C. andreana* to form a South American group of allied species (Wilson, Doebley and Duvall, 1992). Nee (1990) suggests that *C. andreana* appears to be ancestral to *C. maxima,* while Wilson, Doebley and Duvall (1992) suggest that *C. ecuadorensis*, *C. maxima* and *C. andreana* are derived from the same ancestor. *Cucurbita ecuadorensis* itself has been regarded as a species subjected to an incipient domestication process which, however, did not continue (Nee, 1990; Sanjur et al., 2002).

Cucurbita moschata

Although a very important vegetable crop in many parts of the world (e.g. Africa), *C. moschata* has been subject to less scientific scrutiny aimed at elucidating taxonomic relationships within the species than *C. argyrosperma* or *C. pepo*. Filov in 1966 classified more than 20 varieties of *C. moschata* into geographical subspecies. This classification reflects several centres of diversity of *C. moschata* such as Columbia, Japan, Mexico, Central America, the western United States, Florida, India and Asia Minor. Gwanama, Labuschagne and Botha (2000), using random amplified polymorphic DNA analysis, elucidated the relationship between 31 landrace genotypes obtained from Malawi and Zambia. That analysis revealed four clusters, with genotypes from Malawi mainly grouping in three clusters, while all genotypes from Zambia and three from Malawi clustered in another group.

That *C. moschata* has a closer affinity to the *Argyrosperma* group taxa than to other *Cucurbita* species has been argued from floral, seed and ecological similarities, and the fact that these species are partially interfertile (Merrick, 1995, 1990). Sanjur et al. (2002) confirmed this relationship through mitochondrial *nad1* gene sequence analysis.

Cucurbita ficifolia

C. ficifolia has been subject to less scientific scrutiny aimed at elucidating taxonomic relationships within the species than other cultivated *Cucurbita*. The species is reproductively incompatible with the other species of the genus and shows far less variability than the other species of the genus. The scant morphological variation of this species is consistent with the limited variability in the pattern of isozymes studies conducted so far (Andres, 1990). It has no associated wild ancestor in studies to date (e.g. Sanjur et al., 2002).

Work by Sajur et al. (2002) examining a sequence of an intron region from the mitochondrial gene *nad1*, and work by Wilson, Doebley and Duvall (1992) using chloroplast restriction fragment length polymorphism, suggest that *C. ficifolia* is basal to all other mesophytic *Cucurbita* species. More work is needed to decipher the relationship of *C. ficifolia* to the xerophytic species of the genus with which it shares some morphological features.

Centres of origin and distribution

All of the *Cucurbita* are native to the Americas. The centres of origin and domestication for cultivated *Cucurbita* species can be identified as various areas in North and South America (Jeffrey, 1990). Table 2.1 lists the natural distribution for both cultivated and wild *Cucurbita* species. At least five species of the genus *Cucurbita* were domesticated before the European contact in the late 15th century, forming important food sources in Native American economies, and some of these species were among the earliest plants taken under cultivation and domestication in the New World (Smith, 2001; Sanjur et al., 2002).

Cucurbita argyrosperma

The natural distribution of *C. argyrosperma* ranges from the southwestern United States through Mexico into Central America (Table 2.1). The area of domestication for *C. argyrosperma* is considered to be from the southwestern United States to the centre-south region of Mexico (Merrick and Bates, 1989). Archaeological remains of *C. argyrosperma* found from southwestern United States to the centre-south region of Mexico suggest that domestication of this species occurred in the region several thousand years ago (Table 2.2). Unlike the other cultivated *Cucurbita* species, data on the distribution of *C. argyrosperma* outside the Americas are scarce and there is some question as to whether this species was cultivated at any time in other parts of the world. Today, it is cultivated primarily in South America, Mexico, and on very limited areas in the United States.

Cucurbita ficifolia

The natural distribution of *C. ficifolia* ranges from the Mexican highlands south to northern Chile and Argentina (Table 2.1). It grows as an annual in temperate climates and can appear to be a perennial in tropical zones. The precise location of the centre of domestication of *Cucurbita ficifolia* is still uncertain. Some have proposed that its centre of origin is Central America or southern Mexico/Central America. Linguistic evidence tends to support this hypothesis because of the wide use of names based on the Nahuatl name "*chilacayohtli*" as far south as Argentina. Others suggest that its centre of domestication is located in South America and, more specifically, in the Andes (Nee, 1990). Andres (1990) compared diverse types of evidence and was not able to determine precisely the probable domestication site of this crop. So, while native names from indigenous Nahua influence like "*chilacayote*" and "*lacayote*" suggest a Mexican origin, systematic evidence has been inconclusive because a strong reproductive incompatibility has been found between *C. ficifolia* and the wild taxa of *Cucurbita* native to Mexico. While archaeological evidence favours a South American origin since the most ancient remains have been found in Peru (Table 2.2), it has not been possible to support this evidence by means of systematic studies, including those involving the two wild South American taxa (*C. maxima* ssp. *andreana* and *C. ecuadorensis*) as both prosper in habitats different from those in which *C. ficifolia* is cultivated (Nee, 1990). Also, the bee *Peponapis atrata*, until recently considered a pollinator specific to *C. ficifolia*, has not been found in South America. *Cucurbita ficifolia* differs in its karyotype from all other taxa of the genus and is always clearly separated in analyses with various molecular markers (Weiling, 1959; Andres, 1990; Wilson, Doebley and Duvall, 1992; Sanjur et al., 2002). All of these data have led to the proposition that the wild ancestor of *C. ficifolia* could be an undiscovered (or possibly extinct) species from South America (Andres, 1990; Nee, 1990; Sanjur et al., 2002). Although archaeological records show that it was the most cultivated *Cucurbita* species in the Americas prior to the European contact (Bisognin, 2002), currently *C. ficifolia* is the least cultivated of the *Cucurbita* species in the Americas. In the 16th and 17th centuries, Europeans introduced *C. ficifolia* to the Mediterranean countries (specifically France and Portugal) and India. It was also spread to other parts of the world and picked up other names, e.g. shark skin melon in Asia, Malabar gourd in India, and pie melon in Australia and New Zealand.

Table 2.2. **Synthesis of archaeological records of cultivated species of *Cucurbita***

C. argyrosperma	*C. ficifolia*	*C. maxima*	*C. moschata*	*C. pepo*
– Valley of Tehuacán, Puebla, Mexico (before 5200 BCE) – Ocampo, Tamaulipas, Mexico (200-900 CE) – Southwest United States (380-1340 CE) – Zape River, Durango, Mexico (700 CE)	– Huaca Prieta, Peru (2700-300 BCE) – Casma Valley, Peru (2000-1500 BCE) – North of Belize and Tikal, Guatemala (remains of doubtful identification) (2000 BCE-850 CE) – Valley of Oaxaca, Mexico (700 CE)	– Casma Valley, Peru (2000-1500 BCE) – Viru Valley, Peru (1800 BCE-600 CE) – Pampa Grande North of Argentina (500-1000 CE) – Ica, Peru (500-1400 CE) – San Nicolás, Peru (1200 CE)	– Ocampo, Tamaulipas and Valley of Tehuacán, Puebla, Mexico (4900-3500 BCE) – Huaca Prieta, Peru (2700-300 BCE) – North of Belize and Tikal, Guatemala (2000 BCE-850 CE) – Valle Casma, Peru (500 BCE) – Southwest United States (300 BCE) – Viru Valley, Peru (600-1100 CE) – Chincha, Peru (1310-1530 CE) – Great Plains, west of the Mississippi River, United States (1700 CE)	– Valley of Oaxaca (8750 BCE-700 CE) – Ocampo, Tamaulipas (7000-5000 BCE) – West-Center, Illinois (5000-2000 BCE) – Ozark Highlands, Missouri, United States (4000 BCE) – West of Kentucky, United States (3000-600 BCE) – Philips Spring, Missouri, United States (2300 BCE) – Valley of Tehuacan, Mexico (2000-1000 BCE) – Southwest United States (300 BCE) – Zape River, Durango, Mexico (600-700 CE) – Edzna, Campeche, Mexico (850 CE) – Great Plains, west of the Mississippi River, United States (1400 CE)

Source: Adapted from Lira, Andres and Nee (1995).

Cucurbita maxima

The natural distribution of *Cucurbita maxima* is comprised of Argentina, Bolivia and Chile (Table 2.1). *C. maxima* was domesticated in South America. Historical chronicles indicate that during the time of the conquest of Río de La Plata (16th century), this species was one of the main crops of the Guaraní people living in what is now northeastern Argentina and Paraguay (Parodi, 1935). Then, as now, there were many variants under cultivation in the Andean valleys (Cárdenas, 1944). Archaeological evidence for *C. maxima* uses and domestication has been found from Peru to northern Argentina as shown in Table 2.2. The wild species that have shown higher genetic affinity or otherwise are more similar to this crop (Puchalski and Robinson, 1990; Wilson, Doebley and Duvall, 1992; Sanjur et al., 2002) are *C. maxima* ssp. *andreana*, endemic to Argentina and Uruguay (Martínez-Crovetto, 1974; 1965; 1954), and *C. ecuadorensis*, known only from the coast of Ecuador (Cutler and Whitaker, 1969; Nee, 1990). *C. maxima* ssp. *andreana* is the most probable wild ancestor of this crop (Millán, 1945; Fursa and Filov, 1982; Nee, 1990; Sanjur et al., 2002).

From the 16th century, several types of *C. maxima*, such as the turban type, were transported directly from South America to Europe. Many other cultivars reached Australia, Africa and Asia, where local landraces evolved. In the 19th century, several cultivars were introduced into the United States from South America (Decker-Walters and Walters, 2000). Secondary centres of diversity include Bangladesh, Burma, India and the southern Appalachians of the United States, e.g. the landrace "Candy Roaster" was originally developed by the Cherokee people in the southern Appalachians. These findings suggest that for *C. maxima*, in addition to the regions of South America mentioned above, multiple centres of diversity, primarily composed of landraces, exist around the world.

Cucurbita moschata

The natural distribution of *Cucurbita moschata* is from the lowlands of Mexico into Central America (Table 2.1). *Cucurbita moschata* was domesticated in Latin America (Whitaker, 1947) but there is no consensus as to the precise area where domestication likely occurred. It has been proposed that *C. moschata* was domesticated in Mesoamerica (Whitaker and Davis, 1962) or alternatively in South America, more specifically in what is now Colombia. However, available evidence has been difficult to interpret and the centre of origin/domestication question is still open to debate. The oldest archaeological remains of *C. moschata* (4900-3500 BCE) have been recovered from the Ocampo caves, in the state of Tamaulipas, in northeast Mexico; however, very early dates have been registered for several localities in Central America (2000 BCE-850 CE) and South America (2700-300 BCE; Table 2.2). One of the main arguments against the South American origin hypothesis is that *C. moschata* is capable of producing highly fertile hybrids with the wild taxa of the *C. argyrosperma* group (Merrick, 1990), which has an identified centre of origin from the southwestern United States to the centre-south of Mexico. Morphological and ecological studies as well as comparative mitochondrial, ribosomal and chloroplastic analyses (Wilson, Doebley and Duvall, 1992; Jobst, King and Hemleben, 1998; Sanjur et al., 2002) suggest that the ancestor of *C. moschata* might have derived from a wild taxon of *C. argyrosperma* ssp. *sororia*. However, *C. moschata* and *C. argyrosperma* have different isoenzymatic patterns (Sanjur et al., 2002). Although studies by Merrick (1990, 1991) and Sanjur et al. (2002) support a high level of relatedness between *C. moschata* and members of the *Argyrosperma* group, these authors do not support the possibility of the ancestor being the subspecies *sororia*. *C. lundelliana* has also been proposed as the *C. moschata* ancestor, and that wild taxon is indigenous to the Yucatan Peninsula in Central America (Whitaker, 1974). However, there are several morphological differences between *C. lundelliana* and *C. moschata* (i.e. the greenish-grayish-blue seed colour in *C. lundelliana* has not been seen in *C. moschata*). Results from molecular biology studies (Puchalski and Robinson, 1990; Wilson, Doebley and Duvall, 1992) have also lent evidence to exclude *Cucurbita lundelliana* as a possible ancestor. Some characteristics associated with *C. moschata* have been identified in landraces from Bolivia, Columbia and Panama (e.g. dark-coloured seeds, small fruits, a lignified and warty rind; Wessel-Beaver, 2000b), suggesting hybridisation between *C. moschata* and wild local species in Columbia (Nee, 1990). Based on these sets of information, some authors have suggested the existence of two independent domestications, one in Mexico and the other in northern South America. This hypothesis is supported by linguistic evidence – i.e. *C. moschata* is known by native names specific to the language used in each region (Lira, Andres and Nee, 1995; Robinson and Decker-Walters, 1997; Decker-Walters and Walters, 2000).

After its domestication, it is likely that *C. moschata* spread through the Caribbean Islands, giving rise to various native cultivars (Robinson and Decker-Walters, 1997; Piperno, Andres and Stothert, 2000). After the European contact, *C. moschata* was spread rapidly to other continents, adapting to different ecological conditions. *C. moschata* was being cultivated at the end of the 17th century in western Mississippi (United States), and in the 19th century in Angola, India, Java, Northern Africa and Japan, where the species diversified (Sauer, 1993; Lira, Andres and Nee, 1995; Decker-Walters and Walters, 2000). Wu et al. (2011) report the existence of 1 032 landrace accessions of *C. moschata* in the People's Republic of China, hereafter "China"), reflecting a large diversity in landraces developed in China. *C. moschata* accessions are among the best represented *Cucurbita* accessions at the genebank of the Center for the Conservation and Breeding of Agricultural Diversity (COMAV) at the Polytechnic University at Valencia, Spain with a strong representation of landraces still cultivated under traditional cropping systems (Esteras et al., 2008). In Spain the cultivation of this species is mainly based on landraces maintained for centuries. Ferriol et al. (2004) examined 47 COMAV accessions from the Spanish peninsula, the Canary Islands, and Central and South America. The morphological characterisation showed considerable variability. Molecular analysis using amplified fragment length polymorphism and sequence related amplified polymorphism markers showed a genetic variability concordant with the morphological variability (Ferriol et al., 2004). In addition, *C. moschata* is the best represented *Cucurbita* in gene banks of the Americas (e.g. Center for International Food and Agricultural Policy [CIFAP] in Mexico). *C. moschata* is also one of the most important vegetables cultivated in Malawi, Zambia and tropical Africa, where its cultivation is mainly based on landraces with a high degree of autogamy (Gwanama, Labuschagne and Botha, 2000). These landraces have adapted to a variety of climatic conditions and soil types (Gwanana, Labuschagne and Botha, 2000). In general, *C. moschata* is the most widely cultivated *Cucurbita* in the tropics, where it is primarily grown on a small-scale basis for local consumption (Andres, 2004). These findings indicate that for *C. moschata*, in addition to the diversity seen in the Americas, multiple centres of diversity, primarily composed of landraces, exist around the world.

Cucurbita pepo

The natural distribution of *C. pepo* ranges from the eastern United States north into the state of Illinois through the Mississippi Valley, through the state of Texas and south into Mexico (Table 2.1). The archeological record, summarised in Table 2.2, shows that *C. pepo* initially became a fundamental element of agriculture in the Americas in the highlands of Mexico thousands of years ago (Smith, 1986). The most ancient remains of this species have been found in the Valley of Oaxaca (8750 BCE-700 CE) in the Ocampo caves, in the state of Tamaulipas in northeast Mexico (Table 2.2). The appearance of *C. pepo* in the United States as a domesticated crop also dates back thousands of years (King, 1985) as supported by the archaeological record which indicates *C. pepo* was being used by native populations thousands of years ago (7000-5000 BCE; Table 2.2) in what is now west central Illinois (United States). According to archaeological records, *C. pepo* appears to be one of the first domesticated species of *Cucurbita*, with a domestication even older than that attributed to other important domesticated crops:

C. pepo domestication appears to have predated that of other American crops such as maize (*Zea maize*) and common beans (*Phaseolus vulgaris*) by some 2 000 to 5 000 years (Smith, 2006). The archeological record shows that *C. pepo* was cultivated in different ecological niches in the Americas from the high plains to fluvial systems.

The archaeological record suggests that *C. pepo* was domesticated on at least two occasions and in two different regions of North America: in Mexico and in the United States (Decker, 1988, 1986; Andres, 1987a). The hypothesis of two domestication events is supported by the grouping of allozymic patterns of the cultivated forms (Decker, 1985). The hypothesis is also supported by additional allozymic assays, comparisons of chloroplastic and mitochondrial DNA, as well as random amplified polymorphic DNA markers (Kirkpatrick and Wilson, 1988; Wilson, Doebley and Duvall, 1992; Decker-Walters et al., 1993, 2002; Sanjur et al., 2002). As evidence has accumulated, it has become clear that *C. pepo* is comprised of two molecularly divergent groups that had already differentiated through geographical isolation long before humans domesticated them (Decker-Walters et al., 1993, 2002; Smith, 2006). These two divergent groups are classified as two subspecies: *C. pepo* ssp. *pepo* and *C. pepo* ssp. *ovifera*. *C. pepo* ssp. *pepo* includes pumpkin, zucchini and other marrow squashes, Mexican landraces and a few ornamental gourds. *C. pepo* ssp. *ovifera* comprises both domesticated and free-living populations, and is further divided into three taxonomic varieties: *C. pepo* ssp. *ovifera* var. *ovifera* which includes some cultivated cultivars (e.g. acorn, crookneck and scallop squash) and most ornamental gourds, and the free-living populations in the United States, which represent two molecularly distinct populations: *C. pepo* ssp. *ovifera* var. *texana* and *C. pepo* ssp. *ovifera* var. *ozarkana* (Decker-Walters et al., 1993, 2002). *C. pepo* ssp. *ovifera* appears to have been domesticated in what is now the United States, while *C. pepo* ssp. *pepo* appears to have been domesticated in Mexico (Sanjur et al., 2002).

With regard to the likely ancestor(s) of the cultivated *C. pepo*, the three most closely related wild relatives are: *C. pepo* ssp. *fraterna* which is found in Mexico, and *C. pepo* ssp. *ovifera* var. *texana* and var. *ozarkana* which are found in the United States. *C. pepo* ssp. *fraterna* is known only from a few localities in the states of Tamaulipas and Nuevo León in the northeastern region of Mexico (Bailey, 1943; Andres 1987a; Nee, 1990; Rodríguez and Lira, 1992; Wilson, Lira and Rodríguez, 1994). *C. pepo* var. *texana* and var. *ozarkana* have distribution patterns associated with drainage patterns of the river systems that lead from the central United States to the Gulf of Mexico (Smith, 1992; Asch and Sidell, 1992; Wilson, 1998; US Department of Agriculture, 2011a).

With regard to the probable ancestor of *C. pepo* ssp. *ovifera*, Sanjur et al. (2002), on the basis of sequence analysis of and intron of the mitochondrial gene *nad1*, suggest that *C. pepo* ssp. *fraterna* is the most probable ancestor of *C. pepo* ssp. *ovifera*. In the Sanjur et al. studies (2002), *C. pepo* ssp. *fraterna*, *C. pepo* ssp *ovifera*, *C. pepo* ssp. *ovifera* var. *texana* and var. *ozarkana* form a group and share the same haplotype. Smith (2006), however, posits that although the analysis of the *nad1* mitochondrial gene in the studies by Sanjur et al. (2002) could not distinguish between these taxa, random amplified polymorphic DNA data (Decker-Walters et al., 2002; 1993) was able to separate ssp. *fraterna* from ssp *ovifera* var. *texana* and var. *ozarkana* and effectively excluded ssp. *fraterna* from the cluster that includes all cultivated and wild varieties of ssp. *ovifera*. Further support for excluding ssp. *fraterna* from potential contention as an ancestor of ssp. *ovifera* is provided by a recent genetic study employing amplified fragment length polymorphism, inter simple sequence repeat and simple sequence repeat markers, in which ssp. *fraterna* was placed at a greater genetic distance from the domesticates of ssp. *ovifera* than the eastern North American wild gourds (Paris et al.,

2003). Smith (2006) and others (Emshwiller, 2006; Rieseberg and Harter, 2006) therefore argue that *C. pepo* ssp. *ovifera* var. *ozarkana* is the most likely ancestor of cultivated *C. pepo* ssp. *ovifera*.

With regard to the possible wild ancestor of *C. pepo* ssp. *pepo*, some authors consider it unknown or extinct (Wilson, Doebley and Duvall, 1992; Decker-Walters et al., 2002). That *C. pepo* ssp. *fraterna* is found on the muddy or volcanic plains of northeastern Mexico supports the suggestion that it might be the ancestor. Sanjur et al. (2002) suggest that *C. pepo* ssp. *fraterna* could have existed in the past in small and half-isolated populations which were genetically divergent, and some *C. pepo* ssp. *fraterna* population still not collected could be the possible ancestor of *C. pepo* ssp. *pepo*. Morphometric and molecular research, as well as studies on artificial and spontaneous hybridisation support *C. pepo* ssp. *fraterna* as a closely related wild relative. Nonetheless, at this time, no *C. pepo* ssp. *pepo* ancestor has been identified.

Today, populations composed of *C. pepo* wild relatives, *C. pepo* ssp. *fraterna*, *C. pepo* var. *texana* and *C. pepo* var. *ozarkana*, range from northeastern Mexico through the state of Texas, east to the state of Alabama and north through the Mississippi Valley to the state of Illinois in the United States. They occupy a diversity of environments and ecological niches – from upland seasonally dry thornscrub habitat in northeastern Mexico to primarily riverbanks and moist thickets in Texas, to a variety of riparian and other disturbed lowland habitats (e.g. agricultural fields, railroad tracks and highway embankments) throughout the Mississippi Valley. Different morphological and physiological adaptations have evolved in these areas, including early fruit abscission from the peduncle in response to riverine dispersal in the state of Texas, as well as relatively quick seed germination in response to a shorter growing season in the more northerly populations (Decker-Walters et al., 1993). In North America, *C. pepo* is a morphologically and ecologically diverse species composed of genetically distinct groups of cultivars and free-living populations (i.e. self-sustaining wild populations), all of which are interfertile.

Following its domestication, *C. pepo* experienced great diversification in the Americas and subsequent to the European contact, in Europe and Asia (Decker, 1988). That several cultivars were known in the Americas prior to the European contact is demonstrated by the great variability found in the first European herbaria (Whitaker, 1947; Paris, 2001, 1989). The various forms of *C. pepo* which were geographically isolated in the Americas were brought together and cultivated together in European gardens where hybridisation unavoidably occurred to produce new recombinants (Paris, 1989). *C. pepo* accessions are, for example, among the best represented *Cucurbita* accessions at the genebank of the COMAV at the Polytechnic University at Valencia Spain. Landraces still cultivated in Europe under traditional cropping systems (Esteras et al., 2008) are well represented (see also Aliu et al. [2011] for a description of *C. pepo* landrace diversity in the Balkans). The Newe Ya'ar Research Center in Israel maintains a *C. pepo* collection with 320 accessions collected almost entirely from North America, Europe and Asia (Paris, 2011). *C. pepo* is planted in all countries of Africa on a limited scale, even though it is less tolerant of tropical conditions than *C. moschata* (Grubben, 2004). Landraces or traditional varieties are maintained in traditional cropping systems in Mexico and the northern region of Central America, and the variation of this species in rural communities, at least in Mexico and Central America (mainly Guatemala), is also large. It includes variants which are cultivated at altitudes above 2000 m (during the rainy season or even the dry season on land that remains wet), and still others that can grow near the sea and in even more extreme conditions (i.e. those found on the Yucatan Peninsula). On the other hand, it is common for small farmers in Latin American to manage, in common plots, a fair representation of races or local varieties with morphological features similar to those that have been considered specific to some groups of commercial cultivars. *C. pepo* accessions are well represented in genebanks of the Americas (e.g. CIFAP). As with *C. maxima* and *C. moschata*, in addition to the diversity seen in the Americas, multiple centres of diversity exist for *C. pepo*, primarily as landraces, around the world.

Morphological characters

Cucurbita is a genus of dicotyledonous flowering plants in the family *Cucurbitaceae*. The wild and cultivated species of *Cucurbita* are trailing or climbing vines, or subshrubs (bushes) in some cultivated varieties. The cultivated and wild mesophytic species are annuals or such long-lived annuals to be seen as short-lived perennials. The xerophytic species are perennials, persisting for years due to their long tap roots. Mature stems are approximately 1 cm thick, but considerably thicker at the base of the plant. Cultivated forms have internodes that are usually 15-25 cm long, with petioles having a similar range of length and leaf laminae that are generally pentagonal in outline, ranging to 30 cm diameter or more. Wild forms have more diminutive vegetative parts. Emerging from the leaf axil are solitary branched tendrils 20 cm long and solitary flowers. Wild and cultivated species of *Cucurbita* are mostly monoecious, i.e. both male and female flowers are produced by a single plant, the exception being *C. foetidissima*, where some varieties are gynomonoecious. The large flowers are gamopetalous with tubular-campanulated corollas, very showy, light yellow or bright yellow-orange in colour. Rooting commonly occurs at the stem nodes. Some varieties produce tendrils that help secure vines, limit wind damage and improve vine growth across weedy and uneven ground.

For purposes of identification, the five domesticated species are differentiated by the following morphological characteristics: habit of growing, stems, leaves (in distal nodes), indument (petioles and primary veins in the lower surface of blades), receptacle (in staminate flowers), corolla, filaments, peduncle of fruit, size and shape of fruit, surface of fruit, colour of fruit, pulp of fruit, seeds (shape and size), and colour of seeds. Table 2.3 shows only the most outstanding features of the cultivated species. A complete treatment (descriptions, synonyms, illustrations, distribution maps, etc.) of *Cucurbita* is available in Lira, Andres and Nee (1995). Note, however, that cultivated species of *Cucurbita* are very diverse locally, regionally and worldwide (e.g. Jeffrey, 1990; Nee, 1990; Lira, Andres and Nee, 1995; Sanjur et al., 2002), having a wealth of innumerable strains, landraces and varieties that defy facile description (Esquinas-Alcazar and Gulick, 1983; Lira, Andres and Nee, 1995). As more plants are considered, the array of variations and combinations of notable traits increases and the differentiations become less distinct and meaningful.

This section and Annex 2.A2 offer examples of cultivars available in the retail seed market ("commercial" cultivars) for four of the five cultivated *Cucurbita* species; commercially available cultivars for *C. ficifolia* are scarcely found, but some are available online, e.g. from heirloom seed vendors.

Table 2.3. **Principal morphological characters that generally differentiate the domesticated species of *Cucurbita***

	C. argyrosperma	*C. ficifolia*	*C. maxima*	*C. moschata*	*C. pepo*
Stems	Rigid, angulated and sulcated	Rigid, angulated and only slightly sulcated	Soft, rounded or slightly angulated in lateral view, not sulcated	Rigid, angulated and only slightly sulcated	Rigid, angulated and sulcated
Leaves (in distal nodes)	Generally deeply lobed, but not palmatifid with the terminal lobe pinnatifid	Moderately lobed, rounded	Not lobed or only occasionally very slightly so	Never or very rarely with lobed acute or rounded lobes	Generally deeply lobed almost palmatifid with the terminal lobe pinnatifid, rounded or acute
Indument (petioles and primary veins in the lower surface of blades)	Soft and shortly pubescent to pilose	With short glandular hairs that stain dark brown or black	Hirsute to hispid, aculeate, but not conspicuously speculate	Soft, short to long pubescent or villous	Sometimes spiculate with persistent, conic bases of the hairs, strongly aculeate
Peduncle of fruit	Rigid, angulated to non-angulated except where it joins the fruit, smooth or costate, frequently very thickened, becoming cylindrical, claviform or subglobous	Rigid, angulate with obtuse ribs, without obtuse lobes and moderately widened where it joins the fruit, the ribs not having to extend or spread towards the apex of the fruit	Rigid or soft, cylindrical, not angulate, with irregular suberous strias, commonly not widening where it joins the fruit	Rigid, angulate with obtuse ribs, with obtuse lobes and notably widened where it joins the fruit, the ribs not tending to extend or spread towards the apex of the fruit	Rigid, angulate, with acute ribs, widened where it joins the fruit, the ribs sometimes more or less tending to extend or spread towards the apex of the fruit
Colour of the seeds	Usually white, sometimes dull white, tan; margins the same colour as the centre of the seed or a little darker, sometimes yellowish to golden, not fibrillous to very slightly so, or greenish, grayish, blue	Black or brownish-black, less commonly dull white or tan when mature; margins the same colour as the centre of the seed, sometimes lighter particularly near the apex	White to tan; margins usually of different colour than the centre of the seed (never greenish, grayish, blue), except in the white ones	White, dull white to tan or dark brown; margins usually of different colour than the centre of the seed, but never greenish-grayish-blue, usually yellow to golden and fibrillous or fimbriate	Dull white to tan; margins the same color as the centre of the seed

Source: Adapted from Lira, Andres and Nee (1995).

Cucurbita argyrosperma

There is less variation within *C. argyrosperma* as compared to other cultivated species of the genus such as *C. pepo, C. maxima* or *C. moschata*, as *C. argyrosperma* is limited for the most part to races or local varieties that are cultivated in the southwestern United States, Mexico and Central America, plus several commercial cultivars which have been developed mainly in the United States, such as "green striped cushaw", "white cushaw", "magdalena striped", "papago", "Japanese pie", "hopi", "taos", "parral cushaw", "Veracruz pepita" and "silver seed gourd". The most important variations observed between these cultivated varieties correspond to the dimensions, shape and colour patterns of fruits and seeds.

Cucurbita ficifolia

Cucurbita ficifolia is by far the least variable species of the cultivated *Cucurbita.* Morphologically, the most important variation is in colour patterns (white to green fruits with colour patterns: spots or white stripes, and tan to dark brown or black seeds), and dimensions of fruits and seeds. Its relatively low morphological variation coincides with that observed regarding isoenzymatic patterns (Andres, 1990). However, including

South American populations in the survey may uncover wider genetic diversity (e.g. collections with differential resistance to viral diseases have been identified). Field observations have shown that some average-sized fruits contain more than 500 seeds, and that each plant can produce numerous fruits (Lira, Andres and Nee, 1995).

Cucurbita maxima

Cucurbita maxima is one of the most diversified cultivated species of the genus. Its variation includes numerous races and local varieties, and many commercial cultivars, some edible and some ornamental, with trailing and subshrub habits, and with fruits and seeds that are sometimes unique in dimension, shape and colour. Some varieties have also proven to have various degrees of resistance to several viral diseases (Tapley, Enzie and Van Eseltine, 1937; Millán, 1947; Martínez-Crovetto, 1974, 1965; Rochelle, 1974; Maluf and Souza, 1984). A total of 52 cultivars of this species were described by Tapley, Enzie and Van Eseltine in 1937. These cultivars differ not only in morphological characteristics, but in aspects of agronomic interest, such as duration of the life cycle, productivity and, in some cases, in their ability to adapt to limiting ecological conditions.

Castetter (1925) and others (Whitaker and Davis, 1962; Robinson and Decker-Walters, 1997; Decker-Walters and Walters, 2000) have addressed the diversity displayed by *C. maxima* forms, and several types have been identified. The "Banana" type includes long fruits which are pointed at both ends, with a soft rind and brown seeds. The fruits of the "Delicious" type are turbinate-shaped and have shallow ribs, a hard rind and white seeds and a high-quality flesh. The "Hubbard" type includes oval fruits, with pointed or curved ends, a very hard rind and white seeds. Fruits of the "Show" type are wide and orange coloured with a smooth rind and white seeds. A wide variability in this type is seen in India. The "Turban" type contains turban shaped fruits. Most *C. maxima* fruits are eaten at maturity as the fruit is of high quality. However, fruits of the "Marrow" type are picked a few days post-anthesis for consumption, have a soft skin, are oval or pear shaped, bulbous and have white seeds (see Annex 2.A2 for additional information on horticultural types).

Of particular interest among *C. maxima* are varieties of the Mammoth group in the "Show" type ("Mammoth Chilli" or "Mammoth Whale") whose fruits reach gigantic dimensions. Some *C. maxima* fruits exceeding 450 kg in weight have been documented. Also impressive are the turbinate fruits like "French Turban" and "Turks Cap".

Cucurbita moschata

Cucurbita moschata also displays striking morphological variation: the variation of its fruits (colours, shapes and widths, and the durability of the fruits, shell, etc.) and seeds, the number of variants with life cycles of different length, the existence of numerous cultivars developed in various parts of the world, and the existence of races and local varieties with outstanding agronomic characteristics, suggest the extensive genetic variation of this species. Some examples of interesting regional variants are found in the Yucatan Peninsula (and possibly in other regions of Latin America), with two life cycles of different length (Lira, 1988; 1985), as well as the variants grown in some Mexican states (e.g. Chiapas, Guanajuato), which have been found to be resistant to certain viral diseases, and which have been used in improvement programmes (Garzón-Tiznado, Montes-Hernández and Becerra, 1993; Gallegos, 1990). The most variation of fruit and seeds of *C. moschata* occur in Columbia (Wessel-Beaver, 2000b). During the 500 years of cultivation in China, *C. moschata* has adapted to different agro-ecological conditions, and now displays high variability for many agronomic characteristics, such as fruit shapes and colours, flowering habits, leaf characters, etc. (Du et al., 2011).

C. moschata fruits are generally harvested at maturity as the fruits possess a high-quality flesh. *C. moschata* is one of the most popular species for the large number and size of seeds. A few types have been identified. The "Cheese" type includes fruits that are variable in shape, mostly oblong, and have a leather-coloured rind. Fruits of the "Crookneck" type have a rounded distal end and a long curved or straight neck. The "Bell" type presents fruits with a shape that ranges from flared to almost cylindrical. Because *C. moschata* grows well in tropical areas, it is known as "Tropical pumpkin" (see Annex 2.A2 for information on horticultural types).

Cucurbita pepo

C. pepo also displays a very high level of striking morphological variation. *C. pepo* is perhaps the most variable species for fruit characteristics in the plant kingdom. Its fruit range in size to over 20 kg; in shape from round to flat-scalloped to long bulbous cylindrical over 75 cm long; exterior colour is based on hues of green, orange and yellow, with colour intensity ranging from very pale to intense, and a grey contribution ranging from none to very dark. Variegation, including striping and bicolour, can result in as many as four colours on the surface of the same fruit. Fruit mesocarp can be relatively thin or thick, and its colour varies in the range from greenish-white to white, yellow, light orange and intense orange. Fruit rinds can be lignified or non-lignified, and smooth, warted, wrinkled or netted. *C. pepo* not only includes commercial cultivars and edible races with diverse morphological and phenological characteristics, but also several cultivars that are used as ornamentals, e.g. "Orange Ball", "Miniature Ball" and "Striped Pear". Tapley, Enzie and Van Eseltine (1937) recognise 60 edible cultivars, including some with subshrub habits such as "Black Zucchini", "Fordhook Bush" and "Boston Greek", as well as others of trailing to climbing habits and edible fruits, e.g. "Connecticut Field" and "Table Queen". Several authors have proposed organisational groupings to reflect the possible origin and evolution of the high level of morphological variation seen in *C. pepo* (Castetter, 1925; Bailey, 1929; Paris, 1989). Paris (1986, 1989, 2001) grouped the cultivated types into eight morphotypes.

"Pumpkin" (*C. pepo* L. var. *pepo* L. Bailey) includes cultivars of creeping plants which produce spherical, ovoid or oblate fruit that is rounded or flat at the ends. The fruit of this group is grown to be eaten when ripe and is sometimes used as fodder. "Scallop" (*C. pepo* L. var. *clypeata* Alefield) has a semi-shrubby habit, the fruit ranges from flat to almost discoidal, with undulations or equatorial margins, and it is eaten before maturity. "Acorn" (*C. pepo* L. var. *turbinata* Paris) is both a shrubby and a creeping plant with fruit that is obovoid or conical, pointed at that apex and longitudinally costate-grooved. The rind is soft, hence the fruit can be eaten in the ripe state. "Crookneck" (*C. pepo* L. var. *torticollia* Alefield) is a shrubby type with yellow, golden or white fruit which is claviform and curved at the distal or apical end and generally has a verrucose rind. It is eaten unripe as the rind and fruit harden when ripe. "Straightneck (*C. pepo* L. var. *recticollis* Paris) is a shrubby plant with yellow or golden fruit and a verrocose rind similar to var. *torticolla*. "Vegetable marrow" (*C. pepo* L. var. *fastigata* Paris) has creeper characteristics as semi-shrub and has short cylindrical fruit that is slightly broader at the apex, with a smooth rind which hardens and thickens on ripening and which varies in colour from cream to dark green. "Cocozelle" (*C. pepo* L. var. *longa* Paris) has cylindrical, long fruit that is slender and slightly bulbous at the apex, it is eaten in the unripe immature state. "Zucchini" (*C. pepo* L. var. *cylindrica* Paris) is the most commonly cultivated group of cultivars at present. Like the previous group, the zucchini group has a strong affinity with the vegetable marrow and its origin is also recent (19th century). Its plants are generally semi-shrubby and its cylindrical fruit does not broaden or broadens only slightly. It is eaten as a vegetable in the unripe state (see Annex 2.A2 for information on horticultural types.) Taxonomically, the morphotypes "pumpkin", "vegetable marrow", "cocozelle" and "zucchini" are subspecies *pepo.* Some authors have categorised the morphotypes "Scallop", "Acorn", "Crookneck" and "Straightneck" as subspecies *ovifera* (also known as *C. pepo* ssp. *texana*, see e.g. Paris et al., 2003); however, this chapter, which is based on the nomenclature used by Lira, Andres and Nee (1995) and Lira Saade, Eguiarte Fruns and Montes Hernández (2009) and Gong et al. (2012), would group these morphotypes in *C. pepo* ssp. *pepo*. (Both groups are shown as *C. pepo* L. in Table 2.1).

Because of their small, hard shells, ornamental gourds are typically thought of as a distinct grouping within *C. pepo*. Isozymic evidence, however, shows this not to be true, with cultivars having originated in both subspecies *pepo* and *ovifera* and possibly in subspecies *fraterna* (Decker-Walters et al., 1993). What many of these cultivars do share in common are characteristics often ascribed to free-living populations, e.g. tough pericarps and bitter flesh, which ward off predation in the wild.

Reproductive biology

Cucurbita is a genus of dicotyledonous flowering plants in the family *Cucurbitaceae.* The cultivated *Cucurbita* are annual plants, long running and climbing, or short and bushy. The flowers are monoecious and numerous and very showy. Flowers open very early in the morning, and the predominant pollinators of these flowers are bees. In the Americas, the most efficient pollinators are the solitary bees of the genera *Peponapis* and *Xenoglossa* (Hurd and Linsley, 1970, 1976, 1964; Hurd, Linsley and Whitaker, 1971), but the flowers can also be pollinated by other bees such as the honey bee, *Apis mellifera.* Fruits are of the pepo type: a berry with numerous seeds surrounded by a fleshy wall that does not open at maturity. Production of flowers, fruits and seeds varies between species (Quesada et al., 1991).

Reproductive organs

Flower

Unisexual flowers are characteristic of the Cucurbitaceae. The *Cucurbita* species are primarily monoecious with both male and female flowers on the same plant (Whitaker and Robinson, 1986; Lira, Andres and Nee, 1995; Rzedowski and Rzedowski, 2001), although some varieties of *C. foetidissima* are gynomonoecious (Whitaker and Robinson, 1986).

Cucurbita flowers are large, gamopetalous with tubular-campanulated corollas, and showy, with a cream coloured or light yellow or bright-yellow orange corolla. Flowers grow from the axil of a leaf. Male flowers have column-like stamens, with free or more or less connivent filaments, and the anthers are joined together forming a cylindrical or narrowly pyramidal structure. Female flowers have an inferior ovary with numerous horizontally positioned ovules, the styles are fused in almost their entire length or are only shortly free in the apex; stigmas are large, fleshy or more or less sunken or lobulated, and slight modifications can be seen in the structure of the perianth regarding the staminate ones, mainly corresponding to differences in size of one or some of its parts (e.g. the receptacle is always much more reduced). After pollination, fruit develops from the pre-formed ovary at the base of the female flower. The shape of the ovary prior to pollination is indicative of the mature fruit shape.

Cross-pollination is favoured by the monoecious nature of the plants, and in some cases the male flowers are slightly larger than the female ones on the same plant. *Cucurbita* can exhibit wide variation in the proportion of male to female flowers on a plant (Janick and Paull, 2007). Zomlefer (1994) reported that production of female flowers is frequently less than that of male flowers. In *C. pepo*, Nepi and Pacini (1993) found a 16.5:1 relation between the number of male and female flowers. Temperature and light influence the production of male and female flowers in several of the species (Whitaker and Davis, 1962). More male flowers are produced on long and very hot days, whereas short and cold days induce the development of more female flowers (Robinson and Decker-Walters, 1997). The first flowers on the vine are male, after which three or four female flowers appear. Although female flowers differentiate later in plant development, females develop faster than the males, resulting in near synchronization at anthesis of the flowers of both sexes (Janick and Paull, 2007). Flowers open early in the morning and close around noon of the same day, never to reopen (Nepi and Pacini, 1993). Flowering time both in male and female flowers of *C. pepo* varies depending on the time of year in which the plants develop: male flowers begin to open 15 minutes before the female flowers when the days are longer. On days further along in the year this difference is not significant (Nepi and Pacini, 1993). Scheerens et al. (1987) found similarities on the daily opening time of flowers of *C. foetidissima* of both sexes. Whitaker and Robinson (1986) observed that in some genotypes, a short photoperiod is needed for flowering to begin, and flowers develop only when the days are short.

Flower development in Cucurbitaceae, apart from being regulated by genetic and environmental mechanisms such as temperature and the duration of days, can be modulated by chemical regulators – substances such as gibberellins and ethylene (Rudich, 1990). For example, ethylene is involved in the regulation of fruit ripening and sex expression and in the plant's response to herbivore damage.

Pollen

Cucurbita pollen grains are large and sticky, and well suited to transport by insects. Wind does not pollinate *Cucurbita* species. Ovules are fertile only during the period of flowering, or the day prior. Good fruit and set development requires 500-1 000 live pollen grains on the stigma of the female flower (Stephenson, Devlin and Horton, 1988; Vidal et al., 2010). Pollen viability in a newly opened male flower is about 92%, but by the time it closes that same morning the viability will have dropped to 75%, and by the next day will be only 10% (Nepi and Pacini, 1993). Environmental conditions at the time of anthesis are important. High or low temperature can result in a more rapid decrease in pollen viability. In addition, in windy, dry conditions, pollen can lose viability rapidly.

Fruit

Fruits of *Cucurbita* are of the pepo type: a berry with numerous seeds surrounded by a fleshy wall that does not open at maturity. Fruits have a thin and soft, or rigid and woody, shell that emerges from the outer layer of the ovary (exocarp), whereas the pulp around the seed is derived from the ovarian internal layers (mesocarp and endocarp). In cultivated plants, the fruits are produced in a great variety of shapes, sizes, colours and types of surface, whereas in wild plants they are all relatively small and relatively uniform regarding shape (globous, ovoid or rarely pyriform), surface (generally smooth or without ornaments) and colour (white, yellowish or green with or without spots and/or fringes; Lira, Andres and Nee, 1995). Some of them are among the largest in the plant kingdom.

Production of fruits varies between species. While producing 20-30 male flowers and 8-12 female flowers, cultivated plants of *C. pepo* generally produce 2-6 fruits (Quesada et al., 1991). Lira, Andres and Nee (1995) report that each plant of some domesticated varieties of *C. ficifolia* can produce numerous fruits. The wild species *C. pepo* var. *texana* commonly produces approximately 50 mature fruits by the end of the growing season (Avila-Sakar, Krupnick and Stephenson, 2001).

Flower, fruit and seed production, as well as the offspring's performance can be affected by environmental and genetic factors, and by paternal and maternal conditions. Work on the paternal effects on the offspring of cultivated plants of *Cucurbita pepo* has demonstrated that ovule fertilisation and seed production is non-random and depends on the origin of the paternal genotype (Quesada et al., 1991). Similarly, the strongest offspring are obtained from the stylar region of the fruit, where the ovules are fertilised by the most vigorous pollen grains. From research on the effects of pollen competition on the performance of the offspring using hybrids of cultivated *C. pepo* and *C. pepo* var. *texana*, Quesada, Winsor and Stephenson (1996, 1993), concluded that the offspring resulting from large amounts of pollen reaching styles are more vigorous than those produced when smaller amounts reach the styles. Apparently, competition between pollen grains leads to more successful seeds, progeny and their future flower production. The percentage of success in pollination (experimentally) is highest directly after the flowers have opened, and diminishes gradually as midday approaches (Whitaker and Robinson, 1986).

Increase in the size of the ovaries is noticeable within 24 hours of anthesis. Not all pistillate flowers develop into fruit, however, most often because many more flowers are produced than the plant can support nutritionally. Competition is strongest during the first week after anthesis. *Cucurbita* that are grown for consumption of the young fruit are harvested several days past anthesis. The time from anthesis to a fully mature fruit varies considerably among various *Cucurbita* species. Table 2.4 lists the phenology and life cycle in Mexico of 20 *Cucurbita* taxa.

Plants continue to flower and produce fruits consistently until killed by frost. In some papers, *Cucurbita ficifolia* is considered a perennial species (Dane, 1983), but Lira, Andres and Nee (1995) indicate that *C. ficifolia* is an annual species, which, depending on certain environmental conditions (i.e. not too severe frost) can live longer, giving the impression that it is a short-lived perennial. A similar phenomenon has been seen in *C. lundelliana* and *C. moschata*, which can keep on producing flowers and fruits for an extended period of time given appropriate conditions. On the other hand, because of frost, in some areas the perennial species behave as facultative annuals, dying in their first year (Whitaker and Robinson, 1986).

Fruit dispersal

The routes by which seeds are dispersed is determined largely by the size, shape and character of the seed coat or the persisting structures of the fruit. In the case of the *Cucurbita*, the persisting structure of the mature fruit, i.e. the gourd, can be buoyant in water. Hence, water represents a potential means of *Cucurbita* seed dispersal. In addition, the pepo type fruits may represent an adaptation for dispersal by animals, and animals also represent a means by which *Cucurbita* seeds are dispersed. Generally, this occurs through ingestion of fruits with seeds; with the seeds then being excreted at a distance from the initial point of ingestion. In many areas of the world, *Cucurbita* fruits are used as fodder. For example, in tropical regions, domestic animals such as donkeys and horses will consume *Cucurbita* L. fruits and vines when fodder is scarce at the end of the rainy season (Mariano and Dirzo, 2002).

Seed and germination

The seeds are surrounded by several layers of tissue. The most external layer derives from the internal epidermis of the carpel (endocarp) and generally swells with water; the internal layer derives from the testa (seed coat) and develops from the ovule's integument and nucellus (Zomlefer, 1994).

Table 2.4. **Phenology and life cycles of 20 *Cucurbita* taxa**

	Taxa	Phenological notes in Mexico
Annuals	*C. argyrosperma argyrosperma*	
	C. argyrosperma ssp. *sororia*	Flowers from June to January and begins to fructify in August, but it is more common to find mature fruits from November or December
	C. ficifolia	
	C. maxima	
	C. maxima ssp. *andreana*	Produces fruits in September and from February to April
	C. moschata	
	C. pepo ssp. *pepo*	
	C. pepo ssp. *fraterna*	
	C. pepo var. *texana*	Flowers and fruits between July and November
	C. ecuadorensis	It is possible to find male flowers and fruits between February and July
	C. lundelliana	It is possible to find populations with flowers of both sexes. Mature fruits are more surely found between November and February
	C. okeechobeensis ssp. *okeechobeensis*	Flowers and fruits between May-June and January, and the mature fruits from October onwards
	C. okeechobeensis ssp. *martinezii*	Flowers and fruits between June and January or February, and mature fruits can be found from October onwards
Perennials	*C. digitata*	Populations with flowers and fruits occur all year round
	C. cordata	Populations with flowers and fruits occur all year round
	C. palmata	Populations with flowers and fruits occur all year round
	C. foetidissima	Populations with flowers and fruits occur all year round
	C. pedatifolia	Flowers and fruits between August and November and mature fruits from October onwards
	C. radicans	Flowers and fruits between May-June and December, and the mature fruits from October onwards
	C. scabridifolia	Flowers and fruits from June-July and until December, and the fruits from August or September

Seeds are oval, oval-elliptical or oval-lanceolate, compressed or more or less tumescent or inflated, with a smooth or sometimes scarified or scarred centre which can be white, cream-colored, greenish-grayish-blue or black, with or without marked or differentiated margins, and these of the same or different colour and/or texture as the centre of the seed.

The number of seeds produced per fruit varies among the species. Lira, Andres and Nee (1995) report that some domesticated varieties of *C. ficifolia* have average-sized fruits with more than 500 seeds. *C. argyrosperma* produces more than 250 seeds per fruit (Merrick, 1990). The wild species *C. pepo* var. *texana* also can produce more than 250 seeds per fruit (Avila-Sakar, Krupnick and Stephenson, 2001).

Few studies deal with seed banks, germination and establishment of seedlings; nevertheless, in general terms, germination is high and rapid, and there is no seed bank or it is limited for most species. *Cucurbita* such as *C. argyrosperma* ssp. *sororia* can be opportunists, adapted to rapidly colonise available open spaces at the beginning of the rainy season. The colonisation strategy seems to be based more on the germination speed of the seeds produced during the last season than on the development of a seed bank in the soil. Under experimental conditions, 86-100% of the seeds of *C. argyrosperma* ssp. *sororia* germinate two to six days after they begin to hydrate. This germination is synchronic and begins with the first main rains that can keep the soil damp for more than one day (Mariano, 2001).

Sexual reproduction

Pollination

Among the *Cucurbita*, some agent – usually bees – is necessary to transfer pollen from the male to female flower; as the pollen is large (80 to 150 µm diameter) and sticky, the species of the genera are not wind pollinated. In the Americas, the solitary bees of the genera *Peponapis* and *Xenoglossa* (Hurd, Linsley and Whitaker, 1971; Canto-Aguilar and Parra-Tabla, 2000) have developed a close relationship with wild and cultivated *Cucurbita* plants – both adults and larvae feed almost exclusively on the nectar and pollen of the plants. Indeed, the bees are dependent on pollen and nectar produced by *Cucurbita* flowers for their survival (Hurd, Linsley and Whitaker, 1971), and appear to have co-evolved with the *Cucurbita* (Hurd, Linsley and Whitaker, 1971). These bees display some behaviours that appear to be adaptations to their interaction with the *Cucurbita*, e.g. an ability to fly at low temperatures, with low light intensity and certain modifications that allow for an adequate extraction and transportation of pollen. These bees often fly from flower to flower while still dark to see which flowers are open, apparently oriented by olfactory cues emitted by some of the species, and probably also with the help of visual and/or hearing sensors. Both the *Peponapis* and *Xenoglossa* are very efficient pollinators of the *Cucurbita* (Hurd, Linsley and Whitaker, 1971). The efficiency and specificity of these bees makes them responsible for moving larger amounts of pollen between wild and cultivated *Cucurbita* than any other group of pollinators.

Pollinators collect large amounts of nectar from the female flowers, and pollen and nectar from the male flowers. Nectar is secreted from a ring of tissue surrounding the style and just inside the perianth tube. When a bee forages in a masculine flower in search of nectar, the pollen adheres to the bee's body and will then be transferred to the stigmas when it visits female flowers (Zomlefer, 1994). The most active period for the bees coincides with the beginning of the plant's floral opening, just before daybreak, and this high level of activity is maintained for several hours (Mariano and Dirzo, 2002).

The hind legs of the bees of the genera *Peponapis* and *Xenoglossa* are adapted to the collection and manipulation of the pollen grains of this genus. However, the pollen grains vary in size and structure between the species and the pollen-collecting devices of the bees vary also (Hurd, Linsley and Whitaker, 1971). This variation between the bees has been shown to be species specific (Hurd and Lindsley, 1970), and apparently has profoundly influenced the ability of the different species of bees to collect and utilise pollens of the various *Cucurbita*, both wild and domestic (Hurd, Linsley and Whitaker, 1971; see Table 2.5).

In addition to *Peponapis* and *Xenoglossa* genera, other species of bees can pollinate plants of the genus *Cucurbita*. One of the most frequent pollinators in cultivation systems around the world is the domestic honeybee, *Apis mellifera* (Canto-Aguilar and Parra-Tabla, 2000; Mariano and Dirzo, 2002). While the bees of the genera *Peponapis* and *Xenoglossa* make more contact with the reproductive parts of the flower, work faster and work earlier in the morning, they are no more efficient than honey bees (*Apis mellifera*) at setting fruit (Tepedino, 1981). However, honey bees, in contrast to the solitary bees of the genera *Peponapis* and *Xenoglossa*, are generalists and readily move to any competing bloom that offers richer rewards. Also, early bee activity can be important on hot days when the flowers close early, and bees of the genera *Peponapis* and *Xenoglossa* are adapted to working early in the morning. In general, where *Peponapis* and/or *Xenoglossa* populations are strong, the use of honey bee colonies is superfluous as most pollination is accomplished by *Peponapis* and *Xenoglossa* bees before the honey bees arrive (Tepedino, 1981).

In addition to the bees of the *Peponapis* and *Xenoglossa* genera, bees of the *Bombus* genus, e.g. *Bombus impatiens*, have been identified in the United States as pollinators of *Cucurbita pepo* (Julier and Roulston, 2009).

Table 2.5 indicates the known species of pollinators that have been identified in wild and cultivated *Cucurbita* populations in Mexico and the species they are known to pollinate. *Peponapis* species are found from Ontario, Canada through the United States through Mexico to northern Argentina in South America (Michener, 2007). *Xenoglossa* species are found through the United States through Mesoamerica. The ranges of *Peponapis* and *Xenoglossa* species are similar in tropical regions of the Americas. Some species of *Peponapis* are also capable of surviving in temperate regions (thus their distribution extends from Canada to Argentina). In contrast, species of *Xenoglossa* have a more restricted distribution toward lower latitudes, but are capable of establishing in more desertic environments. *Megalopta* species are found in Mexico through the mid section of South America (Discover Life, 2011). *Melitoma* species are found in the central and eastern United States through Mexico to northern Chile and Argentina (Discover Life, 2011). Although most common in the northern hemisphere, *Bombus* species can be found around the world (Discover Life, 2011).

Hurd (1966) identified several insects other than bees that might play a minor role in pollination of *Cucurbita*: cucumber, scarab and meloid beetles, as well as flies and moths.

Asexual reproduction

The *Cucurbita* species propagate in nature through sexual reproduction as there is essentially no asexual propagation by means of runners or stolons, or apomixes in nature.

Genetics

The basic chromosome number of the *Cucurbita* is 2n = 2x = 40. Karyotypes suggest that these species are of allopolyploid origins. Results from electrophoretic analyses also helped confirm this genus' polyploidy (Kirkpatrick, Decker and Wilson, 1985), or more specifically, allotetraploid origin (Weeden, 1984). Weeden (1984) and Singh (1990) suggested that the *Cucurbita* are ancestral tetraploid, derived from an ancestor with a haploid chromosome number of ten. Although these authors suggest an apparent homogeneity, Weiling (1959) suggested that the genome in *Cucurbita ficifolia* is AACC (each letter refers to a different ancestral plant genome), whereas in the four remaining domesticated species it is AABB. A recent sequence analysis of an intron from the mitochondrial gene *nad1* indicated that *C. ficifolia* was basal to all other taxa in this group (Sanjur et al., 2002). Wilson, Doebley and Duvall (1992) determined that the annual *Cucurbita* species evolved from the perennial species.

Table 2.5. **Bee species pollinating *Cucurbita* species in Mexico**

Pollinators	References	*Cucurbita* species
Apis mellifera Linnaeus	Mariano and Dirzo (2002); Canto-Aguilar and Parra-Tabla (2000)	*C. argyrosperma, C. pepo, C. moschata, C. ficifolia*
Peponapis michelbacherorum Hurd & Linsley	Hurd, Linsley and Whitaker (1971)	
Peponapis utahensis Cockerell	Mariano and Dirzo (2002); Hurd, Linsley and Whitaker (1971)	*C. argyrosperma* ssp. *sororia, C. pepo, C. ficifolia, C. moschata*
Peponapis melonis Friese	Hurd, Linsley and Whitaker (1971)	
Peponapis fervens Smith	Hurd, Linsley and Whitaker (1971)	
Peponapis citrullina Cockerell	Hurd, Linsley and Whitaker (1971)	
Peponapis limitaris Cockerell	Hurd, Linsley and Whitaker (1971)	*C. argyrosperma, C. moschata*
Peponapis pruinosa Say	Hurd, Linsley and Whitaker (1971)	*C. argyrosperma* ssp. *argyrosperma, C. argyrosperma* ssp. *sororia, C. ficifolia, C. pepo, C. foetidissima, C. lundelliana, C. moschata, C. maxima*
Peponapis azteca Hurd & Linsley	Hurd, Linsley and Whitaker (1971)	*C. argyrosperma* ssp. *argyrosperma, C. argyrosperma* ssp. *sororia, C. pepo, C. foetidissima, C. ficifolia, C. lundelliana, C. moschata, C. maxima*
Peponapis smithi Hurd & Linsley	Hurd, Linsley and Whitaker (1971)	*C. ficifolia, C. argyrosperma, C. moschata*
Peponapis apiculata Cresson	Hurd, Linsley and Whitaker (1971)	*C. ficifolia*
Peponapis atrata Smith	Hurd, Linsley and Whitaker (1971;) Andres (1990)	*C. argyrosperma* ssp. *argyrosperma, C. ficifolia, C. pepo, C. moschata, C. maxima*
Peponapis timberlakei Hurd & Linsley	Hurd, Linsley and Whitaker (1971)	
Peponapis crassidentata Cockerell	Hurd, Linsley and Whitaker (1971)	*C. argyrosperma, C. pepo, C. moschata*
Peponapis angelica Cockerell	Hurd, Linsley and Whitaker (1971)	
Xenoglossa kansensis Cockerell	Hurd, Linsley and Whitaker (1971)	
Xenoglossa strenua Cresson	Hurd, Linsley and Whitaker (1971)	*C. pepo*
Xenoglossa angustior Cockerell	Hurd, Linsley and Whitaker (1971)	
Xenoglossa mustelina Fox	Hurd, Linsley and Whitaker (1971)	
Xenoglossa patricia Cockerell	Hurd, Linsley and Whitaker (1971)	*C. argyrosperma* ssp. *argyrosperma, C. digitata, C. foetidissima, C. pepo, C. moschata, C. maxima*
Xenoglossa fulva Smith	Hurd, Linsley and Whitaker (1971)	*C. argyrosperma* ssp. *argyrosperma, C. ficifolia, C. epo, C. moschata, C. maxima*
Xenoglossa gabbii Cresson	Mariano and Dirzo (2002); Hurd, Linsley and Whitaker (1971)	*C. argyrosperma* ssp. *sororia, C. ficifolia*
Megalopta sp.	Mariano and Dirzo (2002)	*C. argyrosperma* ssp. *sororia*
Melitoma marginella Cresson	Mariano and Dirzo (2002)	*C. argyrosperma* ssp. *sororia*

The morphology of the chromosomes is not well characterised as they are small and not easily differentiated (Weeden, 1984; Weeden and Robinson, 1990). Using flow cytometry, Arumuganathan and Earle (1991) determined that the haploid genome of zucchini (*C. pepo* ssp. *pepo*) is approximately 500 million base pairs long. A typical nucleus (2n) contains 1.04-1.08 picograms of DNA. Most morphological traits appear to be unlinked, and many markers are required to adequately map the genome. Havey et al. (1998) used restricted fragment length polymorphisms to study the transmission of the chloroplast and mitochondrial genomes in cucurbits. They concluded that both organelle genomes were maternally transmitted in *Cucurbita*.

Some work on self-incompatibility and inbreeding depression in the genus has been performed. Some authors have seen little evidence of inbreeding depression in members of the genus (Borghi, 1976). Others, however, have observed indications of inbreeding depression. Mahzabin, Parvez and Alam (2008) indicate that *Cucurbita maxima* shows abrupt inbreeding depression after two generations of selfing. Cardoso (2004) observed inbreeding depression affecting certain traits in a *Cucurbita moschata* variety after four successive self-pollination generations. Cardoso (2004) and Hayes, Winsor and Stephenson (2005a, 2005b) have studied inbreeding depression after four successive generations of self-pollination in *Cucurbita pepo* var. *texana*. In general, inbreeding depression seems to be intense, which suggests a level of genetic variation at least as recessive deleterious genes. The selfing rate showed a range from 0.16 to 0.54, but this might vary among characters, years and conditions (Hayes, Winsor and Stephenson, 2005a, 2005b). Whitaker and Robinson (1986) suggest these different observations might represent the response of different species or varieties of *Cucurbita* to inbreeding.

Molecular markers

A number of different genetic markers have been used to analyse the phylogeny of the genus. Wilson, Doebley and Duvall (1992) studied 15 species using chloroplast restricted fragment length polymorphism analysis and analysed the relationships between different cultivars of *C. pepo*. The cultivated species, with the exception of *C. ficifolia*, form a cluster, and the relationships among *C. moschata*, *C. argyrosperma* ssp. *argyrosperma* and *C. argyrosperma* ssp. *sororia* were not resolved. Jobst, King and Hemleben (1998) analysed the internal transcribed spacers of ribosomal DNA for 11 species of the genus, but the alleles between species are shared, possibly because they are species of very recent origin and/or because there is gene flow between species. Using ISSRs (inter-simple sequence repeats), Katzir et al. (2000) analysed *C. pepo* and discovered a clear differentiation between *C. pepo* ssp. *pepo* and *C. pepo* ssp. *ovifera* (as found in other studies) with *C. fraterna* clustering with *C. pepo* ssp. *ovifera*.

King, Jobst and Hemleben (1995) studied nuclear microsatellites in nine species of the genus. They suggested that *Cucurbita* (like most of the Cucurbitaceae) contains a large amount of satellite DNA. In particular, species of the genus *Cucurbita* contain interesting specific satellite DNA with individual variations among some species. Within *Cucurbita*, the genes coding for the ribosomal 18S, 5.8S and 25S rRNA are present in high copy numbers and appear highly methylated (Hemleben et al., 1988; King et al., 1993; Torres-Ruíz and Hemleben, 1994). Mitochondrial DNA is also specially structured and larger than in other angiosperms. Sanjur et al. (2002) analysed 65 individuals from 14 taxa (8 species of *Cucurbita* and 2 outgroups) with a mitochondrial *nad1* gene. They found 16 haplotypes. Four groups can be defined in the phylogenetic analysis: a basal group including *C. ficifolia*; a group composed of *C. foetidissima*; a group formed by *C. maxima*, *C. andreana* and *C. ecuadorensis*; and a large group with *C. okeechobeensis* ssp. *martinezii* at the base, including *C. pepo*, *C. argyrosperma* ssp. *sororia*, *C. argyropserma* ssp. *argyrosperma* and *C. moschata*. Six independent origins of domestication can be inferred based on this phylogeny. Other phylogenetic studies have been carried out, both at wider levels, analysing for instance the relationships of *Cucurbita* with other plant groups (an example within the family is offered by Chung, Decker-Walters and Staub, 2003), and within a given species (as within *C. pepo* as in Paris et al., 2003).

The levels of genetic variation and the differentiation (genetic structure) of *C. argyrosperma* and *C. moschata* and their relatives have been described in various studies. A close relationship between the populations of *C. argyrosperma* ssp. *argyrosperma* (average D (Nei's genetic distance) = 0.02 [range 0.00-0.06]) and *C. argyrosperma* ssp. *sororia* (D = 0.01 [0.00-0.06]) was reported by Decker (1986). Populations of *C. moschata* showed a greater genetic distance (D = 0.24 [0.16-0.32]; Wilson, 1989; Wilson, Doebley and Duvall, 1992). On the other hand, data on the genetic diversity show a close kinship between *C. argyrosperma* ssp. *argyrosperma* and *C. argyrosperma* ssp. *sororia* (average D = 0.03), and a large differentiation between *C. argyrosperma* ssp. *argyrosperma* and *C. moschata* [average D = 0.22] (Wilson, 1989; Merrick, 1991). Another study on the degrees of genetic diversity in *Cucurbita* revealed that *C. moschata* has a greater genetic diversity (mean expected heterozygosity, H = 0.052) than *C. argyrosperma* (0.039), although the number of individuals studied was small in both species (Decker-Walters et al., 1990).

Montes-Hernández and Eguiarte (2002) studied cultivated populations of *Cucurbita argyrosperma* ssp. *argyrosperma and C. moschata*, together with adjacent wild populations of *C. argyrosperma* ssp. *Sororia*, and found similar high degrees of genetic variations in the three taxa (P= 0.96, mean allelic diversity of 2.08, average expected heterozygosity (He) = 0.407) and little genetic differentiation among conspecific populations (D = 0.081: Fst = 0.087; Nm = 5.22). These findings indicate that *Cucurbita* possesses a high pollen dispersal potential, and a UPGMA (unweighted pair group method with arithmetic mean) analysis indicated the existence of at least two distinct groups of populations: one consisting of both subspecies of *C. argyrosperma* and another consisting of *C. moschata*. In *C. moschata* in Africa, Gwanama, Labuschagne and Botha (2000) used 39 random amplified polymorphic DNA (RAPDs) markers, generating 144 fragments, 23% of which were polymorphic; 4 clusters were found to be associated to the geographical origin of the samples. Ferriol et al. (2004), using 156 amplified fragment length polymorphism (AFLP) fragments in *C. moschata*, found 86% to be polymorphic; and using 148 repetitive fragments, found 66% to be polymorphic.

Ferriol, Picó and F. Nuez (2004) analysed genetic variability and differentiation (genetic structure) of *C. maxima* with AFLP, where 55% were polymorphic, and, with sequence-related amplified polymorphism (SRAP) markers where 57% were polymorphic.

In the *C. pepo* complex, genetic diversity and its heterozygosity are moderately high (D = 0.17 and H = 0.089; Decker and Wilson, 1987) and alleles typical of the cultivated species have been found in wild populations (Kirkpatrick and Wilson, 1988). This has been interpreted as evidence of gene flow between wild and cultivated populations (Decker and Wilson, 1987; Kirkpatrick and Wilson, 1988; Wilson, 1990). Decker-Walters et al. (2002) analysed with RAPDs, 37 wild populations and 16 cultivated varieties. Twenty-six primers yielded 70 scorable and variable markers. Their data also suggested gene flow between wild and cultivated populations. The results of Ferriol et al. (2003a), in a study with 69 cultivated variants of *C. pepo*, including the 2 subspecies ssp. *pepo* and ssp. *ovifera*, using AFLP markers comprising 476 fragments, showed 53% were polymorphic, with an average genetic diversity of 0.18; and, with SRAP markers and 88 fragments, found a polymorphism of 73%, with an average gene diversity of 0.25. With the SRAP analyses, the percentage of polymorphic fragments and the gene diversity were higher in ssp. *pepo* than in ssp. *ovifera* (0.19 and 0.16 respectively), and with the AFLP analyses were 0.12 for ssp. *ovifera* and 0.10 for ssp. *pepo*. Kwon et al. (2004) analysed 16 varieties, including *C. maxima*, *C. moschata and C. pepo* with RAPDs markers and simple sequence repeats of microsatellites. The degrees of genetic variation were high and these markers permitted the clear identification of the varieties and the species.

In addition, in *C. pepo* different characteristics associated with resistance to different pathogens have been described (Lebeda and Křístová, 1996; Lebeda, Křístková and Doležal, 1999; Paris and Cohen, 2000; Provvidenti and Tricoli, 2002; Cohen, Hanan and Paris, 2003; de Oliveira et al., 2003), as well as the genetic bases of other characteristics, such as the banding patterns and color of fruits (Paris, 2003; 2002; 2000), characteristics of the seed (Teppner, 2000), their yield (Paris, 1997; Mohanty, Mohanty and Mishra, 1999) and fat content (Murkovic, Hillebrand and Winkler, 1996).

Advances in genetic mapping include a study by Brown and Myers (2002) of a cross between *C. pepo* with *C. moschata*, using 148 RAPDs markers found in 28 linkage groups, where quantitative trait loci related to the shape of the fruit and leaves were identified. Using RAPD, AFLP, simple sequence repeats and morphological traits, genetic maps for *C. pepo* have been constructed (Zraidi et al., 2007).

Hybridisation and introgression

A wide range of factors that control the incidence and direction of gene flow and introgression within the *Cucurbita* genus has been identified (Merrick, 1990), including spatial and temporal separation, behaviour of pollinators, genetic compatibility factors, physiological differences and environmental adaptation. Numerous attempts at interspecific hybridisation within *Cucurbita* have been conducted over the years and there has been a wide range of success (Singh, 1990; Lebeda et al., 2006).

In *Cucurbita,* all attempts at crossing the xerophytic species, those adapted to arid environments (*C. digitata, C. foetidissima, C. pedatifolia* and *C. radicans*), with the mesophytic species, those adapted to moist environments (*C. argyrosperma, C. ecuadorensis, C. ficifolia, C. lundelliana, C. maxima, C. moschata, C. okeechobeensis* and *C. pepo*), have failed to produce fertile hybrids (Lebeda et al., 2006).

The genetic compatibility relations between the five cultivated, and with the other mesophytic species of the genus *Cucurbita*, have been widely studied (Whitaker, 1951; Whitaker and Bemis, 1965; Merrick, 1990; Lira, Andres and Nee, 1995). In general, the cultivated *Cucurbita* species are reproductively isolated from one another. The primary gene pools of each species are represented by their landraces and commercial cultivars as well as by their intraspecific taxa (see Table 2.6). Although experimental interspecific crosses can be made among the cultivated species, these frequently result in hybrids that are only partially fertile, while others result in no fruit set (Merrick, 1995). Spontaneous crosses between the cultivated *Cucurbita* are uncommon, but have been reported occasionally between certain of the various species' landraces, mostly in Mexico (Decker-Walters et al., 1990; Merrick 1991, 1990). Given the experimental results, these are also likely to be hybrids that are only partially fertile or result in no fruit set. Nevertheless, none of the genus' species is completely reproductively isolated from the others in terms of barriers to hybridisation.

Table 2.6 displays the cross-compatibility of the cultivated *Cucurbita* species with regard to the primary gene pool, the secondary gene pool and the tertiary gene pool. The cultivated *Cucurbita* species of interest, i.e. those listed in the leftmost column, cross readily with plants within their primary gene pool. The secondary gene pool includes species that when crossed experimentally with the cultivated species in the leftmost column can yield at least partially fertile F1 on hybridisation. Although genes can be moved in breeding between the cultivated species and plants in their secondary gene pool, the F1 are usually sterile or sparingly fertile. Species listed as being in the tertiary gene pool of the cultivated species represent the outer limit of potential genetic resources for breeding: Pre-zygotic and post-zygotic barriers can cause partial or complete hybridisation failure, inhibiting introgression between the cultivated species and plants in the tertiary gene pool (Lebeda et al., 2006). Crosses between a cultivated *Cucurbita* species and other cultivated species in the secondary or tertiary gene pools (Table 2.6), present a more complicated picture; the use of techniques such as embryo culture, which are used to bypass hybrid sterility barriers, may be required. Hybrids obtained from such crosses are frequently sterile or exhibit reduced fertility (Whitaker and Robinson, 1986). Among the *Cucurbita*, success in crossing frequently depends on the genotypes used as parentals.

C. ficifolia is the least compatible species, not only with the other cultivated species, but with all the remaining species of the genus (Whitaker, 1951; Whitaker and Davis, 1962; Whitaker and Bemis, 1965; Merrick, 1990; Lira, Andres and Nee, 1995; Robinson and Decker-Walters, 1997). Some interspecific hybrids have been obtained from crosses between *C. ficifolia* and *C. pedantifolia*, *C. foetidissima*, or *C. lundelliana*, but they often lack the capacity to produce an F2 generation (Lira, Andres and Nee, 1995).

Among the cultivated species, *C. moschata* has the best crossability. Among the cultivated species, it is easiest to cross *C. moschata* with *C. argyrosperma*. Hybridisation experiments (and some field observations) have revealed that *C. moschata* has the highest degree of compatibility with *C. argyrosperma*, placing *C. argyrosperma* into the *C. moschata* secondary gene pool (Table 2.6; see Lebeda et al., 2006). The *C. moschata* tertiary gene pool is formed by *C. lundelliana* and some taxa of the groups Maxima and Pepo (Lira, Andres and Nee, 1995).

Conversely, hybridisation experiments (and some field observations) place *C. moschata* into the *C. argyrosperma* secondary gene pool. The next level of *C. argyrosperma* cross-compatibility involves the wild and cultivated species of *C. pepo*, some cultivars of *C. maxima*, and the wild perennial species *C. foetidessima*, which collectively represent the *C. argyrosperma* tertiary gene pool (Lebeda et al., 2006).

The primary gene pool of *C. maxima* includes *C. andreana*, which some authors classify as a *C. maxima* subspecies (Systax Database, 2011; see also Annex 2.A1). The secondary gene pool of *C. maxima* is represented by *C. ecuadorensis*; and its tertiary gene pool includes *C. lundelliana*, *C. argyrosperma*, *C. ficifolia* and *C. pepo* (Lira, Andres and Nee, 1995; Lebeda et al., 2006).

The primary gene pool of *C. pepo* is formed by its various edible and ornamental cultivars, as well as populations of the wild taxa, ssp. *fraterna*, and ssp. *ovifera* var. *texana* and var. *ozarkana*; until recently these wild taxa were identified as distinct species (Singh, 1990). There are a great many *C. pepo* cultivars with particular characteristics that, together with local landraces (grown mostly in Mexico), constitute an extraordinary genetic stock. Populations that could be considered as part of the *C. pepo* secondary gene pool are scarce; most attempts at hybridising *C. pepo* with other wild or cultivated *Cucurbita* species have required the use of special techniques such as embryo culture (Lebeda et al., 2006).

The wild mesophytic annual taxa *C. lundelliana, C. okeechobeensis* and *C. ecuadorensis* have shown some possibilities of introgression through breeding hybridisation with cultivated species and/or with one or more of these species' ancestors.

Of the wild species, *C. lundelliana* is generally the most crossable with the other mesophytic species, being in the tertiary gene pool of *C. ficifolia*, *C. maxima*, *C. moschata* and *C. pepo*.

Table 2.6. **Cross-compatibility of cultivated *Cucurbita* species: Gene pools**

Species	Primary gene pool	Secondary gene pool	Tertiary gene pool
C. argyrosperma	*C. argyrosperma* ssp. *soraria* *C. argyrosperma* ssp. *argyrosperma*	*C. moschata*	*C. pepo* *C. maxima* *C. foetidessima*
C. ficifolia	*C. ficifolia*	*C. pedatifolia* *C. foetidissima*	*C. lundelliana* *C. maxima* *C. pepo*
C. maxima	*C. maxima* ssp. *maxima* *C.maxima* ssp. *andreana*	*C. ecuadorensis*	*C. lundelliana* *C. argyrosperma* *C. ficifolia*
C. moschata	*C. moschata*	*C. argyrosperma*	*C. lundelliana* *C. maxima* *C. pepo*
C. pepo	*C. pepo* ssp. *pepo* *C. pepo* ssp. *ovifera* *C. pepo* ssp. *ovifera* var. *texana* *C. pepo* ssp. *ovifera* var. *ozarkana* *C. pepo* ssp. *fraterna*	*C. argyrosperma* *C. okeechobeensis* *C. moschata* *C. ecuadorensis*	*C. lundelliana* *C. ficifolia* *C. maxima*

Sources: Adapted from Lira, Andres and Nee (1995); and Lebeda et al. (2006).

Examples of breeding crosses performed to obtain specific introgressions

Interspecific crosses are an important mechanism for the introduction of valuable traits that are not available, or cannot be found, within the gene pool of a crop species. However, such crosses are often only achieved with difficulty as there are many natural barriers, both pre- and post-fertilisation, that protect the integrity of a species. Even if a cross between the parental plants produces hybrid offspring, the alien gene must introgress into the genome, including successful chromosome pairing in the target species.

Diverse studies have analysed hybridisation in *Cucurbita* (Whitaker and Bohn, 1950; Whitaker and Bemis, 1965; Merrick, 1991, 1990). In spite of hybridisation barriers, desirable traits have been successfully introgressed among species of the *Cucurbita*. In most cases, success in crossing between cultivated species depends on the genotypes used, with some attempts more successful than others (Whitaker and Davis, 1962; Robinson and Decker-Walters, 1997; Lebeda et al., 2006). For example, although *C. moschata* is in the secondary gene pool of *C. argyrosperma* and the hybridisation possibilities between the members of the subspecies of *C. argyrosperma* (i.e. *C. argyrosperma* ssp. *argyrosperma* and ssp. *sororia*) and *C. moschata* are good (Wilson, 1990; Wilson, Doebley and Duvall, 1992), there are reports of a decrease in the level of compatibility when *C. moschata* is used as the female parent (Merrick, 1991, 1990; Wessel-Beaver, 2000a).

Information related to hybridisation among *Cucurbita* species and techniques to overcome crossing barriers and hybrid sterility has been summarised by Lira, Andres and Nee (1995; see also Sisko, Ivancic and Bohanec, 2003). The breeding of *Cucurbita* has primarily focused on improving the production and quality of the fruits by attempting to increase resistance to pathogens and diseases, and by modifying plant architecture and sex expression (Lebeda et al., 2006). Interspecific hybrids have been made to identify diverse resistance sources, primarily to diseases caused by viruses and fungi. Resistance to zucchini yellow mosaic virus (ZYMV) and watermelon mosaic virus (WMV), which *C. moschata* was reported to display, has been incorporated into cultivars of *C. pepo* by crosses with *C. moschata* (Garzón-Tiznado, Montes-Hernández and Becerra, 1993; Gilbert-Albertini et al., 1993). Wild species of *Cucurbita* including *C. ecuadorensis* and *C. foetidissima* have been found to be resistant to a number of viruses (Provvidenti, 1990), and have been used as sources of resistance to these diseases. It is difficult to hybridise *Cucurbita foetidissima* with other members of the genus because it is phylogentically distant from the cultivated species; nevertheless, its virus-resistant alleles can be introduced into the extended *Cucurbita* gene pool for use in genetic improvement of the cultivated species as it is a member of the tertiary gene pool of *C. argyrosperma* (see Table 2.6).

In terms of intraspecific crosses being useful in increasing resistance to pathogens and disease, Lebeda and Widrlechner (2004) published the results of screenings on cultivated *C. pepo*, represented by eight groups of morphotypes, for susceptibility or resistance to the fungi *P. cubensis* or *P. xanthii*. The *C. pepo* morphotypes expressed significant differences in resistance/susceptibility to *P. cubensis* or *P. xanthii*. Generally, there was an inverse relationship detected in resistance to the two fungi. While zucchini, cocozelle and vegetable marrow (ssp. *pepo*) were highly resistant to *P. cubensis*, they had relatively high powdery mildew sporulation. Cultivars with the fruit type acorn, straightneck and ornamental gourd (ssp. *ovifera*) were quite susceptible to *P. cubensis*; however, they were considered resistant to *P. xanthii* in laboratory and field evaluations (Lebeda and Křístová, 2000).

Interspecific hybrids have been made to incorporate the gene responsible for the "bush" phenotype of *C. pepo* into *C. moschata* and *C. argyrosperma*, species that are in the secondary gene pool of *C. pepo*, providing these species with the characteristics of a compact plant (Robinson and Decker-Walters, 1997). Bush plants have a more uniform growth and better response to high-density planting compared to vine plants (Loy and Broderick, 1990).

Hybridisation and introgression in the field

The amount and frequency of gene flow between a cultivated plant and its closest wild relatives are affected by several factors, e.g. the existing mating system, similarities in flowering phenology, ease in which the gametes can move and overlapping ecogeographic distribution. Several authors, including Decker (1986) and Decker-Walters et al. (1990), have presented genetic evidence for introgression in the field among various *Cucurbita*.

As noted earlier in this chapter, the *Cucurbita* with limited exception are monoecious, plants may produce flowers over much of their maturity, and the species are insect pollinated. Kirkpatrick and Wilson (1988) examined the potential for gene flow between cultivated *Cucurbita pepo* and its wild relative *C. pepo* var. *texana* by monitoring flower patterns and gene flow among experimental populations. While flowering patterns and pollinator movements tended to maximise self-pollination and local gene exchange, movement of effective pollen was detected up to a distance of 1 300 metres. Hybridisation rates of 5% have been reported (see also Montes, 2002). Spencer and Snow (2001) compared the fitness component of wild *Cucurbita pepo* from Arkansas (United States) with *C. pepo* wild-crop hybrids. Their results suggest that the F1 generation of the wild-crop cross does not present a strong barrier to introgression of crop genes into free-living *C. pepo* populations. Quesada et al. (1991) and Queseda, Winsor and Stephenson (1996) showed that subsequent generations of offspring of such hybrids are viable. Decker and Wilson (1987) and Kirkpatrick and Wilson (1988) have found alleles typical of the cultivated species in wild populations and this has been interpreted as evidence of gene flow between wild and cultivated populations. Allozyme frequency distributions and distinctive patterns of variation in fruit structure, colour and bitterness within populations of free-living *C. pepo* indicate that past hybridisation events have resulted in introgression between cultivated *C. pepo* L. and free-living *C. pepo* ssp. *ovifera* (Decker and Wilson, 1987; Wilson, 1990).

Similarities in flowering phenology can affect the potential for hybridisation among other species of *Cucurbita*. For example, *C. moschata*, *C. pepo* ssp. *fraterna* and both subspecies of *C. argyrosperma* have a very similar flowering phenology in relation to the day and time of opening of male and female flowers (Wilson, Lira and Rodríguez, 1994). Wilson (1990) and Lira (1991) have reported hybrids between *C. argyrosperma* ssp. *sororia* and *C. moschata* in the state of Chiapas, Mexico. Gene flow and introgression between cultivated populations of *C. argyrosperma* ssp. *argyrosperma* and *C. moschata* with adjacent wild populations of *C. argyrosperma* ssp. *sororia* is attributed partly to the plants flowering at the same time, and partly to pollinators visiting plants in these taxa indiscriminately (Montes-Hernández and Eguiarte, 2002). In addition, Mexican farmers permit wild relatives of cultivated *Cucurbita* to grow in the edges of their plots, and inside the plots they sometimes find bitter fruits which indicate hybridisation (Nabhan, 1984; Merrick and Nabhan, 1985; Montes-Hernández, Merrick and Eguiarte, 2005). Wilson, Lira and Rodríguez (1994) noted that a mixed population of *Cucurbita* in Mexico showed an anomalous pattern of fruit bitterness. Some domesticated plants (*C. argyrosperma* and *C. moschata*) expressed bitterness whereas some sympatric free-living plants (*C. pepo* ssp. *fraterna*) produced non-bitter fruits. Wilson hypothesised that this reversal of typical bitterness expression suggested gene flow between crop and wild plants at the site. Using synthetic hybridisation Wilson, Lira and Rodríguez (1994) showed that F1 hybrids can be produced from crosses involving *C. pepo* ssp. *fraterna* as the pistillate parent and *C. argyrosperma* as the staminate parent.

RAPDs, RFLPs and microsatellites, AFLPs and studies involving nuclear DNA, chloroplast DNA (cpDNA) and mitochondrial DNA (mtDNA) (Ferriol et al., 2004; Ferriol, Picó and Nuez, 2004, 2003b) have been applied to study introgression and gene flow in *Cucurbita*. Morphological and isoenzyme analyses have also been used to study introgression between various members of the *Cucurbita* (Bretting, 1990; Decker, 1988; Decker-Walters et al., 1990; Kirkpatrick and Wilson, 1988; Montes-Hernandez and Eguiarte, 2002; Nee, 1990; Wilson, 1990; and Wilson, Lira and Rodríguez, 1994).

Crop production and use

Production statistics

Originally domesticated in the Americas, now disseminated worldwide, the cultivated *Cucurbita* species play a major role in food-production agriculture, as well as in local home gardening throughout tropical, subtropical and temperate regions of the globe. Collectively, *Cucurbita* species rank among the ten most important vegetable crops worldwide (Ferriol and Picó, 2008).

International statistics on production and trade rarely distinguish between the *Cucurbita* species; it is thus difficult to parse out how much of each species is grown in the various regions of the globe. For example, *C. pepo* is the most important commercial species worldwide (Paris, 2001), and in tropical Africa *C. moschata* and *C. maxima* are known to be more important than other species such as *Cucurbita pepo*; however, production figures simply report on "pumpkins, squash and gourds". In addition, international production statistics do not reflect the use of the *Cucurbita* in home gardens or when grown on a small-scale basis for local consumption. Nonetheless, the FAO gathers production statistics worldwide for commercial production of pumpkins, squash and gourds and these numbers offer some indication of the importance of the *Cucurbita* in commercial agriculture. The FAO Statistical Database reports the 2013 world production of pumpkins, squashes and gourds at 24.6 million metric tonnes (mMT) from almost 1.8 million hectares. The People's Republic of China was by far the main producer with 7.1 mMT, followed by India (4.9 mMT), the Russian Federation (1.1 mMT), Iran (897 293 MT), the United States (796 872 MT) and Ukraine (610 800 MT). For Latin America, the main producer was Mexico (544 998 MT), followed by Cuba (412 028 MT), Argentina (302 324 MT) and Peru (232 888 MT). In tropical Africa, substantial production was reported for Rwanda (239 182 MT) and Cameroon (158 801 MT). Also included in the top 25 producers were Egypt (543 334 MT), Spain (533 200 MT), Italy (530 000 MT), Turkey (388 785 MT), Bangladesh (375 000 MT), Indonesia (372 387 MT), Korea (323 364 MT), Algeria (260 913 MT), Pakistan (257 740 MT), Japan (227 303 MT), Morocco (224 314 MT), the Philippines (223 479 MT), Thailand (207 000 MT) and South Africa (181 315 MT) (FAOSTAT, 2013).

Environmental conditions

All of the cultivated *Cucurbita* are warm season crops adapted to monthly mean temperatures of 18-27°C, and are killed by frost. Warm temperatures promote growth and are especially beneficial for germination and development of seedlings. *Cucurbita* species are widely adapted to various types of soils, but prefer good drainage and do not tolerate poorly drained soil.

Cucurbita argyrosperma

Cucurbita argyrosperma ssp. *argyrosperma*, within its native range (southwestern United States to Central America), is cultivated in a wide altitudinal range from sea level to 1 800-1 900 metres, generally in regions with warm and slightly dry climate (with irrigation), or in regions with a well-defined rainy season. *C. argyrosperma* does not tolerate very low temperatures well. *C. argyrosperma* can be found in cultivation in Mexico, and some cultivation can be found in Argentina and Peru and the southwestern United States. Some sporadic cultivation may be found elsewhere.

Cucurbita ficifolia

Cucurbita ficifolia is widely distributed under cultivation from 1 000-3 000 metres, on practically all mountain ranges in Latin America. Cultivation at higher altitudes is a feature that distinguishes *C. ficifolia* from other cultivated species of the genus, which, in general, can be managed in a wider interval of ecological conditions. Studies by Andres (1990) and others have shown that *C. ficifolia* is an annual, which, depending on certain ecological conditions (i.e. not too severe frosts), is capable of surviving for a longer period of time than that corresponding to a species with this type of life cycle.

C. ficifolia requires a rich, well-drained, moisture-retentive soil, and a very warm, sunny and sheltered position, but may be able to tolerate poor, wet and badly drained soils in some instances. Plants are not very frost-tolerant, but they can be grown in temperate climate. *C. ficifolia* can only be grown from seed, and can be used as a rootstock for grafts of other *Cucurbita* due to its hardy root system and virus resistance. *C. ficifolia* may be found in cultivation in high elevations from Mexico to northern Chile and Argentina, and in other parts of the world (e.g. France, Germany, Japan and the Philippines).

Cucurbita maxima

Within its native distribution, there are variants or local races of *Cucurbita maxima* cultivated in places within a wide altitudinal range, from 100 metres (in some Brazilian localities) to 3 000 metres (in Bolivia). A frost tender, annual plant, *C. maxima* is primarily cultivated in regions with temperate climate, and very rarely in warm and damp regions (Robinson and Decker-Walters, 1997). *C. maxima* requires a rich, moisture-retentive well-drained soil and a warm, sheltered, sunny position. Many forms require a temperature range of 20-27°C during the growing season, although there are some forms that tolerate cooler conditions. *C. maxima* is the most tolerant of the cultivated *Cucurbita* of low temperatures. *C. maxima* is cultivated in temperate and subtropical regions worldwide.

Cucurbita moschata

In botanical literature, *C. moschata* is reported as being grown mainly in areas of low altitude with a hot climate and high humidity (Esquinas-Alcazar and Gulick, 1983; Whitaker, 1968). However, while it is true that this species is preferentially grown within these limits, they do not appear to be strictly adhered to, as variants have been found above 2 200 metres, e.g. in Oaxaca, Mexico. For example, Bukasov (1981) fixes 2 200-2 300 metres as the top altitudinal limit for this crop in Colombia and Mexico, and this information has recently been corroborated by means of collections of fruits and seeds of the variants growing above 2 200 metres within the region of the Mixteca Alta in the state of Oaxaca, Mexico (Lira, Andres and Nee, 1995). In general, *C. moschata* is the cultivated *Cucurbita* least tolerant of low temperatures, but is relatively drought tolerant. *C. moschata* is cultivated in subtropical and tropical regions worldwide, but can also be cultivated sporadically elsewhere.

Cucurbita pepo

C. pepo can grow in a variety of ecological conditions. *C. pepo* tolerates a fairly wide range of altitudes ranging from 8-2 300 metres. This species includes variants which are cultivated at altitudes above 2 000 metres (during the rainy season or even during the dry season on land that remains wet), and still others can grow near the sea and in even more extreme conditions (i.e. those found on the Yucatan Peninsula). For example, in Mexico there are native varieties that grow close to sea level in semi-dry climates and limestone soil (i.e. the "*tsol*" in the Yucatan Peninsula), while others are managed at altitudes above 2 000 metres, with colder climatic conditions and sometimes highly eroded soil (i.e. the "*güiches*" in the state of Oaxaca) (Lira, 1991; 1988; 1985). In Guatemala the varieties or native races commonly called "*güicoy*" are grown above 1 800 metres, while those called "*tsol*" are managed in the low and warm-humid areas of the Petén, below 500 metres (Azurdia-Pérez and González-Salán, 1986). *C. pepo* grows best when day temperatures are between 24°C and 28°C and night temperatures between 16°C and 24°C, although it can tolerate monthly average day temperatures of 18-28°C. It needs six to eight hours of sunlight a day and has some level of drought tolerance. Many of the commercial cultivars are widely spread around the world, demonstrating the ability of varieties to adapt to different environments. *Cucurbita pepo* may be grown in temperate, subtropical and tropical regions worldwide.

Agricultural practices

The cultivated *Cucurbita* are frost sensitive and need frost-free growing periods of four to five months. Temperatures of 20-35°C are ideal for growth. The *Curcurbita* can be grown on a wide range of soil types. They prefer a soil pH between 6.0 and 6.5, although they will tolerate both slightly acidic and slightly alkaline soils. As roots can penetrate up to a metre into the soil, a well-drained soil is preferred. *Cucurbita* also have feeder roots close to the surface. Roots can grow to about the same spread as the vines. The *Cucurbita* are sensitive to soil salinity. They are also susceptible to herbicide damage, and this susceptibility would suggest that care should be taken with herbicide use, or that herbicides can be used for control. The cultivated *Cucurbita* are usually established by direct sowing of seed, although seeds can be sprouted in containers and seedlings transplanted to the field when about 10 cm high. Seedling transplant is particularly indicated if the growing season is short. *Cucurbita* are insect pollinated and require bees for pollination. Inadequate pollination results in poor fruit shape and blossom drop.

Cucurbita fruit develop rapidly after the flower closes, and fruit eaten at the immature stage (e.g. zucchini) must be harvested before the rind begins to harden. For those types that are eaten after the fruit fully matures (e.g. pumpkin), three to four months are generally required to mature a crop to this stage. At this stage the fruit is hard and imperious to scratching; the fruit is removed from the vine with a portion of the stem attached. The mature fruit can be kept in long-term storage (e.g. four to six months) if the fruit is properly ripened and cured. Curing hardens the shell, heals superficial wounds, reduces the water content of the fruit and improves the quality of the flesh. Fruits can be cured by leaving them in the field in warm and dry conditions for ten days to two weeks or by keeping them inside at room temperature for a month (OMAFRA, 2011). Table 2.4 offered phenological notes on the various *Cucurbita* species, including the cultivated species, in Mexico.

Cucurbita species can express highly oxygenated, triterpene compounds called cucurbitacins, which taste bitter to humans and can be toxic (US Environmental Protection Agency, 1999). Cultivated *Cucurbita* varieties intended for consumption by humans or domesticated animals have been bred to express low levels of cucurbitacins. However, plants in wild populations express high levels of these substances and if a cultivated plant is visited by bees carrying pollen from plants in a wild population, higher levels of the toxicant can be produced in the fruit. Any resulting seeds would produce plants with bitter fruit as bitterness is a dominant characteristic. Higher cucurbitacin levels can also be expressed by the plant in response to stresses such as drought, high temperatures, low soil fertility and low soil pH. Higher levels can also be expressed in the newly emergent seedling, and by improperly cared for, harvested fruit. In addition to producing inedible fruit, plants producing higher level of cucurbitacins can attract phytophagous Chrysomelidae beetles and attendant pest management problems (see below for additional information on the cucurbitacins and Chrysomelidae beetles).

Cucurbita argyrosperma

The cultivated varieties of *Cucurbita argyrosperma* are used in the traditional heavy rain agricultural systems and are sown at the start of the rainy period (May-June in the northern hemisphere) from 1 000-3 000 metres. Growth of these varieties lasts five to seven months; the young fruit for vegetables is harvested approximately three months after being sown, while the ripe fruit for seed is harvested between October and

December. Unlike other cultivated species of the genus, it is less frequent for varieties of the *argyrosperma* complex to be found in vegetable gardens, plots or in small agricultural holdings, or to be associated with other species (Lira-Saade and Montes-Hernández, 1992). In some areas its rapid growth is used to provide ground cover with the aim of preventing weeds from developing in the field.

Cucurbita ficifolia

Cucurbita ficifolia is a crop grown mainly in traditional heavy rain agricultural systems. It is typically sowed at the beginning of the rainy season, and harvested from the end of September (young fruit and flowers for vegetables) to December or January (ripe fruit for seeds and pulp) in the northern hemisphere. The only form of propagation is the sowing of seed, together with one of the traditional crops of this type of agriculture (maize, bean and other species of *Cucurbita*) or else cultivation in vegetable gardens along with other species or by itself. The ripe fruit is harvested and selected for seed. It can be stored for long periods (18-20 months) and it is frequently seen drying on the roofs of farmers' houses (Lira-Saade and Montes-Hernández, 1992).

Cucurbita maxima

Cucurbita maxima prefers light (sandy), medium (loamy) or heavy soil (clay) which is well-drained. The plant prefers acidic and neutral soils. It can grow in semi-shade or no shade. Dry periods with a relatively low humidity favour the best growth. A frost-sensitive annual plant, it is widely cultivated in the tropical and temperate zones. It is one of the species used in large-scale production agriculture.

Cucurbita moschata

C. moschata variants are grown under traditional, heavy rain agricultural systems. It is possible to find varieties grown in maize fields together with maize, beans and one or two other *Cucurbita*, or in vegetable gardens and other more intensively managed farmland where they are grown alone or with other species. There are some old references to a considerable variation in Colombia, but that has yet to be properly documented and evaluated (Lira-Saade and Montes-Hernández, 1992). The greatest diversity lies in the neotropics where the vines are grown under a wide range of ecological conditions, including under hotter conditions than are tolerated by the other *Cucurbita* species (Andres, 2004). This species is used in large-scale production agriculture.

Cucurbita pepo

In its native area of distribution, *C. pepo* is grown both in maize fields and vegetable gardens as well as in other more intensive systems. In the former case, it is combined with maize, beans and/or with one to three of the other cultivated species of *Cucurbita*, while in the latter system it may be found growing on plots or in small groups, generally combined with other vegetables. Where it is grown commercially, it is generally found as the sole crop, occupying areas of varying size (Lira-Saade and Montes-Hernández, 1992). Although *C. pepo* is grown in several different commercial cropping systems, there may be as much grown in home gardens as grown commercially for sale in local or distant markets, and this may be true even in countries such as the United States where it is frequently grown in larger scale production systems (National Gardening Association, 2009). *C. pepo* is the most important *Cucurbita* species economically. In North America cultivars of both ssp. *pepo* and ssp. *ovifera* are grown. Elsewhere in the world, ssp. *pepo* is the more economically important subspecies.

Management issues

Amongst the *Cucurbita, C. pepo* can present a weed problem in certain agricultural settings; these problems are associated with free-living members of the species in North America. *C. pepo* var. *ozarkana* is considered a weed in the states of Arkansas, Louisiana and Mississippi in the United States in soybean and cotton fields (Boyette, Templeton and Oliver, 1984; Oliver, Harrison and McClelland, 1983). While a perennial problem in Arkansas, reports from Louisiana and Mississippi are based on "outbreaks" that are evidently linked to sporadic flooding events and associated fruit dispersal into cultivated fields. Whereas in wild habitats (i.e. those not directly influenced by human activity), individual plants or small groups of plants are widely dispersed along flood plain corridors, in weedy habitats (i.e. disturbed habitats created by human activities), populations can be very dense and cover agricultural fields. Neither *Cucurbita pepo* var. *texana* nor var. *ozarkana* are found on the United States Department of Agriculture's Federal Noxious Weeds List (USDA, 2011b).

Morphological and isozymic evidence suggests that some free-living *C. pepo* populations in Illinois (Decker and Wilson, 1987; Wilson, 1990), Kentucky (Cowan and Smith, 1993; Decker-Walters et al., 1993) and possibly elsewhere (Asch and Asch, 1992) may have evolved purely as escapes of ornamental gourds, which may or may not have experienced subsequent introgression with other nearby cultivated, weedy or wild material of *C. pepo*. Such wild-habitat populations in northeastern Mexico, Texas and many parts of the Mississippi Valley in the United States have long histories of occupation in their general areas, however, and have been accepted as indigenous (e.g. Smith, Cowan and Hoffman, 1992).

Unlike the wild *C. pepo* which wards off predation by producing small, hard-shelled, tough-pericarped, bitter-fleshed gourds, the edible cultivars under human selection have yielded characteristics that hinder the cultivars' ability to persist in the wild, e.g. large, fleshy, non-bitter fruit. The edible cultivars consequently do not survive as long-lived escaped populations in wild or weedy habitats. *C. moschata*, *C. maxima* and *C. ficifolia* are known to grow outside of cultivation in the United States. The species have been collected from various habitats outside of cultivation: oak-pine woods, agricultural fields, brush and trash heaps, roadsides, ditch banks, vacant lots and disturbed sites. In addition to the US localities, *C. moschata* has been reported as naturalised in the West Indies, Central America (Belize) and South America (Galapagos, Guyana, French Guiana, Surinam). In most cases, these plants are most accurately described as "waifs" as they apparently do not maintain themselves in persistent populations (Nesom, 2011).

The edible cultivars can occur as volunteers in fields and thus present certain management considerations. Because of their rapid germination and large canopy, certain of the *Cucurbita* are used in weed control strategies, e.g. *C. argyrosperma* in traditional growth systems in smaller agricultural holdings (Anaya et al., 1987; Anaya, Ortega and Nava Rodriguez, 1992). Rapid vine growth and large leaves make the *Cucurbita* relatively weed tolerant and these characteristics can be used to reduce weed pressure as seen in traditional native agriculture (Anaya et al., 1987; Anaya, Ortega and Nava Rodriguez, 1992; Radovich, 2011).

General interactions with other organisms (ecology)

This section highlights several interesting interactions of note between the *Cucurbita* and other organisms. It does not attempt to create an exhaustive list of interactions.

Cucurbitacin mediated interactions

Species in the family Cucurbitaceae are characterised by their biosynthesis of a group of secondary compounds that are thought to function as chemical defense compounds (Bar-Nun and Mayer, 1990; Tallamy et al., 1998a) against insects, fungi and herbivores. These compounds are known as cucurbitacins (Rehm et al., 1957) and are responsible for the bitter taste found most obviously in the wild Cucurbitaceae. The cucurbitacins are highly oxygenated tetracyclic triterpene compounds (tetracyclid triterpenoids). These non-volatile compounds possess cytotoxic properties. For example, one form of cucurbitacin antagonises insect steroid responses (Dinan et al., 1997). There are 17 identified cucurbitacin compounds, generally named alphabetically, e.g. A, B, C, D, E, F, I, J, K and L. These compounds are based on the unusual amino acid (-)-3-amino-3-carboxypyrrolidine, and can occur both free and in glycosidic combination. The 17 different members of the cucurbitacin class of natural toxicants can be found as naturally occurring mixtures in species of the *Cucurbita*, primarily in the leaves and seeds. Although originally isolated from species in the Cucurbitaceae, cucurbitacins occur in a variety of plant families (e.g. Brassicaceae, Begoniaceae, Rosaceae) as well as in some mushrooms (e.g. *Russula* and *Hebeloma*). The ability of the *Cucurbita* to produce cucurbitacins influences several aspects of their ecology.

Animals

Humans find almost all the cucurbitacins contained within the fruit of the wild *Cucurbita* to be extremely bitter and the compounds have been found to be toxic to a number of animal species. The most toxic cucurbitacin has an LD_{50} of 5 mg/kg body weight in the mouse. The least toxic has an LD_{50} of 650 mg/kg body weight in the mouse (US Environmental Protection Agency, 1999). In spite of the bitter taste and toxicity which appears to deter most animals, some animals can tolerate at least some of the pulp of wild *Cucurbita*; e.g. coyotes (*Canis latrans*) and porcupines (*Erethizontidae* spp.) eat seeds tainted by the pulp of xerophytic *Cucurbita digitata* (Sowls, 1997). Javelina (*Pecari tajacu*) appears to have even greater tolerance as they have been reported to dig up and eat the bitter tuberous roots of *C. foetidissima* and *C. digitata* (Sowls, 1997).

Phytophagous insects

In general, the cucurbitacins produced by the *Cucurbita* are thought to defend against phytophagous insects (Tallamy et al., 1998a). However, for a group of Chrysomelidae beetles of the tribe Luperini, cucurbitacins act as arrestants and feeding stimulants (Metcalf et al., 1982). The beetles belong to the subtribes Diabroticina (about 900 species distributed in the American continent) and Aulacophorina (about 480 species found in Asia). Diabroticina beetles can detect these compounds in plant tissues and inert substances like silica gel or filter paper at quantities as low as 0.1 ng (Metcalf, Metcalf and Rhodes, 1980). When the beetles encounter bitter plant tissues they compulsively ingest them. Furthermore, they sequester cucurbitacins in hemolymph and elytra as chemical defense against natural enemies and transfer the compounds to their eggs (Ferguson and Metcalf, 1985; Brust and Barbercheck, 1992; Tallamy et al., 1998b). A paper by Nishida, Yokoyama and Fukami (1992) showed, for several members of the

Luperini tribe, that these sequestered cucurbitacins deterred feeding by a bird predator, indicating an allomomal role for these compounds. Interestingly enough, some beetle species which do not rely on cucurbits as a food source still show this behaviour, known as pharmacophagy (Fukami and Nishida, 1990; Eben, Barbercheck and Aluja, 1997). An example of such behaviour is displayed by *Diabrotica virgifera virgifera* which is a specialist on plants of the Poaceae. When it reaches maturity, this beetle leaves the nutritious and toxin-free *Zea maize* in search of cucurbitacin enriched plants (Tallamy et al., 2005). Metcalf (1986) proposed that this behaviour is a relict of a coevolutionary association with cucurbits. Gillespie et al. (2003) argued that phylogenetic analysis within the Luperini tribe supports the theory that this behaviour represents convergent evolution of cucurbitacin feeding. Whatever the origin of the behaviour, compulsive feeding is such a strong and reliable characteristic of Diabroticina beetles that cucurbitacins are used as bait in insecticidal preparations for the control of several pest species within the Diabroticina (Lance and Sutter, 1990).

Micro-organisms

As with many other members of the plant kingdom, *Cucurbita* are attacked by a number of microbial pathogens. The next section provides a listing of those pathogens most commonly found on *Cucurbita* species. The largest diversity of disease-producing organisms on species of *Cucurbita* is found among the fungi (Blancard et al., 1994; Zitter, Hopkins and Thomas, 1996; Davis et al., 2008). The fungi causing the largest economic losses in the *Cucurbita* are those that cause powdery mildew (*Podosphaera xanthii, Erysiphe cichoracearum*). Some research (Bar-Nun and Mayer, 1990) has shown that application of cucurbitacins to plant tissue can reduce the infection rate of a fungus, *Botrytis cinerea*, supporting the hypothesis that cucurbitacins can act as defense compounds against at least some fungi.

One bacterial pathogen, *Erwinia tracheiphila*, is particularly problematic in the *Cucurbita*. It is transmitted to the plant by chrysomelid beetles and, as noted above, these beetles are attracted to plants expressing cucurbitacins.

Other interactions

Insect pollinators

As discussed above, the *Cucurbita* are primarily pollinated by bees, and the most efficient pollinators of the *Cucurbita* are the solitary bees of the genera *Peponapis* and *Xenoglossa*. A coevolutionary relationship exists between the bees of the genera *Peponapis* and *Xenoglossa* and the *Cucurbita*. To the bees, it is a relationship on which their survival depends (Hurd, Linsley and Whitaker, 1971). It also seems to be the chief parameter of the bees' evolution (Hurd, Linsley and Whitaker, 1971). A number of coevolutionary adaptations exist between the bees of the genera *Peponapis* and *Xenoglossa* and the *Cucurbita*. For example, these bees are adapted to collect the large (80-150 µm diameter) and spiny pollen grains and to drink the nectar of *Cucurbita* from which the bees derive the majority of their food (Hurd, Linsley and Whitaker, 1971). Although other plants are occasionally visited, adult females rely solely on plants of the *Cucurbita* for the pollen food used to rear offspring (Hurd and Linsley, 1964). It has been hypothesised that the original ranges of the bees were affected by the spread by humans of *Cucurbita* species through the Americas, with the bees extending their ranges using "pollen avenues" established by these cultivated *Cucurbita* in a coevolutionary facilitation (Hurd, Linsley and Whitaker, 1971). Other bees, e.g. the honey bee (*Apis mellifera*), also pollinate *Cucurbita*.

Phytophagous insects

In general, cultivated plant species are used as a food source by a large number of phytophagous insects (e.g. Hodgekinson and Hughes, 1982; Hendrix, 1988), and the *Cucurbita* are no exception, particularly in an agricultural setting. In addition to chrosomalid beetles, other insects are known as pests of the cultivated *Cucurbita* species; some of these have also been seen feeding from wild plants. These include *Epilachna* spp. (Coleoptera: Coccinellidae) and *Diaphania hyalinata* and *Diaphania nitidalis* (Lepidoptera: Pyralidae) (Mariano and Dirzo, 2002). See below for additional information on common insect pests.

Plants in the genus *Cucurbita* have been shown to respond to herbivory in a number of ways, e.g. in the production of flowers, fruits, pollen and pollen performance. Mariano (2001) has observed such effects in *C. argyrosperma* ssp. *sororia* and *C. pepo* var. *texana*. Ávila-Sakar, Krupnick and Stephenson (2001) have shown that the plants of *Cucurbita pepo* var. *texana* are capable of reassigning resources destined for the production of fruits and seeds to growth and production of staminate flowers as a response to the removal of female flowers. Further, Avila-Sakar, Leist and Stephenson (2003) have shown that *C. pepo* var. *texana* has a high tolerance of simulated herbivory; low to moderate levels of foliar damage significantly affected very few traits. Finally, Theis, Kesler and Adler (2009) in *Cucurbita pepo* var. *texana* showed that simulated leaf damage increased fragrance production in male flowers. Female flowers which were bigger and produced more fragrance than males flowers were unaffected by leaf damage. These results suggest that changes in fragrance following herbivory may mediate interactions between plants, herbivores and pollinators.

Plants

Anaya et al. (1987) and Anaya, Ortega and Nava Rodriguez (1992) suggest that the effectiveness of *Cucurbita* species in weed suppression in traditional American polyculture is due to a combination of competition for light and allelopathy. Qasem and Issa (2005) reported that volatiles from *C. pepo* shoots may be phytotoxic: soil-incorporated *C. pepo* residues prevented seed germination of *P. oleracea* and arrested growth of other weed species tested (Qasem and Issa, 2005). In 2007, Fujiyoshi, Gliessman and Langenheim examined the weed-suppressive properties of *Cucurbita* interplanted with corn (*Zea maize*) by comparing different planting and weeding regimes, and measuring weed biomass, light interception by crop canopy and yield. Shading by the *Cucurbita* appeared to be the major mechanism of weed suppression, but the analysis suggested that other factors, such as allelopathy, might also contribute.

Micro-organisms

Several types of viruses are known to attack the *Cucurbita*. The Mosaic viruses (cucumber mosaic – CMV, watermelon mosaic – WMV, zucchini yellow mosaic virus – ZYMV, and squash mosaic virus – SqMV) are the types most commonly observed in the *Cucurbita*. These viruses are transmitted primarily by insect vectors (aphids) and the primary approach to controlling the incidence of viral disease in cultivated *Cucurbita* is control of the vector. The next section discusses the viruses known to infect the *Cucurbita* in greater detail. The following section then briefly describes newer biotechnological approaches to addressing the economic losses associated with certain of the viruses causing disease in *Cucurbita*.

Animals

Cultivated *Cucurbita* are bred to express only very low levels of cucurbitacins, and are far more palatable to humans and other animals than wild *Cucurbita*. In many regions of the world, for example, fruits of the cultivated *Cucurbita* are used as fodder. In tropical regions, domesticated animals such as donkeys and horses will consume *Cucurbita* fruits and vines when fodder is scarce at the end of the rainy season (Mariano and Dirzo, 2002).

Common pests and pathogens

This section lists some of the common pests and pathogens of *Cucurbita*. It is not an exhaustive list.

Viruses

Although only a dozen problem viral variants have been identified, these variants are serious problems for the crops due to the rate of disease spread, the severity of infection, the potential for large economic losses and the difficulty in controlling the diseases. These viral diseases are particularly important due to the susceptibility of the plants to attacks by virus-transmitting insect vectors such as whiteflies, aphids and chrysomelid beetles.

Cucumber mosaic virus (CMV)

This Cucumovirus has worldwide distribution and the widest host range of any plant virus, including more than 1 200 species in over 100 families of dicotyledonous and monocotyledonous angiosperms. The host range includes cereals, forages, woody and herbaceous ornamental, vegetable and fruit crops such as squash, melons, peppers, beans, tomatoes, carrots, celery, lettuce, spinach and beets, various weeds and many ornamentals and bedding plants. Symptoms seen in infections of the virus include leaf mosaic or mottling, yellowing, ringspots, stunting and leaf, flower and fruit distortion.

CMV can be vectored by 60-80 different aphid species in a non-persistent manner from plant to plant in a stylet-borne fashion. The peach (*Myzus persicae*) and melon (*Aphis gossypii*) aphids are the primary CMV vectors. CMV can also be transmitted in seeds, and by the parasitic weeds, *Cuscuta sp*., as well as mechanically by humans cultivating or touching healthy plants after touching infected plants. It can also be carried by the striped and 12-spotted cucumber beetles but the transmission success rate under field conditions makes these insects minor contributors to CMV infection. Many variants of the virus occur, and it is difficult to identify CMV from symptoms alone. CMV produces a systemic infection in most host plants. Older tissues and organs that developed prior to infection usually are not affected by the virus, but newer cells and tissues that develop after infection may be affected with varying severity. Leaves of infected plants become mottled and vines are stunted. The concentration of the virus increases for several days following inoculation, then decreases until it levels off or the plant dies (Agrios, 1997). The virus can overwinter in perennial weeds, flower and food crops by surviving in the roots.

Papaya ringspot virus Type W (PRSV)

This Potyvirus is distributed worldwide. PRSV is transmitted in a non-persistent manner by various aphids such as the peach aphid *Myzus persicae* (Brunt et al., 1996). It can also be transmitted mechanically by humans. It is not seed transmitted (Brunt et al., 1996). This virus was originally called water melon mosaic virus 1 (WMV1) but today is considered to be the W strain of PRSV. PRSV-W should not be confused with what had been called watermelon mosaic virus 2 (WMV2) but is now simply WMV (Lecoq and Desblez, 2009). PSRV has a different host range, different serological properties and no sequence homology with WMV. As with other mosaic viruses, leaves of infected plants become mottled and vines are stunted (Brunt et al., 1996).

Squash mosaic virus (SqMV)

SqMV is a Comovirus and was first reported in California in 1956 (Brunt et al., 1996). SqMV is probably distributed worldwide. This virus can infect and produce symptoms on several commercially grown cucurbits, including *C. maxima*, *C. moschata* and *C. pepo*. It can also infect some plants in the Leguminosae and the Chenopodiaceae. The virus is insect-transmitted in a non-persistent fashion by several insects (*Acalymma vitatta*, *Acalymma thiemei*, *Diabrotica undecimpunctata*, *Diabrotica bivittula*, *Epilachna chrysomalina*, *Epilachna paenulata*) (Brunt et al., 1996). In nature it is spread principally by the spotted cucumber beetle (*Diabrotica undecipunctata*) and striped cucumber beetle (*Acalymma vittata*). The virus can also be transmitted by seed and by mechanical inoculation.

Watermelon mosaic virus (WMV)

In the 1990s, this Potyvirus was referred to as WMV2 to distinguish it from WMV1. Today, WMV1 is considered to be the W strain of papaya ringspot virus (PRSV), while WMV2 is referred to as WMV (Lecoq and Desblez, 2009). WMV has worldwide distribution and is a major viral pathogen of cucurbit crops (Adlerz et al., 1983; Provvidenti, Gonsalves and Humaydan, 1984; Davis and Mizuki, 1987; Chala, Harrison and Halliwell, 1987). This virus can infect and produce symptoms on all commercially grown cucurbits. It can also infect several leguminous and malvaceous species. The virus is aphid-transmitted in a non-persistent fashion. As the host range for WMV is not limited to cucurbits, overwintering of this virus in several leguminous species such as clover can occur. Mixed infections of cucurbits with CMV and WMV are common. WMV causes mosaic and mottle diseases of cantaloupe, cucumber, pumpkin, squash and watermelon and reduces fruit production and quality in squash and other cucurbits (Thomas, 1971; Greber, 1978). Leaves of infected plants become mottled and vines are stunted.

Zucchini yellow mosaic virus *(ZYMV)*

This Potyvirus is a recently described virus disease of cucurbits, first identified in Europe in 1981. The virus is serologically related to, and has characteristics very similar to, WMV (Brunt et al., 1996). ZYMV is also serologically related to bean yellow mosaic virus (Brunt et al., 1996). Like WMV, the ZYMV host range is not limited to cucurbits. The known host range of ZYMV includes *Cucurbita pepo*, *Cucumis melo*, *Cucumis sativus* and *Citrullus lanatus* (ANU, 2005). ZYMV is transmitted in a non-persistent manner by aphid transmission (Lecoq, Pitrat and Clement, 1981; Lisa et al., 1981; Adlerz et al., 1983; Purcifull et al., 1984; Dodds et al., 1984; Adlerz, 1987). It can also be transmitted vertically through seed. Its effects are severe leaf mosaic, yellowing and eventually shoestring symptoms in the leaves. The fruits are stunted, twisted and deformed by raised protuberances. In cultivated crops, plants cease producing marketable fruits within a week or two of infection. On a given cucurbit host, ZYMV usually causes more severe symptoms than WMV, and there is some indication that WMV may make the plant more susceptible to ZYMV (Xu et al., 2004). Leaves of infected plants become mottled and vines are stunted.

Tobacco ringspot virus (TRSV)

TRSV is a Nepovirus and considered a minor cucurbit virus. It is primarily nematode transmitted (*Xiphinema americanun*) but can also be transmitted nonspecifically by insects such as aphids (*Aphis gossypii*) and mites (*Tetranychus* ssp). Melons and cucumbers are the cucurbits most commonly affected by this virus, but it has been found in the *Cucurbita* (Jossey and Badadoost, 2006). It has been reported to spread in North America and China, and has been reported in Australia, Germany, New Zealand and the United Kingdom (Brunt et al., 1996).

Tomato ringspot virus (ToRSV)

ToRSV is a Nepovirus and is considered a minor cucurbit virus. It causes severe damage to summer and winter squash, but shows only mild symptoms in the other cultivated cucurbits. Like TRSV, ToRSV is nematode transmitted (*Xiphinema americanun*) and can overwinter on many weed species without expressing symptoms (Brunt et al., 1996). It has been reported in Australia, Bulgaria, Chile, China, Germany, Italy, Japan, Korea, New Zealand, North America, Peru, Puerto Rico, Turkey and the former Soviet Union (Brunt et al., 1996).

Clover yellow vein virus (CYVV)

CYVV is a Potyvirus and considered a minor cucurbit virus. It is aphid-transmitted in a non-persistent manner and can infect summer squash. Infected plants mostly show chlorotic or necrotic local lesions. It is probably distributed worldwide (wherever white clover occurs). It was previously considered to be the severe strain of bean yellow mosaic virus (Brunt et al, 1996).

Fungi

The most economically important fungal diseases of the *Cucurbita* are the powdery mildews (OMAFRA, 2011).

Cladosporium cucumerinum

C. cucumerinum causes a disease known as scab or gummosis. The fungus can attack any aboveground portion of the plant, including the leaves, petioles, stems and fruits. Scab produces its greatest damage when infection occurs on the fruit. Infected fruit appears to have small spots or sunken areas similar to insect stings. A sticky substance may ooze from the infected area, especially on fleshy fruit. Soft-rotting bacteria may invade these lesions resulting in foul-smelling decay (Strider and Konsler, 1965; Agrios, 1997; American Phytopathological Society, 2011; OMAFRA, 2011).

Choanephora cucurbitarum

This fungus causes a whisker-like fungal growth that causes blossoms and fruits to rot. The disease is commonly referred to as blossom blight or wet rot (Agrios, 1997; American Phytopathological Society, 2011; OMAFRA, 2011).

Erysiphe cichoracearum

E. cichoracearum causes a disease known as powdery mildew. Whitish, talcum-like, powdery fungal growth develops on both upper and lower leaf surfaces and on petioles and stems. Symptoms usually develop first on older leaves, on shaded lower leaves and on upper leaf surface. Infected leaves usually die, and plants senesce prematurely reducing photosynthesis, thereby reducing yield (Agrios, 1997; Jahn, Munger and McCreight, 2002; American Phytopathological Society, 2011; OMAFRA, 2011).

Fusarium oxysporum

F. oxysporum is soil borne and causes a damping-off disease; i.e. it causes young seedlings to wilt and die or not emerge at all. It is occasionally found in cucurbits (Agrios, 1997; American Phytopathological Society, 2011; OMAFRA, 2011).

Phytophthora capsici

P. capcisi causes a blight resulting in leaf spots and fruit rot, seedling damping-off and possible total crop loss. Stem and leaf petiole lesions appear as light to dark brown, water-soaked and irregular in shape, eventually becoming dry, brittle and papery. Older plants with root infections may suddenly wilt. In fruit, the symptoms begin as small water-soaked lesions in the rind, which enlarge quickly and become a soft sunken area covered with white fungal growth (Agrios, 1997; Lopez, Brune and Henz, 1999; American Phytopathological Society, 2011; OMAFRA 2011).

Plectosporium tabacinum

P. tabacinum, also known as *Microdochium tabinum*, causes a blight characterised by the production of light tan to "bleached" sunken, spindle-shaped lesions, primarily on the main stems, petioles main leaf veins and peduncles and sometimes on leaf blades. On fruit, the fungus causes white, tan or silver russeting on the upper surface. Lesions often coalesce to form a continuous dry, scabby surface (Agrios, 1997; American Phytopathological Society, 2011; OMAFRA, 2011).

Podosphaera xanthii

P. xanthii, also known as *Podosphaera fusca*, is the main causal agent of cucurbit powdery mildew and one of the most important limiting factors for cucurbit production worldwide. Although great efforts have been invested in disease control, many basic aspects of the biology of this pathogen remain unknown. Powdery mildews are characterised by spots or patches of white to grayish, talcum powder-like growth. The disease is most commonly observed on the upper sides of the leaves. It also affects the bottom sides of the leaves, buds, stems, flowers and young fruit. Infected leaves may become distorted, turn yellow with small patches of green, and fall prematurely. Infected buds may fail to open (Agrios, 1997; Jahn, Munger and McCreight, 2002; American Phytopathological Society, 2011; OMAFRA, 2011).

Pseudoperonospora cubensis

The symptoms caused by *P. cubensis* are almost exclusively confined to the leaves, although there are rare reports of sporulation on fruits and floral parts. The first evidence of infection is small, slightly chlorotic to bright yellow areas on the upper leaf surface; the colour is less vivid on the lower leaf surface. As lesions expand, they often coalesce, resulting in necrosis of the infected leaves so that in a few days the entire leaf is dead. This disease is commonly referred to as downy mildew (Agrios, 1997; Lebeda and Wedrlechner, 2004; American Phytopathological Society, 2011; OMAFRA, 2011).

Pythium spp.

These soil-borne micro-organisms can cause damping-off, with young seedlings wilting or not emerging at all (Agrios, 1997; American Phytopathological Society, 2011; OMAFRA, 2011).

Sphaerotheca fuliginea

S. fuliginea causes a powdery mildew wherein whitish, talcum-like, powdery fungal growth develops on both upper and lower leaf surfaces and on petioles and stems. Symptoms usually develop first on older leaves, on shaded lower leaves and on the upper leaf surface. Infected leaves usually die, and plants senesce prematurely reducing photosynthesis, thereby reducing yield (Agrios, 1997; American Phytopathological Society, 2011; OMAFRA, 2011).

Bacteria

Erwinia tracheiphila causes bacterial wilt. It is spread by the stripped cucumber beetle, *Diabrotica undecipunctata*, and the spotted cucumber beetle, *Acalymma vittata*, and controlled by eliminating cucumber beetles. The bacteria live in the digestive tract of the striped and spotted cucumber beetles. The beetles defecate frass as they feed and *E. tracheiphila* invades the plant through the wounds caused by the feeding beetles (Sasu et al., 2010).

Insects

A number of insects can attack *Cucurbita* species. Some of the insects listed below are cosmopolitan and have a worldwide distribution, e.g. *Myzus persicae* and *Aphis gossypii*, while others are more limited in their distribution, e.g. *Anasa tristis*.

Aphididae

Aphid species most commonly found on *Cucurbita* include: *Aphis gossypii*, the melon aphid; *Myzus persicae*, the peach aphid; *Aphis fabae*, the bean aphid; and *Aphis craccivora*, the cowpea aphid.

Aphids extract sap from the terminal leaves and stems of plants. They may also feed on developing pods causing them to shrink or become malformed. Their feeding can result in deformation, wilting or death of the plant depending on populations and size of the plant. Saliva injected during feeding can also cause deformation of plant tissue. While aphids can cause significant damage on their own, they frequently present another concern: the transmission of several plant viruses.

Coleoptera

Beetle species most commonly found on *Cucurbita* species include: the stripped cucumber beetle, *Diabrotica undecipunctata*, and the spotted cucumber beetle, *Acalymma vittata*. Beetles such as the palestriped flea beetle, *Systena blanda*, can also attack plants of the *Cucurbita.*

Cucumber beetles (the stripped cucumber beetle, *Diabrotica undecipunctata*, and the spotted cucumber beetle, *Acalymma vittata*) are common pests on various members of the Cucurbitaceae. The name stems from the tendency of these beetles to be found on cucurbits. These coleopterans are among the first insects to attack cucurbits as the plants emerge. The spotted cucumber beetle is about 0.25 inches long, yellow to greenish-yellow with 12 black spots on its back and a black head. They overwinter in the adult stage near plants and in debris. Some migrate south and have been known to travel 500 miles in 3-4 days. The larvae are yellowish-white with a brown head and a brownish patch on top of the last body segment. The larvae feed on plant roots. When there is ample moisture, they will feed on the flesh of the fruit, especially fruits lying on the soil surface. Whereas larvae are root feeders, adults are primarily pollen feeders and do not damage the leaves of cucurbits to a significant extent (Krysan and Smith, 1987; Eben and Barbercheck, 1996; Gámez-Virués and Eben, 2005). The striped cucumber beetle is pale white-yellow to orange with a black head. Its wings have three black stripes running their entire length. Other than immediate stand loss, and damage to leaves, stems, blossoms and fruit, damage is incurred from the beetles' ability to carry the pathogen *Erwinia tracheiphila*, which is carried in the insects' body and transmitted to the plant as the beetles feed (OMAFRA, 2011).

The palestriped flea beetle (*Systena blanda*) is a general feeder attacking a multitude of plants. Larvae can be found feeding on roots. Adults attack the foliage of plants leaving small round holes.

The squash-ladybird (*Epilachna borealis*) is a black-spotted, yellow hemispherical species of wide geographical distribution. The adult beetles hibernate and lay their eggs on leaves in the spring. The yellow, spiny larvae chew circular holes in the leaves. A closely related species is *Epilachna varvestis*, the Mexican bean beetle. The Mexican bean beetle resembles the ladybird; it is coppery coloured with 16 black dots in 3 rows down its back. Its larvae are orange or yellow, humped-backed and fuzzy. Both feed on the lower surface of leaves, skeletonise the leaf.

Lepidoptera

The term "cutworm" applies to the larvae of various moth species in the Noctuidae family. Cutworms are general feeders and attack a wide range of plants, including the cucurbits. These cucurbit pests include the black cutworm (*Agrotis ipsilon*), the granulate cutworm (*Feltia subterranea*) and the spotted cutworm (*Amathes c-nigrum*). These lepidopterans may injure many types of vegetables and sometimes cereals. Larvae hide under clods or in tracks of the soil by day and feed at night, cutting young plants near the ground or feeding on the foliage. They cause greatest damage to seedlings and newly set plants, resulting in stand loss. Cutworms overwinter as larvae or pupae, depending on the species.

"Melonworm" and "pickleworm" are the common names of the larvae of two moth species in the family Pyralidae, with the name melonworm applying to the species *Diaphania hyalinata* and the name pickleworm applying to the species

Diaphania nitidalis. The larvae of these lepidopterans are restricted to feeding on the cucurbits, with both summer and winter squash being particularly favoured hosts. Melonworms feed mainly on the foliage, being primarily a leaf feeder which seldom feeds on the fruits. The pickleworm, in contrast, does feed on the fruits of squash, and can cause serious damage. Early in the season, pickleworms bore into the stems and terminal buds. Later in the season, pickleworms bore into the fruit from the side next to the ground. After feeding for about two weeks, the larva moves out of the fruit to the leaves, where it will spend seven to ten days as a pupa inside a cocoon. Pickleworm is highly dispersive, e.g. in the United States it overwinters in south Florida, spreading northward each spring. The pickleworm has been reported from Canada southward to South America.

"Squash vine borer" (*Melitta satyriniformis*) is a diurnal species of sesiid moth that attacks wild and cultivated varieties of *Cucurbita*. The moth of this lepidopteran resembles a large wasp without the stinging apparatus. Females deposit eggs near the base of the plant about the time the first planting begins to emerge until bloom. A small larva emerges and enters the stem of the plant. The larva then feeds inside the stem and eventually causes it to die. As the worm feeds, it pushes its excrement out of the entrance hole. The worm will eventually exit the stem and enter the soil to pupate (OMAFRA, 2011).

Hemiptera

"Squash bugs" (*Anasa tristis*) are Hemiptera and colloquially called "squash bugs" in North America because some of the species are pests of squash plants and other cucurbits. Squash bugs are quite mobile and can move easily among plants within a field and later move to late planted fields. The insects spend most of their time within the plant canopy, mainly around the stems and on the underside of the leaves. Both nymphs and adults feed by sucking sap from the plant. The adults often congregate near the base of the plant and young nymphs concentrate on the leaf where they hatch and then migrate to other plant parts. Squash bugs can increase in numbers very rapidly and, in high numbers, can cause plant wilting. This insect injects a toxin into the plant while feeding and this toxin results in wilting (OMAFRA, 2011).

Whitefly (*Bemisia tabaci*) is reported on all continents except Antarctica. Over 900 plant hosts are recorded and it reportedly transmits 111 virus species. Most of these whitefly transmitted diseases are begomoviruses, although whiteflies are also vectors of criniviruses, ipomoviruses (Adkins et al., 2006), potyvirus, torradoviruses and carlaviruses (Markham et al., 1994; Navas-Castillo, Fiallo-Olive and Sanchez-Campos, 2011). Its small size belies its ability to move large distances (Ellsworth and Martinez-Carillo, 2001; ISSG *Global Invasive Species Database*, 2011). *B. tabaci* is phytophagous and has been reported to produce silvering of leaves in *Cucurbita* (Schuster, Kring and Price, 1991).

Diptera

Vegetable leafminer (*Liriomyza* spp): Adult leafminers are small flies with a small wing length. Adult females puncture the upper surfaces of leaves with the ovipositor for feeding and egg laying. Adults feed on fluids that exude from the wounds. Eggs are laid singly in separate leaf punctures and hatch within two to seven days. Larvae feed on the leaf mesophyll for 6-12 days. Full-grown larvae slit the leaf epidermis, exit the leaf, fall to the ground and pupate in the soil. Losses in cucurbits due to these dipterans are difficult to quantify. The mining activity of these insects may cause photosynthetic reduction. High populations of leafminers can cause leaf distortion and premature leaf abscission. Infestation may also predispose the plant to other foliar diseases. Adult leafminers may be able to transmit viruses, because of their feeding habits.

Spider mite

Spider mites (*Tetranychus urticae*) are arachnids. Spider mites feed by sucking the contents from individual leaf cells. The feeding of one mite is not damaging but mites are usually present in huge numbers. Mite populations explode during hot, dry weather as they reproduce very rapidly. A female lays an average of 100 eggs and most eggs hatch within 3 days. Mites can complete a life cycle in 5 days when the temperature is 75°F or above.

Biotechnological developments

Genetic modification

As noted earlier in this chapter, the cultivated *Cucurbita* species are important food sources worldwide. Although some of the wild *Cucurbita* species have been reported to display resistance to viral disease, the cultivated *Cucurbita* display far lower levels of resistance. This is particularly true of the most economically important of the *Cucurbita*, *C. pepo*, and diseases caused by viruses can result in large economic losses (Provvidenti, 1990). The presence of these viruses has been reported in nearly all countries and territories where commercial crops of *C. pepo* are produced: Algeria, Australia, Brazil, Bulgaria, Canada, China, Costa Rica, the Czech Republic, the Dominican Republic, Egypt, England, France, Germany, Greece, Guadeloupe, Guam, Japan, Bailiwick of Jersey, Jordan, Honduras, Islamic Republic of Iran, Israel, Italy, Lebanon, Madagascar, Malaysia, Martinique, Mauritius, Mayotte, Mexico, Morocco, Nepal, the Netherlands, New Caledonia, New Zealand, Nigeria, Pakistan, Portugal, Puerto Rico, Saudi Arabia, Singapore, Spain, Sudan, Swaziland, Syrian Arab Republic, Chinese Taipei, Tunisia, Turkey, the United States, Bolivarian Republic of Venezuela and Yemen (Desbiez and Lecoq, 1997).

The primary means of controlling these diseases is control of the insect vectors, a methodology presenting a less than perfect solution. Due to the importance of the cultivated species affected by these viruses, and the difficulty in controlling spread of the viruses, the use of biotechnological techniques to develop resistant varieties has offered an alternative, successful approach. *Cucurbita pepo* cultivars containing the transgenes ZW20 (OECD Unique Identifier SEM-0ZW20-7) and CZW3 (OECD Unique Identifier SEM-0CZW3-2), have been commercially available since the mid-1990s in the United States. The ZW20 transgene confers resistance to the zucchini yellow mosaic virus (ZYMV) and the watermelon mosaic virus (WMV), both members of the potyvirus group. The CZW3 transgene confers resistance to ZYMV and WMV and to the cucumber mosaic virus (CMV), the type member of the cucumovirus group. Protection against these viruses is provided by insertion of DNA sequences encoding the coat protein gene of the various viruses into the *C. pepo* genome. Analysis shows that for both ZW20 and CZW3, a single copy of the transgene has inserted at a single site in the *C. pepo* genome (USDA, 1994). Although it is now known that protection occurs through interfering RNA mechanism (RNAi), expression of the transgenes was specifically engineered into the *C. pepo* cultivars and is controlled by the 35S promoter of the cauliflower mosaic virus (CaMV) to allow constitutive expression of the various coat proteins.

Other studies involving Cucurbita

Several species of *Cucurbita* have been used in classic studies of plant biochemistry (Frisse, Pimenta and Lange, 2003). *Cucurbita pepo* has long been used in studies of purine synthesis (Lovatt, 1983), amino acid transport and beta oxidation of fatty acids (Bush and Langston-Unkefer, 1988). This type of research has continued and, with use of molecular tools, the expression of the ascorbic acid oxidase has been studied (Lin and Varner, 1991), and cDNA from an anionic peroxidase has been obtained and its expression analysed in different kinds of tissues (Carpin et al., 1999). A chromosomal homologue to the aminocyclopropane-carboxylate synthase has also been cloned and sequenced (Huang et al., 1991).

Modern biotechnology has supported an in-depth study of infection resistance mechanisms by CMV (Havelda and Maule, 2000), and identified the genes that are systemically induced by attacks of the white flies *Bemisia argentifolii* and *B. tabaci* (van de Ven et al., 2000).

In addition, recent research has shown promising results for the use of the certain cucurbitacins or cucurbitacin analogues to arrest the cell cycle in tumor cells and induce apoptosis (Sun et al., 2005; Zhang et al., 2010; Boykin et al., 2011).

References

Adkins, S. et al. (2006), "A novel whitefly transmitted potyvirus isolated from cucurbits in Florida", *Phytopathology*, Vol. 97, S3, pp. 145-154.

Adlerz, W.C. (1987), "Cucurbit potyvirus transmission by alate aphids (Homoptera: Aphididae) trapped alive", *Journal of Economic Entomology*, Vol. 80, No. 1, pp. 87-92, *http://dx.doi.org/10.1093/jee/80.1.87*.

Adlerz, W.C. et al. (1983), "Zucchini yellow mosaic virus: A pathogen of squash and other cucurbits in Florida", *Proceedings of the Florida State Horticultural Society*, Vol. 96, pp. 72-74, *http://fshs.org/proceedings-o/1983-vol-96/72-74%20(ADLERZ).pdf*.

Agrios, G.N. (1997), *Plant Pathology, 4th edition*, Academic Press, New York, New York.

Aliu, S. et al. (2011), "Morphological and nutritive variation in a collection of *Cucurbita pepo* L. growing in Kosovo", *Notulae Scientia Biologicae*, Vol. 3, No. 2, pp. 119-122, *http://notulaebiologicae.ro/index.php/nsb/article/view/6066/5536*.

American Phytopathological Society (2011), APSnet, *www.apsnet.org/Pages/default.aspx* (accessed October 2011).

Anaya, A.L., R.C. Ortega and V. Nava Rodriguez (1992), "Impact of allelopathy in the traditional management of agroecosystems in Mexico", Chapter 17, in: Rizvi, S.J.H. and V. Rizvi (eds), *Allelopathy: Basic and Applied Aspects*, Chapman and Hall, London, pp. 271-301.

Anaya, A.L. et al. (1987), "Perspectives on allelopathy in Mexican traditional agor ecosystems: A case study in Tlaxcala", *Journal of Chemical Ecology*, Vol. 13, No. 11, pp. 2 083-2 101, November, *http://dx.doi.org/10.1007/BF01012873*.

Andrade-Cetto, A. and M. Heinrich (2005), "Mexican plants with hypoglycaemic effects used in the treatment of diabetes", *Journal of Ethnopharmacology*, Vol. 99, No. 3, pp. 325-348, July.

Andres, T.C. (2004), "Progress in cucurbit genetics and breeding research", pp 113-118, in: *Proceedings of Cucurbitaceae 2004*, the 8th EUCARPIA Meeting on Cucurbit Genetics and Breeding, Olomouc, Czech Republic, 12-17 July 2004.

Andres, T.C. (1990), "Biosystematics, theories on the origin, and breeding potential of *Cucurbita ficifolia* (Cucurbitaceae)", in: Bates, D.M., R.W. Robinson and C. Jeffrey, (eds.), *Biology and Utilization of the Cucurbitaceae*, Cornell University Press, Ithaca, New York, pp. 102-118.

Andres, T.C. (1987a), "*Cucurbita fraterna*, the closest wild relative and progenitor of *C. pepo*", *Cucurbit Genetics Cooperative Report*, Vol. 10, Article 36, pp. 69-71, *http://cuke.hort.ncsu.edu/cgc/cgc10/cgc10-36.html*.

Andres, T.C. (1987b), "Hybridization of *Cucurbita foetidissima* with *C. pedatifolia*, *C. radicans*, and *C. ficifolia*", *Cucurbit Genetics Cooperative Report*, Vol. 10, Article 37, pp. 72-73, *http://cuke.hort.ncsu.edu/cgc/cgc10/cgc10-37.html*.

Andres, T.C. (1987c), "Relationship of *Cucurbita scabridifolia* to *C. foetidissima* and *C. pedatifolia*: A case of natural interspecific hybridization", *Cucurbit Genetics Cooperative Report*, Vol. 10, Article 38, pp. 74-75, *http://cuke.hort.ncsu.edu/cgc/cgc10/cgc10-38.html*.

ANU (2005), Research School of Biological Sciences, Australian National University, *http://life.anu.edu.au./viruses/ICTVdB/57010077.htm* (accessed July 2005).

Argueta, A. (1994), *Atlas de las Plantas de la Medicina Tradicional Mexicana. 3 Volumes*, Instituto Nacional Indigenista, Mexico City, Mexico.

Arumuganathan, K. and E.D. Earle (1991), "Nuclear DNA content of some important plant species", *Plant Molecular Biology Reporter*, Vol. 9, No. 3, pp. 208-218.

Asch, D.L. and N.B. Asch (1992), "Archeaobotany", in: Stafford, R. (ed.), *Geoarchaeology of the Ambrose Flick Site*, Res. Series 10, Center for American Archaeology, Kampsville, Illinois.

Asch, D.L. and N.A. Sidell (1992), "Archeobotany", in: Stafford, C.R. (ed.), *Early Woodland Occupations at the Ambrose Flick Site in the Northern Syn Bottom of West Central Illinois*, Research Series 10, Center for American Achaeology, Kampsville, Illinois, pp. 167-263.

Avila-Sakar, G., G.A. Krupnick and A.G. Stephenson (2001), "Growth and resource allocation in *Cucurbita pepo* ssp. *texana*: Effects of fruit removal", *International Journal of Plant Sciences*, Vol. 162, No. 5, pp. 1 089-1 095, *http://dx.doi.org/10.1086/322879*.

Avila-Sakar, G., L.L Leist and A.G. Stephenson (2003), "Effects of the spatial pattern of leaf damage on growth and reproduction: Nodes and branches", *Journal of Ecology*, Vol. 91, No. 5, pp. 867-879, October, *www.jstor.org/stable/3599710*.

Azurdia-Pérez, C.A. and M. González-Salán (1986), *Informe Final del Proyecto de Recolección de Algunos Cultivos Nativos de Guatemala*, Facultad de Agronomía, Universidad de San Carlos de Guatemala (USAC), Instituto de Ciencia y Technología Agrícolas (ICTA) and Consejo Internacional de Recursos Fitogenéticos (CIRF), IBPGR, Guatemala City, pp. 256.

Bailey, L.H. (1948), "Jottings in the cucurbitas", *Gentes Herbarum 7*, pp. 447-477.

Bailey, L.H. (1943), "Species of *Cucurbita*", *Gentes Herbarum 6*, pp. 265-322.

Bailey, L.H. (1929), "The domesticated cucurbitas. I", *Gentes Herbarum 2*, pp. 63-115.

Bar-Nun, N. and A.M. Mayer (1990), "Cucurbitacins protect cucumber tissue against *Botrytis cinerea*", *Phytochem.* Vol. 29 No. 3, pp. 787-791, *http://dx.doi.org/10.1016/0031-9422(90)80019-D*.

Bisognin, A.D. (2002), "Origin and evolution of cultivated Cucurbits", *Ciencia Rural, Santa Maria*, Vol. 32, No. 5, pp. 715-713, *www.scielo.br/pdf/cr/v32n4/a28v32n4.pdf*.

Blancard, D., H. Lecoq and M. Pitrat with M. Javoy (1994), *A Colour Atlas of Cucurbit Diseases: Observation, Identification and Control*, Manson Publishing, London, and Wiley & Sons, N.Y.

Borghi, B. (1976), "Evaluation of necrosis in Cucurbita pepo L.", in: *Heterosis in Plant Breeding*, Proceedings 7th Congress, Eucarpia, Elsevier, Amsterdam, Netherlands.

Boyette, G., E. Templeton and L.R. Oliver (1984), "Texas gourd (*Cucurbita texana*) control", *Weed Science*, Vol. 32, No. 5, pp. 649-655, September, *www.jstor.org/stable/4043984*.

Boykin, C. et al. (2011), "Cucurbitacin IIa: A novel class of anti-drug inducing non-reversible actin aggregation and inhibiting survivin independent of JAK2/STAT3 phosphorylation", *British J. of Cancer*, Vol. 104, No. 5, pp. 781-789, March, *http://dx.doi.org/10.1038/bjc.2011.10*.

Bretting, P.K. (1990), "New perspectives on the origin and evolution of New World domesticated plants: Introduction", *Economic Botany*, Vol. 44, No. 3, Supplement: New Perspectives on the Origin and Evolution of New World Domesticated Plants, pp. 1-5, *www.jstor.org/stable/4255267*.

Brown, R.N. and J.R. Myers (2002), "A genetic map of squash (*Cucurbita* sp.) with randomly amplified polymorphic DNA markers and morphological markers", *J Amer. Soc. for Hort. Science*, Vol. 127 No. 4, pp. 568-575, *http://journal.ashspublications.org/content/127/4/568.full.pdf*.

Brunt, A.A. et al. (eds.) (1996), "Plant viruses online: Descriptions and lists from the VIDE Database", C.A.B. International, U.K., *http://pvo.bio-mirror.cn/refs.htm#vide96*.

Brust, G.E. and M.E. Barbercheck (1992), "Effect of dietary cucurbitacin C on southern corn rootworm (Coleoptera: Chrysomelidae) egg survival", *Environmental Entomology*, Vol. 21, No. 6, pp. 1 466-1 471, *http://dx.doi.org/10.1093/ee/21.6.1466*.

Bukasov, S.M. (1981), *Las Plantas Cultivadas de México, Guatemala y Colombia*, Proyecto CATIE-GTZ de Recursos Geneticos, Turrialba, Costa Rica.

Bush, D.R. and P.J. Langston-Unkefer (1988), "Amino acid transport into membrane vesicles isolated from zucchini", *Plant Physiology*, Vol. 88, No. 2, pp. 487-490, October.

Canto-Aguilar, M.A. and V. Parra-Tabla (2000), "Importance of conserving alternative pollinators: Assessing the pollination efficiency of the squash bee, *Peponapis moschata* (Cucurbitaceae)", *Journal of Insect Conservation*, Vol. 4, No. 3, pp. 203-210, September.

Cárdenas, M. (1944), "Las cucurbitas cultivadas de Bolivia", *Revista Agricola* (*Cochabamba*), Vol. 2, pp. 3-12.

Cardoso, A. (2004), "Depression by inbreeding after four successive self-pollination squash generations", *Scientia Agricola*, Vol. 61, pp. 224-227.

Carpin, S. et al. (1999), "Molecular cloning and tissue-specific expression of an anionic peroxidase in zucchini", *Plant Physiology*, Vol. 120, No. 3, pp. 799-810, July.

Castetter, E.F. (1925), "Horticultural groups of cucurbits", *Proceedings of the American Society for Horticultural Science*, Vol. 22, pp. 338-340.

Chala, V.H., C.W. Harrison and R.S. Halliwell (1987), "Identification of two distinct strains of watermelon mosaic virus 2 affecting cucurbits in Texas", *Plant Disease*, Vol. 71, No. 8, pp. 750-752, *http://dx.doi.org/10.1094/PD-71-0750*.

Chiej, R. (1984), *Encyclopaedia of Medicinal Plants (MacDonald Encyclopedias)*, Little Brown and Company, London.

Chopra, R.N., N. Nayar and I.C. Chopra (1956), *Glossary of Indian Medicinal Plants*, Council of Scientific and Industrial Research, New Delhi, India.

Chou, H.C. and M. Huangfu (1960), "Pumpkin seed (*Cucurbita moschata*) in the treatment of acute schistosomiasis", *Chinese Medical Journal*, Vol. 80, No. 2, pp. 115-120, February.

Chung, S.M., D.S. Decker-Walters and J.E. Staub (2003), "Genetic relationships within the Cucurbitaceae as assessed by consensus chloroplast simple sequence repeats (ccSSR) marker and sequ. analyses", *Can. J. Botany* Vol.81 No.8 pp. 814-832, *http://dx.doi.org/10.1139/b03-074.*

Cohen, R., A. Hanan and H.S. Paris (2003), "Single-gene resistance to powdery mildew in zucchini squash (*Cucurbita pepo*)", *Euphytica*, Vol. 130, No. 3, pp. 433-441, April.

Cowan, C.W. and B.D. Smith (1993), "New perspectives on a wild gourd in eastern North America", *Journal of Ethnobiology*, Vol. 13, No. 1, pp. 17-54, Summer, available at: *https://ethnobiology.org/sites/default/files/pdfs/JoE/13-1/CowanSmith.pdf.*

Cutler, H.C. and T.W. Whitaker (1969), "A new species of *Cucurbita* from Ecuador", *Annals of the Missouri Botanical Garden*, Vol. 55, No. 3, pp. 392-396, *http://dx.doi.org/10.2307/2395132.*

Cutler, H.C. and T.W. Whitaker (1961), "History and distribution of the cultivated cucurbits in the Americas", *American Antiquity* Vol. 26 No. 4, pp. 469-485, *http://dx.doi.org/10.2307/278735.*

Dane, F. (1983), "Cucurbits", in: Tanksley, S.D. and T.J. Orton (eds.), *Isozymes in Plant Genetics and Breeding*, Part B, Elsevier Science Publishers, Amsterdam, Netherlands.

Davis, R.F. and M.K. Mizuki (1987), "Detection of cucurbit viruses in New Jersey", *Plant Disease*, Vol. 71, No. 1, pp. 40-44, *www.apsnet.org/publications/plantdisease/backissues/Documents /1987Articles/PlantDisease71n01_40.pdf.*

Davis, R.M. et al. (2008), "Diseases", in: *UC IPM Pest Management Guidelines: Cucurbits*, University of California, Agriculture and Natural Resources Publication 3445, pp. 45-65, *www.ipm.ucdavis.edu/PMG/selectnewpest.cucurbits.html* (accessed November 2009).

de Oliveira, A.C.B. et al. (2003), "Resistance to papaya ringspot virus in summer squash *Cucurbita pepo* L. introgressed from an interspecific *C. pepo* × *C. moschata* cross", *Euphytica*, Vol. 132, No. 2, pp. 211-215, July.

Decker, D.S. (1988), "Origin(s), evolution, and systematics of *Cucurbita pepo*", *Economic Botany*, Vol. 42, No. 1, pp. 4-15, January-March, *www.jstor.org/stable/4255033.*

Decker, D.S. (1986), "A biosystematic study of *Cucurbita pepo*", Ph.D. Thesis, Texas A&M University, College Station.

Decker, D.S. (1985), "Numerical analysis of allozyme variationin *Cucurbita pepo* L.", *Economic Botany*, Vol. 39, No. 3, pp. 300-309, July.

Decker, D.S. and H.D. Wilson (1987), "Allozyme variation in the *Cucurbita pepo* complex: *C. pepo* var. *ovifera* vs. *C. texana*", *Systematic Botany*, Vol. 12, No. 2, pp. 263-273, April-June, *www.jstor.org/stable/2419320.*

Decker-Walters, D.S. and T.W. Walters (2000), "Squash", in: Kiple, K.F. and K.C. Ornelas (eds.), *The Cambridge World History of Food*, Cambridge University Press, pp. 335-351.

Decker-Walters, D.S. et al. (2002), "Diversity in free-living populations of *Cucurbita pepo* (Cucurbitaceae) as assessed by random amplified polymorphic DNA", *Systematic Botany*, Vol. 27, No. 1, pp. 19-28, January-March, *www.jstor.org/stable/3093892.*

Decker-Walters, D.S. et al. (1993), "Isozymic chacterization of wild populations of *Cucurbita pepo*", *Journal of Ethnobiology*, Vol. 13, No. 1, pp. 55-72, Summer, available at: *https://ethnobiology.org/sites/default/files/pdfs/JoE/13-1/Decker-Walters.et.al.pdf.*

Decker-Walters, D.S. et al. (1990), "Genealogy and gene flow among annual domesticated species of *Cucurbita*", *Can. J. of Botany*, Vol. 68, No. 4, pp. 782-789, *http://dx.doi.org/10.1139/b90-104.*

Desbiez, C. and H. Lecoq (1997), "Zucchini yellow mosaic virus", *Plant Pathology*, Vol. 46, pp. 809-829, *http://onlinelibrary.wiley.com/doi/10.1046/j.1365-3059.1997.d01-87.x/pdf.*

Dinan, L. et al. (1997), "Cucurbitacins are insect steroid hormone antagonists acting at the ecdysteroid receptor", *Biochemical Journal*, Vol. 327, Pt. 3, pp. 643-650, November.

Discover Life (2011), Interactive encyclopedia about the taxonomy, natural history, distribution, abundance and ecology of species, at: *www.discoverlife.org* (accessed September 2011).

Dodds, J.A. et al. (1984), "Aphid- and whitefly-transmitted cucurbit viruses in Imperial County, California", *Phytopathology*, Vol. 72, No. 2, pp. 221-225, *http://dx.doi.org/10.1094/Phyto-74-221*.

Du, X. et al. (2011), "Genetic divergence among inbred lines of *Cucurbita moschata* from China", *Scientia Horticulturae* Vol. 127 No. 3, pp. 207-213, *http://dx.doi.org/10.1016/j.scienta.2010.10.018*.

Duke, J.A. and E.S. Ayensu (1985), *Medicinal Plants of China*, Reference Publications, Inc., Missouri, United States.

Eben, A. and M.E. Barbercheck (1996), "Field observations on host plant associations and natural enemies of diabroticite beetles (Chrysomelidae: Luperini) in Veracruz, Mexico", *Acta Zoológica Mexicana*, Vol. 67, pp. 47-65.

Eben, A., M.E. Barbercheck and M. Aluja (1997), "Mexican diabroticite beetles (Chrysomelidae: Galerucinae): II. Laboratory tests for the preference of cucurbitacins", *Entomologia Experimentalis et Applicata*, Vol. 82, No. 1, pp. 63-72, January, *http://dx.doi.org/10.1046/j.1570-7458.1997.00113.x*.

Ellsworth, P.C. and J.L. Martinez-Carrillo (2001), "IPM for *Bemisia tabaci*: A case study from North America", *Crop Protection*, Vol. 20, pp. 853-869, *www.issg.org/database/species/reference_files/bemtab/Ellsworth%20and%20Martinez-Carrillo,%202001.pdf*.

Emshwiller, E. (2006), "Genetic data and plant domestication", Chapter 8, in: Zeder, M.A. et al. (eds.), *Documenting Domestication: New Genetic and Archaeological Paradigms*, University of California Press, Berkeley and Los Angeles, California, pp. 99-122, *www.botany.wisc.edu/courses/botany_940/06CropEvol/papers/Emshwiller06.pdf*.

Esquinas-Alcazar, J.T. and P.J. Gulick (1983), *Genetic Resources of Cucurbitaceae: A Global Report*, IBPGR, Rome.

Esteras, C. et al. (2008), "Diversity of Spanish landraces of Cucumis Sativus and Cucurbita ssp.", in: Pitrat, M. (ed.), *Cucurbitaceae 2008*, Proceedings of the IX EUCARPIA Meetings on Genetics and Breeding of Cucurbitaceae, INRA, Avignon, France, 21-24 May 2008.

Facciola, S. (1990), *Cornucopia: A Source Book of Edible Plants*, Kampong Publications, Vista, California.

FAOSTAT (2013), FAO Statistics online database, "Production / Crops, – pumpkins, squash and gourds, Year 2013", Food and Agriculture Organization of the United Nations, *http://faostat3.fao.org/home/E* (accessed 5 January 2016).

Ferguson, J.E. and R.L. Metcalf (1985), "Cucurbitacins: Plant derived defense compounds for *Diabroticites* (Coleoptera: Chrysomelidae)", *Journal of Chemical Ecology*, Vol. 11, No. 3, pp. 311-318, March, *http://dx.doi.org/10.1007/BF01411417*.

Ferriol, M. and B. Picó (2008), "Pumpkin and winter squash", in: Prohens-Tomas, J., F. Nuez and M.J. Carena (eds.), *Handbook of Vegetable Breeding*, Springer, Berlin, pp 317-349.

Ferriol, M., B. Picó and F. Nuez (2004), "Morphological and molecular diversity of a collection of *Cucurbita maxima* landraces", *Journal of the American Society for Horticultural Science*, Vol. 129, No. 1, pp. 60-69, January.

Ferriol, M., B. Picó and F. Nuez (2003a), "Genetic diversity of a germplasm collection of *Cucurbita pepo* using SRAP and AFLP markers", *Theoretical and Applied Genetics*, Vol. 107, No. 2, pp. 271-282, July.

Ferriol, M., B. Picó and F. Nuez (2003b), "Genetic diversity of some accessions of *Cucurbita maxima* from Spain using RAPD and SBAP markers", *Genetic Resources and Crop Evolution*, Vol. 50, No. 3, pp. 227-238, May.

Ferriol, M. et al. (2004), "Molecular diversity of a germplasm collection of squash (*Cucurbita moschata*) determined by SRAP and AFLP markers", *Crop Science*, Vol. 44, No. 2, pp. 653-664, *http://dx.doi.org/10.2135/cropsci2004.6530*.

Filov, A.I. (1966), "Ekologija i klassifikatzija tykuy", *Bjulleten Glavnogo Botaniceskogo Sada*, Vol. 63, pp. 33-41 (in Russian).

Frisse, A., M.J. Pimenta and T. Lange (2003), "Expression studies of gibberellin oxidases in developing pumpkin seeds", *Plant Physiology*, Vol. 131, No. 3, pp. 1 220-1 227, March, *http://dx.doi.org/10.1104/pp.015206*.

Fujiyoshi, P.T., S.R. Gliessman and J.H. Langenheim (2007), "Factors in the suppression of weeds by squash interplanted with corn", *Weed Biology and Management*, Vol. 7, No. 2, pp. 105-114, June, *http://dx.doi.org/10.1111/j.1445-6664.2007.00242.x*.

Fukami, H. and R. Nishida (1990), "Sequestration of distasteful compounds by some pharmacophagous insects", *Journal of Chemical Ecology*, Vol. 16, No. 1, pp. 151-164, January, *http://dx.doi.org/10.1007/BF01021276*.

Fursa, T.B. and A.I. Filov (1982), "Cucurbitaceae", in: Korovina, O.N. and T.B. Fursa (eds), *Flora of Cultivated Plants*, Kolos, Moscow, pp. 102-132.

Gallegos, J.J.G. (1990), "Avance generacional de genotipos de calabacita con resistencia a virus mosaico del pepino y de sandía 1 y 2", in : *Informe de Investigación, Ciclo 1989-90*, Proyecto: Estudio y Control de las Enfermedades Virales en el Cultivo de la Calabacita, INIFAP-SARH, Mexico, pp. 1-6.

Gámez-Virués, S. and A. Eben (2005), "Comparison of beetle diversity and incidence of parasitism in Diabroticina (Coleoptera: Chrysomelidae) species collected on cucurbits", *Florida Entomol.* Vol. 88 No. 1 pp. 72-76, *http://journals.fcla.edu/flaent/article/view/75391/73049*.

Garzón-Tiznado, J.A., S. Montes-Hernández and F.A. Becerra (1993), "Fuentes de resistencia a los virus mosaico del pepino y de la sandía-2, en calabaza (*Cucurbita moschata*) en México", *Revista Mexicana de Fitopatología*, Vol. 11, pp. 167-171.

Gilbert-Albertini, F. et al. (1993), "Resistance of *Cucurbita moschata* to watermelon mosaic virus type 2 and its genetic relation to resistance to zucchini yellow mosaic virus", *Euphytica*, Vol. 69, No. 3, pp. 231-237, January.

Gillespie, J.J. et al. (2003), "Convergent evolution of cucurbitacin feeding in spatially isolated rootworm taxa (Coleoptera: Chrysomelidae; Galerucinae, Luperini)", *Molecular Phylogenetics and Evolution*, Vol. 29, No. 1, pp. 161-175, October.

Gong, L. et al. (2012), "Genetic relationships and evolution in *Cucurbita pepo* (pumpkin, squash, gourd) as revealed by simple sequence repeat polymorphisms", *Theoretical and Applied Genetics*, Vol. 124, No. 5, pp. 875-891, March, *http://dx.doi.org/10.1007/s00122-011-1752-z*.

Gossell-Williams, M., A. Davis and N. O'Connor (2006), "Inhibition of testosterone-induced hyperplasia of the prostate of sprague-dawley rats by pumpkin seed oil", *Journal of Medicinal Food*, Vol. 9, No. 2, pp. 284-286, Summer.

Gray, A. (1850), "Cucurbitaceae *Cucurbita texana*", *Boston Journal of Natural History*, Vol. 6. See also: *www.theplantlist.org/tpl/record/kew-2747227*.

Greber, R.S. (1978), "Watermelon mosaic virus 1 and 2 in Queensland cucurbit crops", *Australian Journal of Agricultural Research*, Vol. 29, No. 6, pp. 1 235-1 245, *http://dx.doi.org/10.1071/AR9781235*.

Grubben, G.L.H. (2004), *Plant Resources of Tropical Africa 2: Vegetables*, PROTA Foundation/Backhuys Publishers/CTA Wageningen, Netherlands, *www.ebookdb.org/reading/53283D1E3E6535243D407369/Plant-Resources-Of-Tropical-Africa--Vegetables*.

Gwanama, C., M.T. Labuschagne and A.M. Botha (2000), "Analysis of genetic variation in *Cucurbita moschata* by random amplified polymorphic DNA (RAPD) markers", *Euphytica*, Vol. 113, No. 1, pp. 19-24, May.

Hart, J.P., R.A. Daniels and C.J. Sheviak (2004), "Do *Cucurbita* gourds float fishnets?", *American Antiquity*, Vol. 69, No. 1, pp. 141-148, January, *http://dx.doi.org/10.2307/4128352*.

Havelda, Z. and A.J. Maule (2000), "Complex spatial responses to cucumber mosaic virus infection in susceptible *Cucurbita pepo* cotyledons", *The Plant Cell*, Vol. 12, No. 10, pp. 1 975-1 985, October.

Havey, M.J. et al. (1998), "Differential transmission of the *Cucumis* organellar genomes", *Theoretical Applied Genetics*, Vol. 97, No. 1, pp. 122-128, July.

Hayes, C.N., J.A. Winsor and A.G. Stephenson (2005a), "Environmental variation influences the magnitude of inbreeding depression in *Cucurbita pepo* ssp. *texana* (Cucurbitaceae)", *Journal of Evolutionary Biology*, Vol. 18, No. 1, pp. 147-155, January.

Hayes, C.N., J.A. Winsor and A.G. Stephenson (2005b), "Multigenerational effects of inbreeding in *Cucurbita pepo* ssp. *texana* (Cucurbitaceae)", *Evolution*, Vol. 59, No. 2, pp. 276-286.

Hemleben, V. et al. (1988), "Organization and length heterogeneity of plant ribosomal RNA genes", in: Kahl, G. (ed.), *Architecture of Eukaryotic Genes*, VCH Verlagsgesellschaft, Weinheim, pp. 371-383.

Hendrix, S.D. (1988), "Herbivory and its impact on plant reproduction", in: Lovett Doust, J. and L. Lovett Doust (eds.), *Plant Reproductive Biology*, Oxford University Press, Oxford, pp. 246-263.

Hodgekinson, I.D. and M.K Hughes (1982), *Insect Herbivory*, Chapman and Hall, New York.

Huang, P.L. et al. (1991), "Two genes encoding 1-aminocyclopropane-1-carboxylate synthase in zucchini (*Cucurbita pepo*) are clustered and similar but differentially regulated", *Proceedings of the National Academy of Sciences USA*, Vol. 88, No. 16, pp. 7 021-7 025.

Hurd, P.D. (1966), "The pollination of pumpkins, gourds and squashes (Genus Cucurbita)", in: Second International Symposium on Pollination, 1964, *Bee World*, Vol. 47 (Supp), pp. 97-98.

Hurd, P.D. and E.G. Linsley (1970), "A classification of the squash and gourd bees *Peponapis* and *Xenoglossa*", *California Publications in Entomology*, Vol. 62, pp. 1-39.

Hurd, P.D. and E.G. Linsley (1967), "South American squash and gourd bees of the genus *Peponapis* (Hymenoptera: Apoidea)", *Annals of the Entomology Society of America*, Vol. 60, No. 3, pp. 647-661, *http://dx.doi.org/10.1093/aesa/60.3.647*.

Hurd, P.D. and E.G. Linsley (1964), "The squash and gourd bees – genera *Peponapis* Robertson and *Xenoglossa* Smith – inhabiting America north of Mexico (Hymenoptera: Apoidea)", *Hilgardia*, Vol. 35, No. 15, pp. 375-477, *http://dx.doi.org/10.3733/hilg.v35n15p375*.

Hurd, P.D., E.G. Linsley and T.W. Whitaker (1971), "Squash and gourd bees (*Peponapis*, *Xenoglossa*) and the origin of the cultivated *Cucurbita*", *Evolution*, Vol. 25, No. 1, pp. 218-234, March, *http://dx.doi.org/10.2307/2406514*.

ISSG, *Global Invasive Species Database*, *www.issg.org* (accessed October 2011).

Jahn, M., H.M. Munger and J.D. McCreight (2002), "Breeding cucurbit crops for powdery mildew resistance", in: Berlanger, R.R. et al (eds.), *The Powdery Mildews, A Comprehensive Treatise*, American Phytopathological Society, St. Paul, Minnesota, pp. 239-248.

Janick, J. and R.E. Paull (2007), *The Encyclopedia of Fruits and Nuts*, CAB International, Oxfordshire, United Kingdom.

Jeffrey, C. (1990), "Systematics of the Cucurbitaceae: An overview", in: Bates, D.M., R.W. Robinson and C. Jeffrey (eds.), *Biology and Utilization of the Cucurbitaceae*, Cornell University Press, Ithaca, New York, pp. 3-9.

Jobst, J., K. King and V. Hemleben (1998), "Molecular evolution of the internal transcribed spacers (ITS1 and ITS2) and phylogenetic relationships among species of the family Cucurbitaceae", *Molecular Phylogenetics and Evolution*, Vol. 9, No. 2, pp. 204-219, April.

Jossey, S. and M. Badadoost (2006), "First report of Tobacco ringspot virus in *Cucurbita pepo* in Illinois", *Plant Disease*, Vol. 90, No. 10, p. 1 361, *http://dx.doi.org/10.1094/PD-90-1361B*.

Julier, H.E. and T.H. Roulston (2009), "Wild bee abundance and pollination service in cultivated pumpkins; farm management, nesting behavior, and landscape effects", *Journal of Economic Entomology*, Vol. 102, No. 2, pp. 563-573, April.

Katzir, N. et al. (2000), "Further ISSR and preliminary SSR analysis of relationships among accessions of *Cucurbita pepo*", *Acta Horticulturae*, Vol. 510, No. 69, pp. 433-439, *http://dx.doi.org/10.17660/ActaHortic.2000.510.69*.

King, F.B. (1985), "Early cultivated cucurbits in Eastern North America", in: Ford, R.I. (ed.), *Prehistoric Food Production in North America*, Anthropological Paper No. 75, Museum of Anthropology, University of Michigan, Ann Arbor, Michigan.

King, K., J. Jobst and V. Hemleben (1995), "Differential homogenization and amplification of two satellite DNAs in the genus *Cucurbita* (Cucurbitaceae)", *Journal of Molecular Evolution*, Vol. 41, No. 6, pp. 996-1 005, December.

King, K. et al. (1993), "Molecular evolution of the intergenic spacer in the nuclear ribosomal RNA genes of Cucurbitaceae", *Journal of Molecular Evolution*, Vol. 36, No. 2, pp. 144-152.

Kirkpatrick, K.J. and H.D. Wilson (1988), "Interspecific gene flow in *Cucurbita: C. texana* vs. *C. pepo*", *Americ. J. of Botany*, Vol. 75, No. 4, pp. 519-527, *www.jstor.org/stable/2444217*.

Kirkpatrick, K.J., D.S. Decker and H.D. Wilson (1985), "Allozyme differentiation in the *Cucurbita pepo* complex: *C. pepo* var. *medullosa* vs. *C. texana*", *Economic Botany*, Vol. 39, No. 3, pp. 289-299, July.

Kreuter, M.H. (2000), "Hormone disbalance and urinary bladder disease", in: *Phyto Novum*, Emil Flachsmann AG, Zurich, Switzerland, pp. 2-24.

Krysan, J.L. and R.F. Smith (1987), "Systematics of the *virgifera* species group of *Diabrotica* (Coleoptera: Chrysomelidae: Galerucinae)", *Entomography*, Vol. 5, pp. 375-484.

Kwon, Y.-S. et al. (2004), "A comparative study of the RAPD and SSR markers in establishing a genetic relationship of the various types of *Cucurbita*", *Korean Journal of Genetics*, Vol. 26, No. 2, pp. 115-122.

Lance, D.R. and G.R. Sutter (1990), "Field-cage and laboratory evaluations of semiochemical-based baits for managing western corn rootworm (Coleoptera: Chrysomelidae)", *Journal of Economic Entomology*, Vol. 83, No. 3, pp. 1 985-1 990, *http://dx.doi.org/10.1093/jee/83.3.1085*.

Lebeda, A. and E. Křístová (2000), "Interactions between morphotypes of *Cucurbita pepo* and obligate biotrophs (*Pseudoperonospora cubensis*, *Erysiphe cichoracearum* and *Sphaerotheca fulginea*)", *Acta Hort.* Vol. 510, pp. 210-225, *http://dx.doi.org/10.17660/ActaHortic.2000.510.35*.

Lebeda, A. and E. Křístová (1996), "Genotypic variation in field resistance of *Cucurbita pepo* cultivars to powdery mildew (*Erysiphe cichoracearum*)", *Genetic Resources and Crop Evolution*, Vol. 43, No. 1, pp. 79-84, February.

Lebeda, A. and M.P. Widrlechner (2004), "Response of wild and weedy *Cucurbita* L. to pathotypes of *Pseudoperonospora cubensis* (Berk. & Curt.) Rostov (cucurbit downy mildew)", in: Spencer-Phillips, P.T.N. and M. Jeger (eds.), *Advances in Downy Mildew Research – Volume 2*, Kluwer Academic Publishers, Dordrecht, The Netherlands, pp. 203-210.

Lebeda, A., E. Křístková and K. Doležal (1999), "Peroxidase isozyme polymorphism in *Cucurbita pepo* cultivars with various morphotypes and different level of field resistance to powdery mildew", *Sci. Hort.*, Vol. 81, No. 2, pp. 103-112, *http://dx.doi.org/10.1016/S0304-4238(98)00261-1.*

Lebeda, A. et al. (2006), "Cucurbits (Cucurbitaceae; Cucumis spp., Cucurbita spp., Citrullis spp.)", Chapter 8, in: Singh, R.S. (ed.), *Genetic Resources, Chromosome Engineering and Crop Improvement: Vegetable Crops Volume 3*, CRC Press, Boca Raton, Florida, pp. 271-376, *http://dx.doi.org/10.1201/9781420009569.ch8.*

Lecoq, H. and C. Desblez (2009), "Watermelon mosaic virus and zucchini yellow virus", in: Mahy, B.W.J. and M.H.V. Van Regenmortel (eds.), *Desk Encyclopedia of Plant and Fungal Virology*, Academic Press, Oxford.

Lecoq, H., M. Pitrat and M. Clement (1981), "Identification and characterization of a potyvirus that induces muskmelon yellow stunt", *Agronomie*, Vol. 1, pp. 827-834.

Lin, L.S. and J.E. Varner (1991), "Expression of ascorbic acid oxidase in zucchini squash (*Cucurbita pepo* L.)", *Plant Physiology*, Vol. 96, No. 1, pp. 159-165, May.

Lira, R. (1991), *Estudios Taxonómico y Ecogeográfico de las Cucurbitáceas de Latinoamérica, 2º Reporte Semestral*, International Board for Plant Genetic Resources, Rome.

Lira, R. (1988), *Cucurbitaceae de la Peninsula de Yucatán: Taxonomía y Etnobotánica*, M.S. Thesis, Instituto Nacional de Investigaciones sobre Recursos Bióticos, Xalapa, Veracruz, Mexico.

Lira, R. (1985), "Identidad taxonómica de las calabazas cultivadas (*Cucurbita* spp.) en la Peninsula de Yucatán", *Biótica*, Vol. 10, No. 3, pp. 301-307.

Lira, R., T.C. Andres and M. Nee (1995), "Cucurbita L.", in: Lira, R. (ed.), *Estudios Taxonómicos y Ecogeográficos de las Cucurbitaceae Latinoamericanas de Importancia Económica: Cucurbita, Sechium, Sicana y Cyclanthera*, Systematic and Ecogeographic Studies on Crop Genepools 9, International Plant Genetic Resources Institute, Rome, pp. 1-115.

Lira-Saade, R. and S. Montes-Hernández (1992), "Cucúrbitas (Cucurbita spp.)", in: Hernández-Bermejo, J.E. and J. León (eds.), *Cultivos Marginados: Otra Perspectiva de 1492*, Colección FAO, Producción y Protección Vegetal No. 26, FAO, Rome, pp. 61-75.

Lira-Saade, R., L. Eguiarte Fruns and S. Montes Hernández (2009), *Proyecto Recopilación y Análisis de la Información Existente de las Especies de los Géneros* Cucurbita *y* Sechium *que Crecen y/o se Cultivan en México*, Informe Final, CONABIO, México, D.F, *www.biodiversidad.gob.mx/genes/centrosOrigen/Cucurbita%20y%20Sechium/Informe_Final/Informe%20final%20Cucurbita%20y%20Sechium.pdf.*

Lisa, V. et al. (1981), "Characterization of a potyvirus that causes zucchini yellow mosaic", *Phytopathology*, Vol. 71, No. 7, pp. 667-672, available at: *www.apsnet.org/publications/phytopathology/backissues/Documents/1981Articles/Phyto71n07_667.pdf.*

Lopez, J.F., S. Brune and G.P. Henz (1999), "Metodologia de avaliacao da resistencia de germoplasma de aboboras e morangas a *Phytophthora capsici*", *Horticulura Brasileira*, Vol. 17, Supplement, pp. 30-32 (in Portuguese).

Lovatt, C.J. (1983), "*De novo* purine biosynthesis in intact cells of *Cucurbita pepo*", *Plant Physiology*, Vol. 73, No. 3, pp. 766-772, November.

Loy, J.B. and C.E. Broderick (1990), "Growth, assimilate partitioning, and productivity of bush and vine *Cucurbita maxima*", in: Bates, Robinson and Jeffrey (eds.), *Biology and Utilization of the Cucurbitaceae*, Cornell University, Ithaca, New York and London, pp. 436-447.

Mahzabin, F., S. Parvez and M.F. Alam (2008), "Micropropagation of *Cucurbita maxima* Duch. through shoot tip culture", *Journal Bio-science*, Vol. 16, pp. 59-65.

Maluf, W.R. and E.L.S. Souza (1984), "Resistência ao vírus do mosaico da melancia-1 (WMV-1) em moranga *Cucurbita maxima* Duch", *Horticultura Brasileira*, Vol. 2, pp. 22-25 (in Portuguese).

Mariano, B.N. (2001), "Efectos de la Herbivoría sobre la Adecuación Masculina y Femenina de *Cucurbita argyrosperma* ssp. *Sororia*", PhD Thesis, Instituto de Ecología, Universidad Nacional Autónoma de México, México, D.F.

Mariano, N. and R. Dirzo (2002), "*Cucurbita argyrosperma* ssp. *sororia* (L.H. Bailey) Merrick & Bates (Cucurbitaceae), Agüichichi", in: Noguera-Martínez, F. et al. (eds.), *Historia Natural de Chamela*, Instituto de Biología, Universidad Nacional Autónoma de México, pp. 167-170.

Markham, P.G. et al. (1994), "The transmission of geminiviruses by *Bemisia tabaci*", *Pesticide Science*, Vol. 42, No. 2, pp. 123-128, October, *http://dx.doi.org/10.1002/ps.2780420209*.

Martínez-Crovetto, R. (1974), "*Cucurbitaceae*", in: Burkart, A. (ed.), *Flora Ilustrada de Entre Ríos (Argentina)*, Colección Científica del I.N.T.A., Buenos Aires, pp. 63-94.

Martínez-Crovetto, R. (1965), "*Cucurbitaceae*", in: Cabrera, A.L. (ed.), *Flora de la Provincia de Buenos Aires*, Colección Científica del I.N.T.A., Buenos Aires, pp. 390-407.

Martínez-Crovetto, R. (1954), "Synopsis des Cucurbitacées de l'Uruguay", *Notulae Systematicae (Paris)*, Vol. 15, Nos. 1-4, pp. 47-55.

Merrick, L.C. (1995), "Squashes, pumpkins and gourds", in: Smartt, J. and S.W. Simmons (eds.), *Evolution of Crop Plants*, pp. 97-105, Longman Scientific and Technical, London.

Merrick, L.C. (1991), "Systematics, evolution, and ethnobotany of a domesticated squash, its wild relatives and allied species in the genus Cucurbita", PhD Thesis Cornell Univ., Ithaca, NY.

Merrick, L.C. (1990), "Systematics and evolution of a domesticated squash, *Cucurbita argyosperma*, and its wild and weedy relatives", in: Bates, Robinson and Jeffrey (eds.), *Biology and Utilization of the Cucurbitaceae*, Cornell University Press, Ithaca, New York, pp. 77-95.

Merrick, L.C. and D.M. Bates (1989), "Classification and nomenclature of *Cucurbita argyrosperma* Huber", *Baileya*, Vol. 23, pp. 94-102.

Merrick, L.C. and G.P. Nabhan (1985), "Natural hybridization of wild and cultivated *Cucurbita* in Mexico: Implications for crop evolution", *Ecological Society of America Bulletin*, Vol. 66.

Metcalf, R.L. (1986), "Coevolutionary adaptations of rootworm beetles (Coleoptera: Chrysomelidae) to cucurbitacins", *Journal of Chemical Ecology*, Vol. 12, No. 5, pp. 1 109-1 124, May, *http://dx.doi.org/10.1007/BF01638999*.

Metcalf, R.L., R.A. Metcalf and A.M. Rhodes (1980), "Cucurbitacins as kairomones for diabroticite beetles", *Proceedings of the National Academy of Sciences USA*, Vol. 77, No. 7, pp. 3 769-3 772, July.

Metcalf, R.L. et al. (1982), "Cucurbitacin contents and diabroticite (Coleoptera: Chrysomelidae) feeding upon *Cucurbita* spp", *Environmental Entomology*, Vol. 11, No. 4, pp. 931-937, *http://dx.doi.org/10.1093/ee/11.4.931*.

Michener, C.D. (2007), *The Bees of the World*, 2nd edition, Johns Hopkins University Press, Baltimore, Maryland.

Millán, R. (1947), "Los zapallitos de tronco de Sudamérica extratropical", *Darwiniana*, Vol. 7, No. 3, pp. 333-345, January, *www.jstor.org/stable/23211608*.

Millán, R. (1945), "Variaciones del zapallito amargo *Cucurbita andreana* y el origen de *Cucurbita maxima*", *Revista Argentina de Agronomía*, Vol. 12, pp. 86-93.

Mohanty, B.K., S.K. Mohanty and R.S. Mishra (1999), "Genetics of yield and yield components in pumpkin (*Cucurbita moschata*)", *Indian J. of Agric. Sciences*, Vol. 69, No. 11, pp. 781-783.

Montes, H.S. (2002), "Flujo Génico en Calabaza (*Cucurbita* spp.) Dentro del Sistema Milpa en el Occidente de México", PhD Thesis, Facultad de Ciencias, Universidad Nacional Autónoma de México, México, D.F.

Montes-Hernández, S. and L.E. Eguiarte (2002), "Genetic structure and indirect estimates of gene flow in three taxa of *Cucurbita* in western Mexico", *American Journal of Botany*, Vol. 89, No. 7, pp. 1 156-1 163, July, *http://dx.doi.org/10.3732/ajb.89.7.1156*.

Montes-Hernández, S., L.C. Merrick and L.E. Eguiarte (2005), "Maintenance of squash (*Cucurbita* spp.) landrace diversity by farmers' activities in Mexico", *Genetic Resources and Crop Evolution*, Vol. 52 No. 6, pp. 697-707, *http://dx.doi.org/10.1007/s10722-003-6018-4*.

Murkovic, M., A. Hillebrand and J. Winkler (1996), "Variability of fatty acid content in pumpkin seeds (*Cucurbita pepo* L)", *Zeitschrift für Lebensmittel-Untersuchung und -Forschung A*, Vol. 203, No. 3, pp. 216-219, May.

Nabhan, G.P. (1984), "Evidence of gene flow between cultivated *Cucurbita mixta* and a field edge population of wild *Cucurbita* at Onavas, Sonora", *Cucurbit Genetics Cooperative Report*, No. 7.

National Gardening Association (2009), *The Impact of Home and Community Gardening in America*. Harris Interactive, Inc.

Navas-Castillo, J., E. Fiallo-Olive and S. Sanchez-Campos (2011), "Emerging virus diseases transmitted by whiteflies", *Annual Review of Phytopathology*, Vol. 49, pp. 219-248, *http://dx.doi.org/10.1146/annurev-phyto-072910-095235*.

Nee, M. (1990), "The domestication of *Cucurbita* (Cucurbitaceae)", *Economic Botany*, Vol. 44, No. 3, pp. 56-58, July.

Nepi, M. and E. Pacini (1993), "Pollination, pollen viability and pistil receptivity in *Cucurbita pepo*", *Annals of Botany* Vol. 72 No. 6, pp. 527-536, *http://dx.doi.org/10.1006/anbo.1993.1141*.

Nesom, G.L. (2011), "New state records for *Citrullus, Cucumis*, and *Cucurbita* (Cucurbitaceae) outside of cultivation in the USA", *Phytoneuron*, Vol. 1, pp. 1-7, *www.phytoneuron.net/PhytoN-CucurbRecords.pdf*.

Newsom, L.A., S.D. Webb and J.S. Dunbar (1993), "History and geographic distribution of *Cucurbita pepo* gourds in Florida", *Journal of Ethnobiology*, Vol. 13, No. 1, pp. 75-97, Summer, *https://ethnobiology.org/sites/default/files/pdfs/JoE/13-1/Newsom.pdf*.

Nishida, R., M. Yokoyama and H. Fukami (1992), "Sequestration of cucubitacin analogs by New and Old World chrysomelid leaf beetles in the tribe Luperini", *Chemoecology*, Vol. 3, No. 1, pp. 19-24, April.

Oliver, L., S. Harrison and M. McClelland (1983), "Germination of Texas gourd (*Cucurbita texana*) and its control in soybeans (*Glycine max*)", *Weed Science*, Vol. 31, No. 5, pp. 700-706, September, *www.jstor.org/stable/4043694*.

OMAFRA, (2011), "Pumpkin and squash production", Ontario Ministry of Agriculture, Food and Rural Affairs, Ontario, *www.omafra.gov.on.ca* (accessed October 2011).

Paris, H.S. (2011), "Characterization of the *Cucurbita pepo* collection at the Newe Ya'ar Research Center, Israel", *PGR Newsletter*, Vol. 126, pp. 41-45.

Paris, H.S. (2003), "Genetic control of irregular striping, a new phenotype in *Cucurbita pepo*", *Euphytica*, Vol. 129, No. 1, pp. 119-126, January.

Paris, H.S. (2002), "Multiple allelism at a major locus affecting fruit coloration in *Cucurbita pepo*", *Euphytica*, Vol. 125, No. 2, pp. 149-153, May.

Paris, H.S. (2001), "History of the cultivar-groups of *Cucurbita pepo*", *Horticultural Reviews*, Vol. 25, pp. 71-170, *http://dx.doi.org/10.1002/9780470650783.ch2.*

Paris, H.S. (2000), "Gene for broad, contiguous dark stripes in cocozelle squash (*Cucurbita pepo*)", *Euphytica*, Vol. 115, No. 3, pp. 191-196, October.

Paris, H.S. (1997), "Genes for developmental fruit coloration of acorn squash", *Journal of Heredity*, Vol. 88, No. 1, pp. 52-56.

Paris, H.S. (1989), "Historical records, origins and development of the edible cultivar groups of *Cucurbita pepo* (Cucurbitaceae)", *Economic Botany*, Vol. 43, No. 4, pp. 423-443, October-December, *www.jstor.org/stable/4255187.*

Paris, H.S. (1986), "A proposed subclassification for *Cucurbita pepo*", *Phytologia*, Vol. 61, pp. 133-138.

Paris, H.S. and S. Cohen (2000), "Oligogenic inheritance for resistance to zucchini yellow mosaic virus in *Cucurbita pepo*", *Annals of Applied Biology*, Vol. 136, No. 3, pp. 209-214, June, *http://dx.doi.org/10.1111/j.1744-7348.2000.tb00027.x.*

Paris, H.S. et al. (2003), "Assessment of genetic relationships in *Cucurbita pepo* (Cucurbitaceae) using DNA markers", *Theoretical Applied Genetics*, Vol. 106, No. 6, pp. 971-978, April, *http://dx.doi.org/10.1007/s00122-002-1157-0.*

Parodi, L.R. (1935), "Relaciones de la agricultura prehispánica con la agricultura argentina actual", *Anales de la Academia Nacional de Agronomía y Veterinaria*, Buenos Aires, Vol. 1, pp. 115-167.

Piperno, D.R., T.C. Andres and K.E. Stothert (2000), "Phytolits in *Cucurbita* and other neotropical Cucurbitaceae and their occurrence in early archaeological sites from the lowland American tropics", *Journal of Archaeological Science*, Vol. 27, No. 3, pp. 193-208, March, *http://dx.doi.org/10.1006/jasc.1999.0443.*

Provvidenti, R. (1990), "Viral diseases and genetic sources of resistance in Cucurbita species", in: Bates, D.M., R.W. Robinson and C. Jeffrey (eds.), *Biology and Utilization of the Cucurbitaceae*, Cornell University Press, Ithaca, New York, pp. 427-435.

Provvidenti, R. and D.M. Tricoli (2002), "Inheritance of resistance to squash mosaic virus in a squash transformed with the coat protein gene of pathotype 1", *HortScience*, Vol. 37, No. 2, pp. 575-577, *http://hortsci.ashspublications.org/content/37/3/575.full.pdf.*

Provvidenti, R., D. Gonsalves and H.S. Humaydan (1984), "Occurrence of zucchini yellow mosaic virus in cucurbits from Connecticut, New York, Florida, and California", *Plant Disease*, Vol. 68, pp. 443-446, *www.apsnet.org/publications/plantdisease/backissues/Documents/1984Articles/PlantDisease68n05_443.PDF.*

Puchalski, J.T. and R.W. Robinson (1990), "Electrophoretic analysis of isozymes in *Cucurbita* and *Cucumis* and its application for phylogenetic studies", in: Bates, D.M., R.W. Robinson and C. Jeffrey (eds), *Biology and Utilization of the Cucurbitaceae*, Cornell University Press, Ithaca, New York, pp. 60-76.

Purcifull, D.E. et al. (1984), "Serological relationships and partial characterization of zucchini yellow mosaic virus isolated from squash in Florida", *Plant Disease*, Vol. 68, No. 3, pp. 230-233, *http://dx.doi.org/10.1094/PD-69-230.*

Qasem, J.R. and N. Issa (2005), "Allelopathic effects of squash (*Cucurbita pepo* L cv Scarlette) on certain common weed species in Jordan", in: Harper, J.D. et al. (eds.), *Proceedings of the 4th World Congress on Allelopathy: "Establishing the Scientific Base"*, Wagga Wagga, New South Wales, Australia, 21-26 August 2005, pp. 258-262.

Quesada, M., J.A. Winsor and A.G. Stephenson (1996), "Effects of pollen competition on the reproductive performance in cucurbit hybrids (Cucurbitaceae): F1 and backcross generations", *Canadian J. of Botany*, Vol. 74, No. 7, pp. 1 113-1 118, *http://dx.doi.org/10.1139/b96-136.*

Quesada, M., J.A. Winsor and A.G. Stephenson (1993), "Effects of pollen competition on progeny performance in a heterozygous cucurbit", *American Naturalist*, Vol. 142, No. 4, pp. 694-706, October, *http://dx.doi.org/10.1086/285564.*

Quesada, M. et al. (1991), "Effects of genotype on pollen performance in *Cucurbita pepo*", *Sexual Plant Reproduction*, Vol. 4, No. 3, pp. 208-214, July.

Radovich, T. (2011), "Farm and forestry production and marketing profile for pumpkin and squash", Specialty crops for Pacific Island agroforestry at: *www.ctahr.hawaii.edu.*

Rehm, S. et al. (1957), "Bitter principles of the Cucurbitaceae. VII. The distribution of bitter principles in this plant family", *Journal of the Science of Food and Agriculture*, Vol. 8, No. 12, pp. 679-686, December, *http://dx.doi.org/10.1002/jsfa.2740081203.*

Rieseberg, L.H. and A.V. Harter (2006), "Molecular evidence and the evolutionary history of the domesticated sunflower", in: Motley, T. (ed), *Darwin's Harvest: New Approaches to the Origins, Evolution, and Conservation of Crops*, Columbia Univ. Press, New York, pp. 31-48.

Robinson, R.W. and D.S. Decker-Walters (1997), *Cucurbits. Crop Production Science in Horticulture*, No. 6, CAB International, Cambridge, United Kingdom.

Rochelle, L.A. (1974), "Descriçao taxonômica de cultivares de *Cucurbita maxima* Duch. e *C. pepo* L", *Anais de Escuela Superior de Agricultura "Luiz de Queiroz*, Vol. 31, pp. 667-700, *http://dx.doi.org/10.1590/S0071-12761974000100051.*

Rodríguez, I. and R. Lira (1992), "Diversidad, biología reproductiva y relaciones de Cucurbita fraterna L.H. Bailey, con especies cultivadas del género", in: *Resúmenes Simposio Etnobotánica 92*, Jardines Botánicos de Córdoba, Spain.

Rudich, J. (1990), "Biochemical aspects of hormonal regulation of sex expression in cucurbits", in: Bates, D.M., R.W. Robinson and C. Jeffrey (eds.), *Biology and Utilization of the Cucurbitaceae*, Cornell University Press, Ithaca, New York, pp. 127-134.

Rzedowski, G.C. and J. Rzedowski (2001), *Flora Fanerogámica del Valle de México*, 2nd edition, Instituto de Ecología y Comisión Nacional para el Conocimiento y Uso de la Biodiversidad, Pátzcuaro, Michoacán, Mexico.

Sanjur, O.I. et al. (2002), "Phylogenetic relationships among domesticated and wild species of *Cucurbita* (Cucurbitaceae) inferred from a mitochondrial gene: Implications for crop plant evolution and areas of origin", *Proceedings of the National Academy of Sciences USA*, Vol. 99, No. 1, pp. 535-540, January.

Sasu, M.A. et al. (2010), "Floral transmission of *Erwinia tracheiphila* by cucumber beetles in a wild *Cucurbita pepo*", *Environmental Entymology*, Vol. 39, No. 1, pp. 140-148, February, *http://dx.doi.org/10.1603/EN09190.*

Sauer, J.D. (1993), *Historical Geography of Crop Plants*, CRC Press, Boca Raton, Florida.

Schabort, J.C. (1978), "Cucurbitacin 19-hydroxylase in *Cucurbita maxima*", *Phytochemistry*, Vol. 17, No. 6, pp. 1 062-1 064, *http://dx.doi.org/10.1016/S0031-9422(00)94283-3.*

Scheerens, J.C. et al. (1987), "Floral development, flowering patterns, and growth rate of monoecious and gynoecious buffalo gourd", *Journal of the American Society for Horticultural Science*, Vol. 112, No. 3, pp. 574-578, May.

Schuster, D.J., J.B. Kring and J.F. Price (1991), "Association of the Sweetpotato whitefly with a silver leaf disorder of squash", *HortScience*, Vol. 26, No. 2, pp. 155- 156, available at: *http://hortsci.ashspublications.org/content/26/2/155.full.pdf.*

See, E.F., W.A. Wan Nadiah and A.A. Noor Aziah (2007), "Physico-chemical and sensory evaluation of breads supplemented with pumpkin flour", *ASEAN Food Journal*, Vol. 14, No. 2, pp. 123-130.

Singh, A.K. (1990), "Cytogenetics and evolution in the Cucurbitaceae", in: Bates, D.M., R.W. Robinson and C. Jeffrey (eds.), *Biology and Utilization of the Cucurbitaceae*, Cornell University Press, Ithaca, New York, pp. 10-28.

Sisko, M., A. Ivancic and B. Bohanec (2003), "Genome size analysis in the genus *Cucurbita* and its use for determination of interspecific hybrids obtained using the embryo-rescue technique", *Plant Science*, Vol. 165, No. 3, pp. 663-669, *http://dx.doi.org/10.1016/S0168-9452(03)00256-5*.

Smith, B.D. (2006), "Eastern North America as an independent center of plant domestication", *Proceedings of the National Academy of Sciences USA*, Vol. 103, No. 33, pp. 12 225-12 226, *http://dx.doi.org/10.1073/pnas.0604335103*.

Smith, B.D. (2001), "Documenting plant domestication: The consilience of biological and archaeological approaches", *Proceedings of the National Academy of Sciences USA*, Vol. 98, No. 4, pp. 1 324-1 326, *http://dx.doi.org/10.1073/pnas.98.4.1324*.

Smith, B.D. (1992), *Rivers of Change: Essays on Early Agriculture in Eastern North America*, Smithsonian Institution Press, Washington, DC.

Smith, C.E. (1986), "Preceramic plant remains from Guila Naquitz", in: Flannery, K.V. (ed.), *Guila Naquitz, Archaic Foraging and Early Agriculture in Oaxaca, Mexico*, Academic Press, New York, pp. 265-274.

Smith, B.D., C.W. Cowan and M.P. Hoffman (1992), "Is it an indigene or a foreigner?", in: Smith, B.D. (ed.), *Rivers of Change: Essays on the Origin of Agriculture in Eastern North America*, Smithsonian Institution Press, Washington, DC, pp. 67-100.

Sowls, L.K. (1997), *Javelinas and Other Peccaries: Their Biology, Management, and Use*, Texas A&M University Press, College Station, Texas.

Spencer, L.J. and A.A. Snow (2001), "Fecundity of transgenic wild-crop hybrids of *Cucurbita pepo* (Cucurbitaceae): Implications for crop-to-wild gene flow", *Heredity*, Vol. 86, Pt. 6, pp. 694-702, June.

Stephenson, A.G., B. Devlin and J.B. Horton (1988), "The effects of seed number and prior fruit dominance on the pattern of fruit production in *Cucurbita pepo* (zucchini squash)", *Annals of Botany*, Vol. 62, No. 6, pp. 653-661.

Strider, D.L. and T.R. Konsler (1965), "An evaluation of the *Cucurbita* for scab resistance", *Plant Disease Reporter*, Vol. 49, pp. 388-391.

Sun, J. et al. (2005), "Cucurbitacin Q: A selective STAT3 activation inhibitor with potent antitumor activity", *Oncogene*, Vol. 24, No. 20, pp. 3 236-3 245, May.

Systax Database at the University of Ulm, *www.biologie.uni-ulm.de/systax* (accessed October 2011).

Tallamy, D.W. et al. (2005), "Western corn rootworm, curcubits and cucurbitacins",in: Vidal, S. et al. (eds.), *Western Corn Rootworm: Ecology and Management*, pp. 67-93, CAB International, Oxfordshire, United Kingdom, *http://web.missouri.edu/~hibbardb/CV_files/Tallamy%20et%20al%20chap04%20proof.pdf*.

Tallamy, D.W. et al. (1998a), "Cucurbitacins as feeding and oviposition deterrents to insects", *Environmental Entomology*, Vol. 26, No. 3, pp. 678-683, *http://dx.doi.org/10.1093/ee26.3.678*.

Tallamy, D.W. et al. (1998b), "Sequestered cucurbitacins and pathogenicity of *Metarhizium anisopliae* (Moniliales: Moniliaceae) on spotted cucumber beetle eggs and larvae (Coleoptera: Chrysomelidae)", *Environmental Entomology*, Vol. 27, No. 2, pp. 366-372, *http://dx.doi.org/10.1093/ee/27.2.366*.

Tapley, W.T., W.D. Enzie and G.P. Van Eseltine (1937), *The Vegetables of New York, Vol. 1, Part 4: The Cucurbits*, New York Agricultural Experiment Station, Geneva, New York.

Tepedino, V.J. (1981), "The pollination efficiency of the squash bee (*Peponapis pruinosa*) and the honey bee (*Apis mellifera*) on summer squash (*Cucurbita pepo*)", *Journal of the Kansas Entomological Society*, Vol. 54, No. 2, pp. 359-377, April, *www.jstor.org/stable/25084168.*

Teppner, H. (2000), "*Cucurbita pepo* (Cucurbitaceae): History, seed coat types, thin-coated seeds and their genetics", *Phyton – Annales Rei Botanicae*, Vol. 40, Fasc. 1, pp. 1-42, *www.landesmuseum.at/pdf_frei_remote/PHY_40_1_0001-0042.pdf.*

Theis, N., K. Kesler and L.S. Adler (2009), "Leaf herbivory increases floral fragrance in male but not female *Cucurbita pepo* subsp. *texana* (Cucurbitaceae) flowers", *American Journal of Botany*, Vol. 96, No. 5, pp. 897-903, May, *http://dx.doi.org/10.3732/ajb.0800300.*

Thomas, A. (1971), "The incidence and economic importance of watermelon mosaic virus", *New Zealand Journal Agricultural Research*, Vol. 14, No. 1, pp. 242-247, *http://dx.doi.org/10.1080/00288233.1971.10421320.*

Torres-Ruiz, R.A. and V. Hemleben (1994), "Pattern and degree of methylation in ribosomal RNA genes of *Cucurbita pepo* L", *Plant Molecular Biology*, Vol. 26, No. 4, pp. 1 167-1 179.

USDA (2011a), *PLANTS Database*, United States Department of Agriculture, *http://plants.usda.gov* (accessed September 2011).

USDA (2011b), *Federal Noxious Weed List*, United States Department of Agriculture, *http://plants.usda.gov/java/noxious* (accessed September 2011).

USDA (1994), "Availability of determination of nonregulated status for virus resistant squash", *Federal Register*, Vol. 59, Issue 238, 13 December, pp. 64 187-64 189.

US Environmental Protection Agency (1999), "Notice of Filing a Pesticide Petition to Establish a Tolerance for Certain Pesticide Chemicals in or on Food", *Federal Register*, Vol. 64, 1 September, pp. 47 788.

van de Ven, W.T.G. et al. (2000), "Local and systemic changes in squash gene expression in response to silverleaf whitefly feeding", *The Plant Cell*, Vol. 12, No. 8, pp. 1 409-1 423.

Vidal, M. et al. (2010), "Pollination and fruit set in pumpkin (*Cucurbita pepo*) by honey bees", *Revista Brasileira de Botanica*, Vol. 33, No. 1, pp. 106-113, January/March, *http://dx.doi.org/10.1590/S0100-84042010000100010.*

Weeden, N.F. (1984), "Isozyme studies indicate that the genus *Cucurbita* is an ancient tetraploid", *Cucurbit Genetics Cooperative Report*, Vol. 7, pp. 84-85, Article 37, available at: *http://cuke.hort.ncsu.edu/cgc/cgc07/cgc7-37.html.*

Weeden, N.F. and R.W. Robinson (1990), "Isozyme studies in Cucurbita", in: Bates, D.M., R.W. Robinson and C. Jeffrey (eds.), *Biology and Utilization of the Cucurbitaceae*, Cornell University Press, Ithaca, New York, pp. 51-59.

Weiling, F. (1959), "Genomanalytische Untersuchungen bei Kürbis (*Cucurbita* L.)", *Zuchter*, Vol. 29, No. 4, pp. 161-179, January.

Wessel-Beaver, L. (2000a), "*Cucurbita aryrosperma* sets fruit in fields where *C. moschata* is only pollen source", *Cucurbit Genetics Cooperative Report*, Vol. 23, pp. 62-63, Article 20, *http://cuke.hort.ncsu.edu/cgc/cgc23/cgc23-20.html.*

Wessel-Beaver, L. (2000b), "Evidence for the center of diversity of *Cucurbita moschata* in Columbia", *Cucurbit Genetics Cooperative Report*, Vol. 23, pp. 54-55, Article 16, *http://cuke.hort.ncsu.edu/cgc/cgc23/cgc23-16.html.*

Whitaker, T.W. (1974), "Cucurbita", in: King, R.C. (ed.), *Handbook of Genetics*, Plenum Press, New York, pp. 135-144.

Whitaker, T.W. (1968), "Ecological aspects of the cultivated *Cucurbita*", *HortScience*, Vol. 3, No. 1, pp. 9-11.

Whitaker, T.W. (1951), "A species cross in *Cucurbita*", *J. of Heredity*, Vol. 42, No. 2, pp. 65-69.

Whitaker, T.W. (1947), "American origin of the cultivated cucurbits", *Annals of the Missouri Botanical Garden*, Vol. 34, No. 2, pp. 102-111, May, *www.jstor.org/stable/2394459.*

Whitaker, T.W. and W.P. Bemis (1965), "Evolution in the genus *Cucurbita*", *Evolution*, Vol. 18, No. 4, pp. 553-559, December, *http://dx.doi.org/10.2307/2406209.*

Whitaker, T.W. and G.W. Bohn (1950), "The taxonomy, genetics, production and uses of the cultivated species of *Cucurbita*", *Economic Botany*, Vol. 4, No. 1, pp. 54-81, January-March, *www.jstor.org/stable/4251961.*

Whitaker, T.W. and G.N. Davis (1962), *Cucurbits: Botany, Cultivation, and Utilization*, Interscience Publishers, New York.

Whitaker, T.W. and R.W. Robinson (1986), "Squash breeding", in: Bassett, M.J. (ed.), *Breeding Vegetable Crops*, AVI Publishing, Westport, Connecticut, pp. 209-242.

Wilson, H.D. (1998), "Free-living *Cucurbita pepo* in the United States: Viral resistance, gene flow and risk assessment", *http://botany.csdl.tamu.edu/FLORA/flcp/flcp1.htm.*

Wilson, H.D. (1990), "Gene flow in squash species: Domesticated *Cucurbita* species may not represent closed genetic systems", *BioScience*, Vol. 40, No. 6, pp. 449-455, June, *http://dx.doi.org/10.2307/1311392.*

Wilson, H.D. (1989), "Discordant patterns of allozyme and morphological variation in Mexican *Cucurbita*", *Systematic Botany*, Vol. 14, No. 4, pp. 612-623, October-December, *http://dx.doi.org/10.2307/2419006.*

Wilson, H.D., J. Doebley and M. Duvall (1992), "Chloroplast DNA diversity among wild and cultivated members of *Cucurbita* (Cucurbitaceae)", *Theoretical and Applied Genetics*, Vol. 84, No. 7, pp. 859-865, September.

Wilson, H.D., R. Lira and I. Rodríguez (1994), "Crop/weed gene flow: *Cucurbita argyrosperma* Huber and *C. fraterna* L.H. Bailey (Cucurbitaceae)", *Economic Botany*, Vol. 48, No. 3, pp. 293-300, July-September, *www.jstor.org/stable/4255643.*

Wu, J. et al. (2011), "Molecular diversity of Chinese *Cucurbita moschata* germplasm collections detected by AFLP markers", *Scientia Horticulturae*, Vol. 128, No. 1, pp. 7-13, February, *http://dx.doi.org/10.1016/j.scienta.2010.12.006.*

Xia, T. and Q. Wang (2006), "D-*chiro*-inositol found in *Cucurbita ficifolia* (Cucurbitaceae) fruit extracts plays the hypoglycemic role in streptozocin-diabetic rats", *Journal of Pharmacy and Pharmacology*, Vol. 58, No. 11, pp. 1 527-1 532, November.

Xu, Y. et al. (2004), "Inheritance of resistance to zucchini yellow mosaic virus and watermelon mosaic virus in watermelon", *Journal of Heredity*, Vol. 95, No. 6, pp. 498-502, December.

Zhang, M. et al. (2010), "Inhibition of pancreatic cancer cell growth by cucurbitacin B through the modulation of signal transducer and activator of transcription 3 signaling", *Pancreas*, Vol. 39, No. 6, pp. 923-929, August, *http://dx.doi.org/10.1097/MPA.0b013e3181ce719e.*

Zitter, T.A., D.L. Hopkins and C.E. Thomas (1996), *Compendium of Cucurbit Diseases*, APS (American Phytopathological Society) Press, St. Paul, Minnesota.

Zomlefer, W.B. (1994), *Guide to Flowering Plant Families*, University of North Carolina Press, Chapel Hill and London.

Zraidi, A. et al. (2007), "A consensus map for *Cucurbita pepo*", *Molecular Breeding*, Vol. 20, No. 4, pp. 375-388, November, *http://dx.doi.org/10.1007/s11032-007-9098-6.*

Annex 2.A1
List of taxonomic names of various *Cucurbita*

The primary source of the information in Table 2.A1.1 is "The Plant List" at: *www.theplantlist.org*.

The Plant List is the result of collaboration between the Royal Botanic Gardens, Kew and Missouri Botanical Garden. The collaboration between these groups enabled the creation of the list by combining multiple checklist data sets held by these institutions and other collaborators. Based on information from these sources, The Plant List provides the name agreed by the collaborators to be an "accepted" name, as well as synonyms by which that species has been known. It also contains "unresolved" names for which the contributing data sources did not contain sufficient information to decide whether they were "accepted" names or synonyms.

Other sources are also available, and information from these sources has also been incorporated: e.g. USDA-ARS Germplasm Resources Information Network at: *www.ars-grin.gov*.

The rightmost column of Table 2.A1.1 enumerates the names of cultivars offered as examples in the chapter in relation to the cultivated *Cucurbita* species.

The reader should be aware that botanists over time have applied some 400 names at various taxonomic ranks to the huge range of diversity observed in the *Cucurbita* (Nee, 1990).

Table 2.A1.1. **List of taxonomic names of various *Cucurbita***

Names used preferentially in text	Synonyms	Associated names found in literature	Varietal names in text
C. argyrosperma Huber *C. argyrosperma* ssp. *argyrosperma*	*C. argyrosperma* var. *callicarpa* *C. argyrosperma* var. *palmeri* *C. argyrosperma* ssp. *sororia* *C. argyrosperma* var. *stenosperma*	*C. palmeri* *C. sororia* *C. kellyana*	Green striped cushaw White cushaw Magdalena striped Papago Japanese pie Silver seed gourd
C. digitata	None recorded	*C. cordata* *C. palmata* *C. californica* *C. cylindrata*	
C. ecuardorensis	None recorded		
C. ficifolia Bouche	*C. ficifolia* f. *leucosperma* *C. ficifolia* f. *melanosperma* *C. ficifolia* var. *mexicana*	*C. melanosperma*	
C. foetidissima Kunth	*C. foetidissima* var. *foetidissima* *C. foetidissima* var. *scabridifolia*	*C. scabridifolia*	
C. galeottii			
C. lundelliana			
C. maxima Duchesne	*C. maxima* ssp. *maxima* *C. maxima* var. *triloba* *C. maxima* var. *turgida* *C. maxima* var. *zapallito* *C. maxima* var. *zipinka*	*C. andreana*	Delicious Hubbard Buttercup Mammoth whale French turban

Table 2.A1.1. **List of taxonomic names of various *Cucurbita*** (*cont.*)

Names used preferentially in text	Synonyms	Associated names found in literature	Varietal names in text
C. moschata Duchesne	*C. moschata* var. *argyrosperma* *C. moschata* var. *columbiana* *C. moschata* var. *meloniformis* *C. moschata* f. *yokohamana*		Butternut squash Golden cushaw
C. okeechobeensis (Small) L.H. Bailey *C. okeechobeensis* ssp. *martinezii* (L.H. Bailey) T.C. Andres & G.P. Nabhan	*C. okeechobeensis* ssp. *martinezii*	*C. martinezii*	
C. pedatifolia L.H. Bailey	*C. moorei*	*C. moorei*	
C. pepo L. *C. pepo* ssp. *ovifera* (L.) D.S. Decker *C. pepo* ssp. *pepo* *C. pepo* ssp. *ovifera* var. *texana* (Scheele) D.S. Decker *C. pepo* ssp. *ovifera* var. *ozarkana* D.S. Decker	*C. pepo* var. *akoda* *C. pepo* var. *americana* *C. pepo* var. *condensa* *C. pepo* var. *fibropulposa* *C. pepo* var. *flogra* *C. pepo* ssp. *fraterna* *C. pepo* var. *georgica* *C. pepo* ssp. *gumala* *C. pepo* var. *kintogwa* *C. pepo* var. *maxima* *C. pepo* var. *medullosa* *C. pepo* var. *melopepo* *C. pepo* var. *moschata* *C. pepo* var. *ovifera* *C. pepo* var. *ozarkana* *C. pepo* var. *texana* *C. pepo* var. *sororia* *C. pepo* ssp. *texana* *C. pepo* var. *toonas* *C. pepo* var. *torticollis*	*C. fraterna* *C. texana* *C. pepo* L. var. *cylindrica* *C. pepo* L. var. *clypeata* *C. pepo* L. var. *fastigata* *C. pepo* L. var. *longa* *C. pepo* L. var. *recticollis* *C. pepo* L. var. *turbinata*	Black zucchini Fordhook bush Connecticutt field Table queen Cherokee roaster Orange ball Miniature ball Striped pear
C. radicans		*C. gracilior*	

Source: Adapted from "The Plant List" (2011).

References

Nee, M. (1990), "The domestication of *Cucurbita* (Cucurbitaceae)", *Economic Botany*, Vol. 44, No. 3, pp. 56-58, July.

The Plant List (2011), A working list of all known plant species created through a collaboration between the Royal Botanic Gardens, Kew and Missouri Botanic Gardens, *www.theplantlist.org* (accessed October 2011).

USDA-ARS (n.d.), Germplasm Resources Information Network (GRIN), United States Department of Agriculture, Agricultural Research Service, *www.ars-grin.gov* (accessed October 2011).

Annex 2.A2
Horticultural types in *Cucurbita* species

Table 2.A2.1. **Horticultural types in *Cucurbita* species**

Species	Type	Description	Typical cultivars
C. argyrosperma	Cushaw	Striped, green or white hard rind. Pear shaped or with a straight or curved neck.	Green striped cushaw; Japanese pie; Tennessee sweet potato
C. maxima	Banana	Elongated fruit pointed at the ends. Orange or pink moderately hard rind.	Banana; Pink banana
	Delicious	Top shaped. Orange or green hard rind.	Delicious; Golden delicious
	Hubbard	Round at the middle tapering at each end. Blue, orange or green, hard warty rind.	Hubbard; Blue hubbard; Golden hubbard
	Marrow	Lemon-shaped with orange hard rind.	Boston marrow
	Show	Very large globular, sutured, light orange fruit. Moderately hard rind.	Atlantic giant; Big Max
	Turban	Turban shaped with a large button. Hard rind.	Turks turban; Warren; Turks cap
C. moschata	Tropical pumpkin	Round, oblate or irregular shape. Green, buff, yellow or piebald hard rind.	La Primera; Seminole; Solar
	Cheese	Variable shape, smooth, hard, buff-coloured hard rind.	Dickinson; Kentucky field
	Crookneck	Long, curved or straight neck. Smooth, hard rind, usually buff.	Golden crookneck; Walthan butternut; Zenith
	Bell	Bell-shaped. Orange flesh. Tan hard rind.	Seminole; Upper ground sweet potato
C. pepo	Acorn	Acorn-shaped grooved fruit. Dark green, orange or white hard rind.	Heart of gold; Table ace; Tay belle
	Cocozelle	Long, cylindrical, bulbous blossom end. Striped or variegated green soft rind.	Cocozelle; Long cocozelle
	Crookneck	Elongated with narrow, curved neck. Yellow soft rind.	Dixie; Yellow summer crookneck; Supersett
	Ornamental gourd	Variously shaped and coloured. Smooth or warty hard rind.	Orange ball; Crown of thorns
	Pumpkin	Large, round, oval oblate shape. Mostly orange, sometimes white relatively soft rind.	Connecticutt field; Howden Jack-be-little; Small sugar
	Scallop	Flattened with scalloped margins. White, yellow, green or bicoloured soft rind.	Peter pan; Sunburst; White bush scallop
	Straightneck	Long, cylindrical, yellow soft rind.	Enterprise; Goldbar; Multipic
	Vegetable marrow	Short, tapered, cylindrical. Light green.	Clarita; Goya; Zahra
	Zucchini	Uniformly cylindrical. Green or yellow to gray soft rind.	Dividend; Revenue; Spineless beauty

Some of the types listed in Table 2.A2.1 are not grown in production agriculture. For example, "Show" pumpkins are grown for competition in the heaviest fruit contests held in various parts of the United States. The 2000 winner weighed in at 517 kilograms. Other types that are regionally important and of historical interest are certain cushaw and vegetable marrow squash. These cultivars can be bought commercially at concerns dedicated to the preservation of heirloom varieties (e.g. *www.sandhillpreservation.com*).

Also, it should be noted that this annex offers examples only as a means of illustrating the types and varieties that can be associated with the various *Cucurbita* species. Paris (1989), for example, notes that for *C. pepo* alone, hundreds or perhaps thousands of named cultivars exist.

Reference

Paris, H.S. (1989), "Historical records, origins and development of the edible cultivar groups of *Cucurbita pepo* (Cucurbitaceae)", *Economic Botany*, Vol. 43, No. 4, pp. 423-443, October-December, *www.jstor.org/stable/4255187*.

Chapter 3.

Brassica crops (*Brassica* spp.)

This chapter deals with the biology of Brassica *species which comprise oilseed rape, turnip rape, mustards, cabbages and other oilseed crops. The chapter contains information for use during the risk/safety regulatory assessment of genetically engineered varieties intended to be grown in the environment (biosafety). It includes elements of taxonomy for a range of* Brassica *species, their centres of origin and distribution, reproductive biology, genetics, hybridisation and introgression, crop production, interactions with other organisms, pests and pathogens, breeding methods and biotechnological developments, and an annex on common pathogens and pests.*

The OECD gratefully acknowledges the contribution of Dr. R.K. Downey (Canada), the primary author, without whom this chapter could not have been written.

The chapter was prepared by the OECD Working Group on the Harmonisation of Regulatory Oversight in Biotechnology, with Canada as the lead country. It updates and completes the original publication on the biology of *Brassica napus* issued in 1997, and was initially issued in December 2012. Data from USDA Foreign Agricultural Service and FAOSTAT have been updated.

Introduction

The plants within the family Brassicaceae constitute one of the world's most economically important plant groups. They range from noxious weeds to leaf and root vegetables to oilseed and condiment crops. The cole vegetables are perhaps the best known group. Indeed, the *Brassica* vegetables are a dietary staple in every part of the world with the possible exception of the tropics. The Food and Agriculture Organization of the United Nations estimates that world commercial production of cabbages, cauliflowers, broccoli and other *Brassica* vegetables in 2013 was over 93 million tonnes from about 3.7 million hectares, with a 2013 farm gate value of some USD 31 billion (FAOSTAT, 2013). These figures do not include the root vegetables or the production from kitchen gardens.

Less well known are the *Brassica* oilseed crops that annually occupy over 34 million hectares of the world's agricultural lands (FAOSTAT, 2013). Because of their ability to survive and grow at relatively low temperatures, they are one of the few edible oil sources that can be successfully produced in cool temperate regions. This characteristic makes them well adapted to cultivation at high elevations and as winter crops in the subtropics. In temperate regions, oilseed rape (*Brassica napus*)[1] and turnip rape (*Brassica rapa*) predominate, while in the subtropics of Asia, Indian mustard or rai (*Brassica juncea*) is the major oil source. Among all the commodities moving in world trade, only petroleum has a greater value than vegetable oils (United States Census Bureau, n.d.; United Nations, n.d.). In total, *Brassica* oilseeds provide 15% of the world's edible vegetable oil and are the third most important source of edible oil after soybean and palm (Table 3.1).

Table 3.1. **World production of edible vegetable oils, averages 1996-2000 to 2011-15**

Millions of tonnes

Crop	1996-2000 (MMt)	2001-05 (MMt)	2006-10 (MMt)	2011-15[2] (MMt)
Soybean	22.4	29.8	36.8	46.2
Palm	18.3	28.2	40.8	58.5
Rape/mustard	*11.8*	*13.7*	*19.1*	*25.7*
Sunflower	8.9	8.4	11.0	14.7
Groundnut	4.3	4.9	4.8	5.5
Cottonseed	3.7	4.1	4.9	5.1
Others[1]	7.9	9.5	11.2	13.1
Total	**77.3**	**98.6**	**128.6**	**168.7**

Notes: 1. Others include olive, coconut and palm kernel. 2. This column was added in January 2016.

Source: After USDA Foreign Agricultural Service (2015).

Species or taxonomic group

Classification and nomenclature

The family Brassicaceae (= Cruciferae) contains over 338 genera and 3 709 species (Al-Shehbaz, Beilstein and Kellogg, 2006; Warwick, Francis and Al-Shehbaz, 2006). The species of greatest interest to those concerned with genetically modified crops are given in Table 3.2 with their chromosome number, genome identification and common English name(s).

Table 3.2. **Nomenclature and genome relationships of cultivated *Brassica* species and related genera**

Species name	Common synonym	Haploid chromosome number and genome	Common name
Brassica rapa L.	*B. campestris* L.	10 AA	
subsp. *campestris* (L.) A.R. Clapham			Summer turnip rape, canola
subsp. *oleifera* (DC.) Metzg.			Winter turnip rape
subsp. *campestris* (L.) A.R. Clapham	subsp.*eu-campestris* (L.) Olsson		Bird or wild turnip rape
subsp. *trilocularis* (Roxb.) Hanelt	subsp. *sarson* (Prain) Denford		Yellow and brown Sarson
subsp. *dichotoma* (Roxb.) Hanelt			Toria
subsp. *chinensis* (L.) Hanelt	*B. chinensis* L. *B. chiensis* var. *parachinsis* (L.H. Bailey)		Pak-choi or bok choy, Chinese mustard, Chinese broccoli, Gai Lan
subsp. *pekinensis* (Lour.) Hanelt	*B. pekinensis* (Lour) Rupr.		Pe-tsai, Chinese cabbage
subsp. *nipposinica* (L.H. Bailey) Hanelt			Curled mustard
subsp. *Rapa*	*B. rapa* L.		Turnip
Brassica tournefortii Gouan		10 TT	Wild turnip
Brassica nigra (L.) W.D.J. Koch		8 BB	Black mustard
Brassica oleracea L.		9 CC	
var. *viridis* L.	var. *acephala* DC.		Kale, collard
var. *botrytis* L.			Cauliflower and broccoli
var. *capitata* L.			Cabbage
var. *gongylodes* L.	var. *caulorapa* Pasq.		Kohlrabi
var. *gemmifera* (DC.) Zenker			Brussels sprouts
var. *italica* Plenck.			Broccoli
var. *oleracea*	subsp. *sylvestris* (L.) Miller		Wild cabbage
subsp. *alboglabra* L.H. Bailey	*B. alboglabra* L.H. Bailey		Chinese kale, kailan
Brassica juncea (L.) Czern.		18 AABB	Brown and oriental mustard, rai
Brassica napus L.		19 AACC	
var. *napus*	subsp. *oleifera* (Delile) Sinskaya		Summer oilseed rape, canola
var. *napus*	*B. napus* f. *biennis* (Schübl. & G. Martens) Thell.		Winter oilseed rape, winter canola
var. *pabularia* (DC.) Rchb.			Rape-kale
var. *napobrassica* (L.) Rchb.	subsp. *rapifera* (Metzg.) Sinskaya		Rutabaga, swede
Brassica carinata A. Braun.		17 BBCC	Abyssinian mustard
Hirschfeldia incana (L.) Lagr.-Foss.	*Brassica adpressa* Boiss.	7 HH	Hoary mustard
Sinapis arvensis L.	*B. kaber* (DC.) L.C. Wheeler	9 SarSar	Wild mustard, charlock
Sinapis alba L.	*B. hirta* Moench	12 SalSal	Yellow or white mustard
Raphanus sativus L.		9 RR	Radish
Raphanus raphanistrum L.		9 RR	Wild radish
Diplotaxis muralis (L.) DC.		21 DD	Annual wall-rocket
Erucastrum gallicum (Willd.) O.E. Schulz		15	Dog mustard
Eruca vesicaria (L.) Cav. subsp. *sativa* (Mill.) Thell.	*Eruca sativa* Mill.	11 EE	Rocket salad

Source: Modified from Yarnell (1956).

Early humans recognised the edible value of many of these species and through selection modified nearly every plant part to suit their needs. Such modifications include the compacting of the leaves to form a head, the root or stem to form a bulb, the inflorescence to form a curd or bunch and the seed to provide both oil and condiment. Species grown as oilseeds include *B. napus*, *B. juncea*, *B. rapa* and *B. carinata*. The vegetable Brassicaceae includes *B. napus* (rutabaga, Siberian kale), *B. rapa* (Chinese cabbage, bok choy, pai-tsai, mizuna, Chinese mustard, broccoli raab and turnip), *B. oleracea* (cabbage, broccoli, cauliflower, Brussels sprouts, kohlrabi, collards and kale), *Raphanus sativus* (radish), *Lepidium sativum* (garden cress) and *Nasturtium officinale* (watercress). The condiment crops include *B. juncea* (brown and oriental mustard), *Sinapis alba* (yellow mustard), *B. nigra* (black mustard, but now little used), *Armoracia rusticana* (horseradish) and *Eutrena japonica* (wasabi). There are a number of other minor potherbs and salad vegetables. There are numerous weedy species, but those of greatest interest with regard to cross-pollination with *B. napus* are *Sinapis arvensis* (wild mustard or charlock), *Raphanus raphanistrum* (wild radish), *B. rapa* (wild or bird rape) and *Hirschfeldia incana* (hoary mustard).

The genus *Brassica* is classified as follows:

Order Brassicales (= Cruciales)

Family Brassicaceae (= Cruciferae)

Tribe Brassiceae

Subtribe Brassicinae

Genus *Brassica* L.

The Brassicaceae family is comprised of 25 tribes with about an additional 5 under study (Al-Shehbaz, Beilstein and Kellogg, 2006). The tribe Brassiceae, which contains the genus *Brassica* and its wild relatives, is made up of 48 genera and approximately 240 species (Warwick and Hall, 2009). Warwick, Francis and Al-Shehbaz (2006) have prepared a checklist and a current taxonomic database for the family on CD-ROM. Also on CD-ROM are chromosome numbers from the literature for 68.6% of the genera and 42.0% of the Brassicaceae species (Warwick and Al-Shebbaz, 2006; Warwick, Francis and Gugel, 2009). The morphological traits that characterise the tribe are conduplicate cotyledons (the radical enclosed by longitudinally folded cotyledons) and/or transversely segmented fruits, that have seeds or rudimentary ovules in both segments and, if present, only simple trichomes or hairs (Warwick and Hall, 2009). Modern molecular studies have reinforced the monophlyetic origin of the tribe.

Taxonomic research on the tribe conducted by Schulz (1936; 1919) established the basic classification that is followed today, although it has been modified and criticised (Al-Shehbaz, 1984). Within the tribe, Schulz (1936; 1919) recognised ten subtribes, with Gómez-Campo (1980) later recommending a reduction to nine. Of the nine subtribes, three are of greatest relevance to those concerned with *Brassica* crops, namely: Brassicinae, Moricandiinae and Raphaninae. Within these subtribes *Brassica*, *Sinapis*, *Diplotaxis*, *Erucastrum*, *Hirschfeldia*, *Eruca* and *Raphanus* are of primary interest.

The association and relationships among species within these subtribes have been studied cytogenetically, chemically and morphologically (reviewed by Prakash and Hinata, 1980; Takahata and Hinata, 1986, 1983) without providing a clear separation of the subtribes and their genera. Recent molecular, morphological and hybridisation data give strong support for a rearrangement of the three subtribes into two clades, namely, the Rapa/Oleracea and the Nigra lineages (see the section on "Centres of origin and ancestors", as well as Warwick and Hall [2009] and references therein). Such a division is also referred to in some publications as the *Brassica* and *Sinapis* lineages. It is expected that the realignment of the species from the three subtribes into the two clades will eventually require renaming of many of the species involved.

The difficulty in clearly separating the genera and species among the *Brassica* and their close relatives has arisen because similar plant forms and morphological traits occur in more than one genus or species. The difficulties encountered by early taxonomists in separating and classifying the various species and forms within the Brassicaceae family are well documented by Hedge (1976) and Prakash and Hinata (1980). As a result, there have been numerous changes and modifications to Schulz's (1919; 1936) original species names and arrangement. The cytological studies by Morinaga (1928; 1929; 1931; 1933; 1934a; 1934b) and his student, U (1935) clarified the broad relationships among the economically important *Brassica* species in which chromosome pairing clearly showed the three species with the higher chromosome number, *B. napus*, *B. juncea* and *B. carinata* are amphidiploids derived from the monogenomic or basic species, *B. nigra*, *B. rapa* and *B. oleracea* (Figure 3.1).

Figure 3.1. **Genome relationships of *Brassica* species and allied genera**

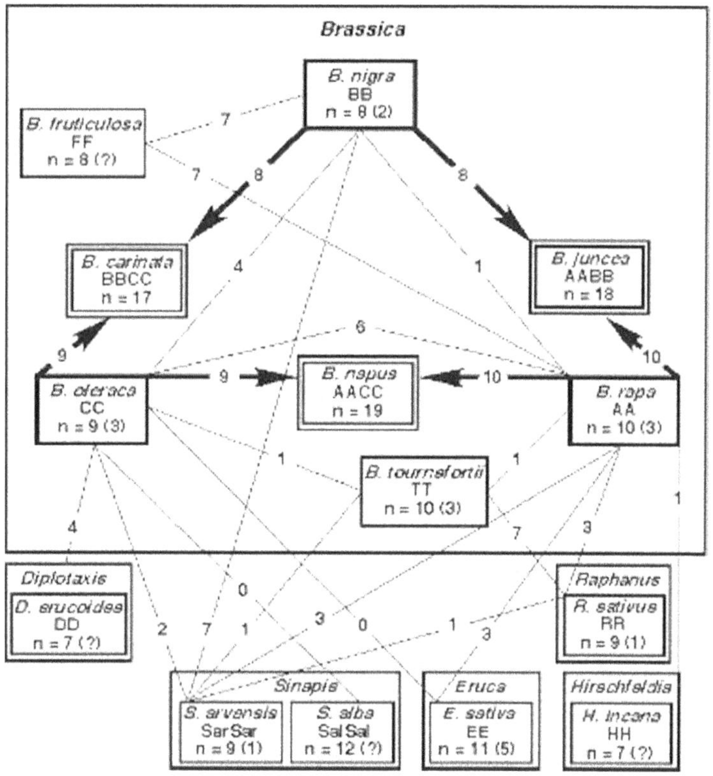

Notes: A, B, C... are the genome symbols. The number in brackets following the haploid chromosome number (n) indicates the maximum possible number of autosyndetic[2] chromosome pairs. The numbers within lines connecting two genomes give the maximum allosyndesis, i.e. the number of bivalents possible between the respective interspecific hybrids (Downey and Röbbelen, 1989).

Source: Modified from Mizushima (1980).

The genome relationships among the amphidiploids were confirmed by resynthesis of the three species from their diploid parents (Frandsen, 1947, 1943; Ramanujam and Srinivasachar, 1943). Further verification of these species' relationships were obtained from studies on phenolic compounds (Dass and Nybom, 1967), protein patterns (Vaughan, 1977), isozymes (Coulthart and Denford, 1982; Chen, Heneen and Simonsen, 1989) and nuclear DNA, restriction fragment length polymorphisms (RFLP; Song, Osborn and Williams, 1988a; 1988b). Additional verification has been achieved through molecular analysis of nuclear and chloroplast DNA and fluorescence *in situ* hybridisation (Snowdon et al., 2003; Snowdon, 2007; Warwick and Sauder, 2005; Lysak et al., 2005).

To further establish the true relationships among the genus and species of the subtribe, Harberd (1976; 1972) proposed grouping them into "cytodemes" based on the crossability of related subspecies with the same chromosome number. Harberd (1976) defined cytodemes as follows: "If two populations have a common chromosome number and are easily crossed to form a hybrid, which is neither obviously weak in vigour nor of low fertility, then they belong in the same cytodeme. By contrast, different cytodemes (which sometimes have the same chromosome number) are (a) difficult to cross, or (b) give a weak hybrid, or (c) have a sterile hybrid, and frequently exhibit all three criteria." Harberd (1976; 1972) also defined the *Brassica* coenospecies as "the group of wild species sufficiently related to the six cultivated species of *Brassica* to be potentially capable of experimental hybridisation with them". On this basis and their chromosome number the coenospecies have been classified into 43 diploid and 13 tetraploid cytodemes (Warwick and Black, 1993: Table 3). This grouping, with the inclusion of *Raphanus* and *Enarthrocarpus* in the subtribe, is supported by both chloroplast and nuclear DNA analysis (Warwick and Black, 1993; Warwick and Hall, 2009).

Cytological analyses of chromosome pairing in interspecific crosses among some of the more important *Brassica* cytodemes by Mizushima (1980) provided information on the maximum possible number of autosyndetic[2] chromosome pairs (Figure 3.1). Harberd and McArthur (1980) extended the study of meiotic chromosome pairing to more than 50 species hybrids. These distant crosses were facilitated using embryo culture.

A chromosome analysis of the monogenomic *Brassica* species by Röbbelen (1960), established that only six chromosomes were distinctly different, the remaining being homologous with one or the other of the basic six. This evidence pointed to the presence, in the evolutionary pathway of the *Brassica* species, of a now-extinct, ancient progenitor with a basic chromosome number of × = 6. The long-standing hypothesis, that the cultivated diploid *Brassica* species are ancient polyploids, has been strongly supported by modern genomic investigations.

The genomes of *Brassica* species are extensively triploid (Lysak et al., 2007, 2005; Rana et al., 2004). In *B. nigra* Lagercrantz and Lydiate (1996) reported that every chromosome region appeared to be present in triplicate and the genomes of *B. oleracea* and *B. rapa* also exhibit tripling (Rana et al., 2004; Mun et al., 2009; Wang, 2010). High density comparative mapping of *Arabidopsis* and *B. napus* also supported the hypothesis of a hexaploid ancestor (Parkin et al., 2005). Indeed, chromosome tripling has been documented for the entire Brassiceae tribe (Lysak et al., 2005). Linkage maps and genome size data (Lysak et al., 2009) indicate that the *B. oleracea* genome, and probably the other monogenomic species which exhibit a range in chromosome number from 7 to 12, increased or reduced their chromosome number through duplication, translocations (Quiros, Ochoa and Douches, 1988; Hosaka et al., 1990; McGrath et al.,1990; Truco and Quiros, 1994), transposition of elements (Zhang and Wessler, 2004) as well as deletions (Hu and Quiros, 1991) and fusions (Lysak et al., 2006).

Table 3.3. **List of 43 diploid cytodemes and 6 amphidiplod taxa in the *Brassica* coenospecies**

N	CYTODEME
	Diploids
10	*Brassica barrelieri* (L.) Janka
7	*Brassica deflexa* Boiss.
11	*Brassica elongata* Ehrh.
8	[a] *Brassica fruticulosa* Cirillo (includes *B. maurorum* Dur., *B. spinescens* Pomel, *Erucastrum littoreum* (Pau & Font Quer) Maire subsp. *glabrum* (Maire) Gómez-Campo (=*Erucastrum laevigatum* subsp. *glabrum* Maire)
10	[a] *Brassica gravinae* Ten.
8	*Brassica nigra* (L.) W.D.J. Koch
9	*Brassica oleracea* L. (includes *B. alboglabra* L.H. Bailey, *B. bourgeaui* (Webb.) Kuntze, *B. cretica* Lam., *B. hilarionis* G.E. Post, *B. incana* Ten., *B. insularis* Moris, *B. macrocarpa* Guss., *B. montana* Pourr., *B. rupestris* Raf., *B. villosa* Biv.)
9	*Brassica oxyrrhina* (Coss.) Willk.
10	*Brassica rapa* L. (= *B. campestris* L.) (includes wild and cultivated varieties)
10	*Brassica repanda* (Willd.) DC. (includes *B. desnottesii* Emb. & Maire, *B. nudicaulis* (Lag.) O.E. Schulz, *B. saxatilis* DC.)
11	*Brassica souliei* (Batt.) Batt. (= *B. amplexicaulis* (Desf.) Pomel)
10	*Brassica tournefortii* Gouan
12	[a] *Coincya* spp. (=*Hutera* =*Rhynchosinapis*) (includes all species in the genus)
11	*Diplotaxis acris* (Forssk.) Boiss.
9	*Diplotaxis assurgens* (Delile) Gren.
9	*Diplotaxis berthautii* Braun-Blanq. & Maire
9	*Diplotaxis catholica* (L.) DC.
7	*Diplotaxis cossoniana* (Reut.) O.E. Schulz
7	*Diplotaxis erucoides* (L.) DC.
13	*Diplotaxis harra* (Forssk.) Boiss. (includes *D. crassifolia* (Raf.) DC., *D. lagascana* DC.)
8	*Diplotaxis siettiana* Maire (includes *D. ibicensis* (Font Quer) Gómez-Campo)
10	*Diplotaxis siifolia* Kunze
11	*Diplotaxis tenuifolia* (L.) DC. (includes *D. cretacea* Kotov., *D.* simplex (Viv.) Spreng., the latter species was incorrectly listed as *D. pitardiana* Maire)
9	*Diplotaxis tenuisiliqua* Delile
10	*Diplotaxis viminea* (L.) DC.
9	*Diplotaxis virgata* (Cav.) DC.
10	*Enarthrocarpus* spp. (includes *E. lyratus* (Forssk.) DC., *E. pterocarpus* (Pers.) DC., *E. strangulatus* Boiss.)
11	*Eruca* spp. (includes *E. vesicaria* (L.) Cav., *E. sativa* Mill., *E. pinnatifida* (Desf..) Pomel)
8	[a] *Erucastrum abyssinicum* R.E. Fr.
9	*Erucastrum canariense* Webb & Berthel. (includes *E. cardaminoides* (Webb) O.E. Schulz)
8	[a] *Erucastrum nasturtiifolium* (Poir.) O.E. Schulz (includes *E. leucanthum* Coss. & Dur.)
8	*Erucastrum strigosum* (Thunb.) O.E. Schulz
7	*Erucastrum varium* (Durieu) Durieu
7	[a] *Erucastrum virgatum* C. Presl
7	*Hirschfeldia incana* (L.) Lagr.-Foss.
9	*Raphanus* spp. (includes *R. raphanistrum* L., *R. sativus* L., *R. caudatus* L., *R. maritimus* Smith and *R. landra* DC.)
10	*Sinapidendron* spp. (includes *S. angustifolium* (DC.) Löwe, *S. frutescens* (Ait.) Löwe, *S. rupestre* Löwe)
12	*Sinapis alba* L. (includes *S. dissecta* Lag.)
9	*Sinapis arvensis* L. (includes *S. allionii* Jacq., *S. turgida* (Pers.) Delile)
7	*Sinapis aucheri* (Boiss.) O.E. Schulz (=*Brassica aucheri* Boiss.)
12	*Sinapis flexuosa* Poir.
9	[a] *Sinapis pubescens* L.
8	*Trachystoma* spp. (includes *T. aphanoneurum* Maire & Weiller, *T. ballii* O.E. Schulz and provisionally *T. labasii* Maire)
	Amphidiploids (with proposed parentage in parentheses)
17	*Brassica carinata* A. Braun (*B. nigra* × *B. oleracea*)
18	*Brassica juncea* (L.) Czern. (*B. rapa* × *B. nigra*)
19	*Brassica napus* L. (*B. rapa* × *B. oleracea*)
21	*Diplotaxis muralis* (L.) DC. (*D. tenuifolia* × *D. viminea*)
15	*Erucastrum gallicum* (Willd.) O.E. Schulz (*E. leucanthum* × ? unknown $n = 7$ taxon)
15	*Erucastrum elatum* (Ball.) O.E. Schulz (*E. littoreum* × ? unknown $n = 7$)

Note: N = haploid chromosome number. Information was obtained from the following sources: Gómez-Campo (1983), Harberd (1976, 1972), Harberd and McArthur (1980, 1972), Leadlay and Heywood (1990), Snogerup, Gustafsson and Von Bothmer (1990), Sobrino-Vesperinas (1988), Takahata and Hinata (1983) and Warwick, Black and Aguinagalde (1992). Nomenclature is based on that in USDA-ARS (The Germplasm Resources Information Network) (2011). a) Allotetraploids (4x) were also indicated for these cytodemes.

Source: Warwick and Black (1993) © Canadian Science Publishing or its licensors.

Based on cotyledonary studies of various taxa in the tribe Brassiceae, Gómez-Campo and Tortosa (1974) proposed that *Brassica* evolved from the Macaronesian plant taxon *Sinapidendron*. This Miocene relic survived several paleo-climatic changes that destroyed most of the Mediterranean Tertiary flora and is put forward as the archetype from which *Brassica* evolved through the *Diplotaxis* and *Erucastrum* complexes. However, the use of such morphometric data to establish evolutionary relationships withinBrassiceae has not always provided results that agreed with those from cytological and molecular studies.

Description

Prakash and Hinata (1980) have summarised the early taxonomic difficulties when only morphological characteristics were used to categorise the numerous and varied forms of the commercially important *Brassica* species. The early proliferation of species' names and misclassifications resulted from the wide array of plant forms that occur among plants within the same genome, plus the mimicking of the same morphological features in plants with a different genetic makeup. Although the application of advanced genetic techniques and chemical investigations has clarified relationships, there is still some disagreement among authorities as to whether a particular form should be considered a species, or a subspecies or variety within a species.

Brassica nigra

Figure 3.2. **Illustration of a *Brassica nigra* plant and its parts**

Source: Koehler's Medical-Plants (1887) provided by Wikipedia, the free encyclopedia.

Sinskaia (1928) identified two major geographic forms of *B. nigra*, a western form grown in Europe, Africa, Asia Minor and Afghanistan and an eastern form grown in India and as far west as Palestine and the Syrian Arab Republic. The early forms were of short season, spreading, with semi-erect growth up to a metre tall but taller, more erect material was selected for commercial production (Hemingway, 1995). The prevalent annual weedy form of today varies in height from 0.6-2.4 metres, depending on the competing vegetation and growing conditions. The plant is lightly covered with soft hairs; the lower leaves are large with upper leaves reduced in size. *B. nigra* can be easily distinguished from the commercial *Brassica* crops in that *B. nigra* does not produce a rosette of basal leaves. A typical plant image, including the tap root, is shown in Figure 3.2. The siliques are short (2-5 cm), hirsute and appressed to the stem of the flowering raceme, with a beak about 0.6 cm long. The small, brown to black seeds exhibit primary dormancy and tend to germinate throughout the growing season.

Brassica rapa

Plants of *B. rapa* species are widely cultivated as leaf and root vegetables, fodder and oilseed crops. In addition, they can be a weed of cultivated land and disturbed sites. The widest array of vegetable forms evolved in the People's Republic of China (hereafter "China") with many of the selected forms corresponding to or mimicking those found in the *B. oleracea* complex. Because the selected forms exhibited significantly different morphological traits, early botanists classified them as separate species. Today they are more correctly classified as subspecies or varieties of *B. rapa.*

Brassica rapa vegetables

The plants in the *B. rapa* subsp. *pekinensis* group of vegetables are biennials that have been classified into three variant forms. The var. *cylindrica* has broad but thin, crinkled and conspicuously veined green leaves with white petioles (Figure 3.3). The leaves are usually tightly wrapped in a cylindrical formation to form a head with a length of 30-60 cm and a diameter about 10-17 cm. The var. *cephalata* forms a flat head similar to a drum-head cabbage (Figure 3.3) while the var. *laxa* forms a loose heart. In the second year of growth bolting occurs and the flowering stem is quickly thrust upwards reaching a height of 1.5 metres and bearing the characteristic raceme with typical *Brassica* yellow flowers. Common names for this group include pe-tsai, celery cabbage or Chinese cabbage.

Figure 3.3. ***B. rapa* subsp. *pekinensis***

A. var. cylindrica B. var. cephalata

Source: Courtesy Evergreen Seeds.

The *B. rapa* subsp. *chinensis* group includes both annual and biennial forms. Bailey (1930) described the subspecies as "a very smooth biennial with large ladle-shaped upstanding radial leaves with thick ivory-white but not wing-margined or toothed petioles." The clasping, entire leaves have prominent veins and resemble leaves of Swiss chard (Figure 3.4). The common name for this plant group is pak choi or bok choy. If the plants are harvested in the early stages of growth they may be called "baby bok choy" or "Shanghai bok choy" (Figure 3.5). The subsp. *parachinensis* is usually included within this group (Figure 3.6). It is grown for its thick stemmed flowering shoots that are cut for market as the first flowers open, allowing for several harvests. The common names for this variant include Gai Lan and Chinese broccoli. Tsen and Lee (1942) include the subsp. *rosularis* and subsp. *narinosa* in this *chinensis* group. The USDA *Germplasm Resources Information Network* (GRIN) database includes *rosularis* in the *chinensis* group, but keeps *narinosa* as a separate subspecies (USDA-ARS, 2011). The plants of the subsp. *narinosa* are stout, low growing, glabrous biennials. The lower leaves are small, puckered and orbicular-ovate with broad white petioles, arranged in short clusters. The upper stem leaves are very broad, entire and clasping. Siliques are about 2 cm long or less with a very short, stout beak about one-half or one-third as long as the pod. Tsen and Lee (1942) also place subsp. *japonica* and subsp. *nipposinica* within the *chinensis* group; however, the USDA keeps both of these subspecies separate (USDA-ARS, 2011). These two subspecies are considered synonyms for this form, exhibiting pencil-thin leaf stems supporting deeply indented feathery leaves (Figure 3.7). The flowering stalks produce siliques about 6 cm long.

Figure 3.4. ***B. rapa*** **subsp.** ***Chinensis*****, Bok choy**

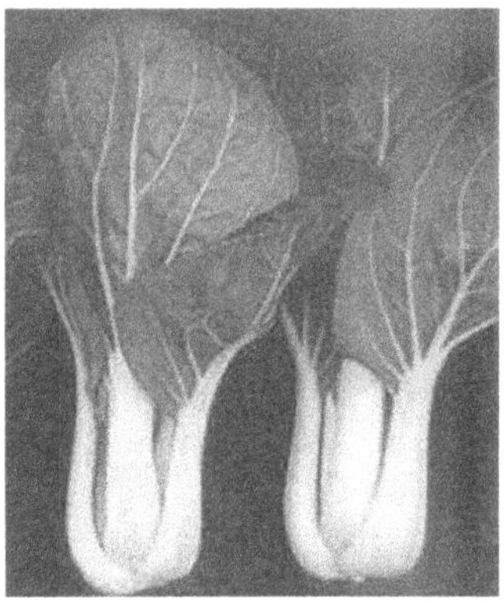

Source: Courtesy Tainong Seeds.

Figure 3.5. ***B. rapa*** **subsp.** ***Chinensis*****, Baby or Shanghai bok choy**

Source: Courtesy Tainong Seeds.

Figure 3.6. ***B. rapa*** **subsp.** ***Parachinensis*****, Gai Lan or Chinese broccoli**

Source: Courtesy Evergreen Seeds.

Figure 3.7. ***B. rapa*** **subsp.** ***Nipposinica***

Source: Courtesy North Carolina State University.

B. rapa subsp. *rapa*, the common turnip, develops a bulbous storage organ in the first year of growth. The top 1-6 cm above ground is an expansion of the hypocotyl that is fused with the expanded root below ground. A narrow tap root extends below the storage organ (Figure 3.8). Most cultivars are white fleshed except for the exposed above-ground portion which, when exposed to sunlight, may turn purple, red or green. Yellow, orange and red fleshed cultivars are also grown. Leaves grow directly from the above-ground shoulder of the expanded hypocotyl and not from a visible crown or neck as occurs in rutabagas (*Brassica napus* var. *napobrassica*). The leaves may be harvested and eaten as "turnip greens". Turnip roots for edible purposes will each weigh about 1 kg but weight will vary with the variety and growing conditions. Cultivars grown for cattle and sheep feed produce much larger roots. The flowering stalk bolts from the overwintered root the following spring, producing a terminal raceme with siliques about 6 cm long.

Figure 3.8. ***B. rapa* subsp. *rapa* the common turnip**

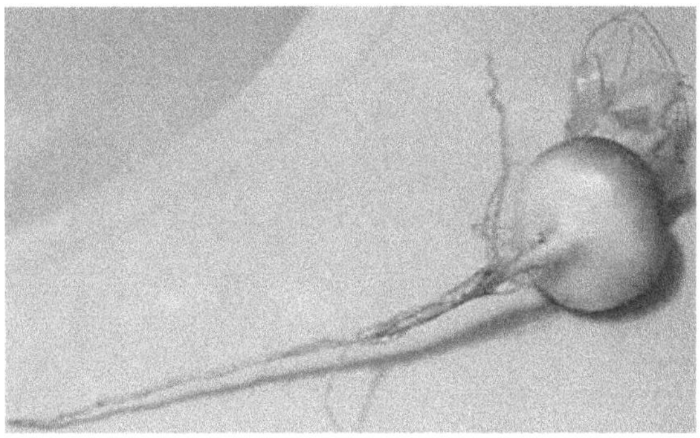

Source: Courtesy Wikipedia, the Free Encyclopedia.

Brassica rapa oilseed and weedy forms

The oilseed form of *B. rapa* subsp. *oleifera* includes both annual and biennial varieties. Both the spring and winter forms of *B. rapa* mature earlier and withstand cold temperatures better than their *B. napus* counterparts. However, the seed and oil yield is normally lower than *B. napus* so production of the winter form is limited to the more rigorous climates of central Sweden, Finland, north-west China and the foothills of the Himalayan mountains. The plant and growth stages of spring *B. rapa* are illustrated in Figure 3.9. Following the emergence of the cotyledons, the plant quickly produces a tap root and a rosette of leaves that shades the surrounding area reducing weed competition. The lower leaves are stalked, lyrate-pinnatifid with a large end lobe exhibiting sparse hairs on the under side. The upper leaves are much smaller and slightly stalked. In the winter form, the plant remains in the rosette stage until exposed to a long vernalization period (40 days) at near freezing temperatures. Day length, and where required vernalization, determine when bolting of the flower stem will occur. Figure 3.9 shows only a single raceme but under field conditions the plant produces many flowering branches and with *B. rapa*, as opposed to *B. napus*, it can be difficult to identify the primary raceme. The plant grows to a height of a meter or less. The position of the flower buds on a raceme, relative to the just opened, self-incompatible flowers, can be used to distinguish plants of *B. rapa* from *B. napus*. In *B. rapa* the flowers over top the buds while the reverse is true for *B. napus*. Siliques, some 6 cm long, contain up to 30 brown to yellow seeds in 2 locules (Figure 3.10).

B. rapa subsp. *campestris* (formerly subsp. *sylvestris*), the weedy form of subsp. *oleifera*, is morphologically indistinguishable from the cultivated spring oilseed *B. rapa*, except that the seed of subsp. *campestris* exhibits primary dormancy, a recessively inherited characteristic.

B. rapa subsp. *dichotoma*, commonly referred to as toria, is an oilseed form grown on the Indian sub-continent. Morphologically it is indistinguishable from the spring form of *B. rapa* subsp. *oleifera*. Other forms grown on the sub-continent are termed yellow and brown sarson (*B. rapa* subsp. *trilocularis*). These forms have broad siliques containing larger seeds than toria. However, yellow sarson is distinguished by its introse anthers, self-compatibility and pure yellow seeds.

Figure 3.9. **Growth stages in turnip rape (*B. rapa*)**

Stages: 0: pre-emergence; 1: seedling; 2: rosette; 3: bud; 4: flower; 5: ripening

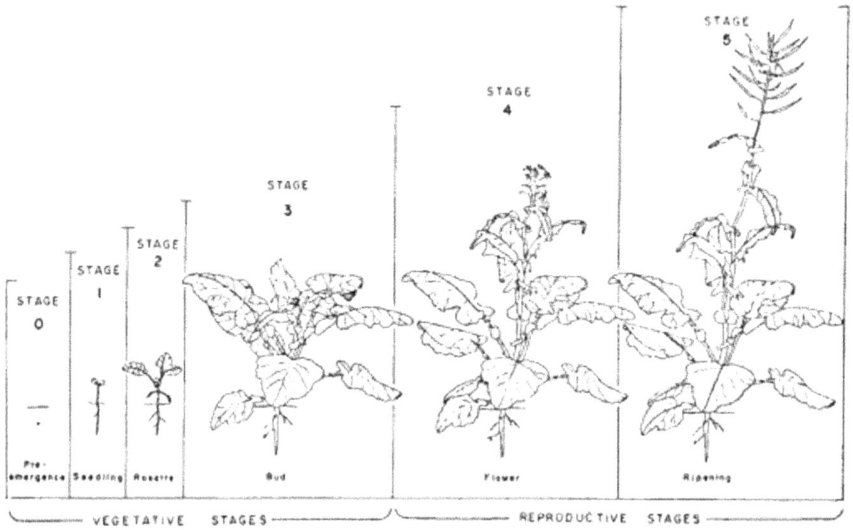

Source: Harper and Berkenkamp (1975). © Canadian Science Publishing or its licensors.

Figure 3.10. **Typical intact and opened siliques of *B. napus* and *B. rapa***

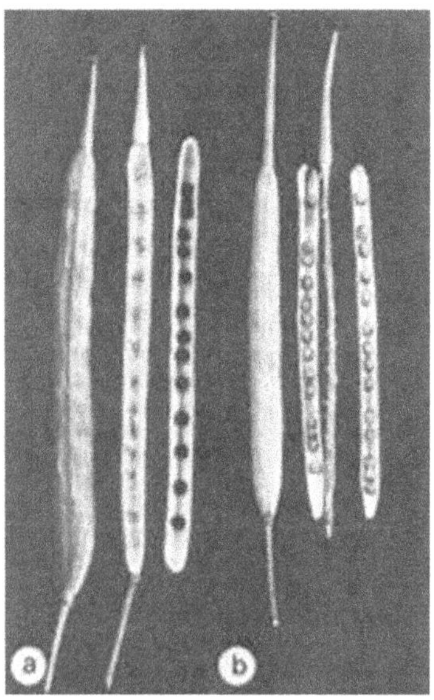

Notes: a) *B. napus* showing intact and opened siliques with seeds of the upper locule exposed, while those of the lower locule are partially obscured by the lamella. b) Intact and open silique of *B. rapa*.

Source: Downey (1983). Courtesy AAFC Reseach Station, Saskatoon (Photographer R.E. Underhood).

Autotetraploid *B. rapa* varieties have been developed for use as leafy vegetables, fodder (turnips) and green manure. Tetraploid plants have larger leaves, thicker stems, greater height and larger seeds than their corresponding diploids (Abel and Becker, 2007). However, tetraploids are not used as oilseed crops as their seed and oil yields are significantly lower than their diploid progenitors (Downey and Armstrong, 1962). The much larger tetraploid pollen also takes significantly longer to germinate than *B. rapa* diploid pollen. Thus, pollen from *B. rapa* diploid plants has a selective advantage resulting in triploid embryos, which abort (Downey and Armstrong, 1962; Håkansson, 1956), providing strong selection pressure against *B. rapa* tetraploid plants growing in *B. rapa* diploid populations. The slower pollen germination of tetraploid plants could predispose them to out-crossing with related species. On the other hand, since tetraploid *B. rapa* crops are normally consumed or ploughed down before flowering they are unlikely to be a significant factor in gene flow.

Brassica oleracea

The *B. oleracea* vegetables are often referred to as the "cole crops" and comprise cabbage, cauliflowers (including broccoli), kales and kohlrabi, but not the *B. rapa* vegetables.

Wild *B. oleracea*

Wild *B. oleracea* var. *oleracea* or wild cabbage is native to the western and southern seaboard of Europe where its tolerance of salt and lime, but its intolerance to competing vegetation, tends to restrict its presence to limestone sea cliffs (Heywood, 1964; Rich, 1991). The plants of this subspecies are biennial or perennial and in the first year produce a rosette of thick, fleshy leaves (Figure 3.11). Following vernalization a flowering stalk 1-2 metres tall arises from the centre of the rosette bearing a raceme of self-compatible, yellow flowers.

Figure 3.11. **Wild *B. oleracea* plants in their first year of growth**

Source: Wikipedia, the Free Encyclopedia.

B. oleracea var. *capitata*, cabbage

The cabbage is a biennial plant that in the first year of growth produces a dense, terminal head of tightly wrapped leaves on a short stout stem. The head is surrounded by a rosette of large fleshy leaves (Figure 3.12A). Three main types of heads – smooth green, red and Savoy – are commercially produced (Figure 3.12B). In the second year, the head splits open and the flowering stalk bolts to 1.5-2.0 metres tall with branches bearing flowering racemes of self-incompatible flowers.

Figure 3.12. **Heads of *B. oleracea* var. *capitata* and Savoy cabbages**

A. Head of cabbage, *B. oleracea* var. *capitata* with its rosette leaves intact

B. Heads of red, smooth green and Savoy cabbage with lower leaves removed

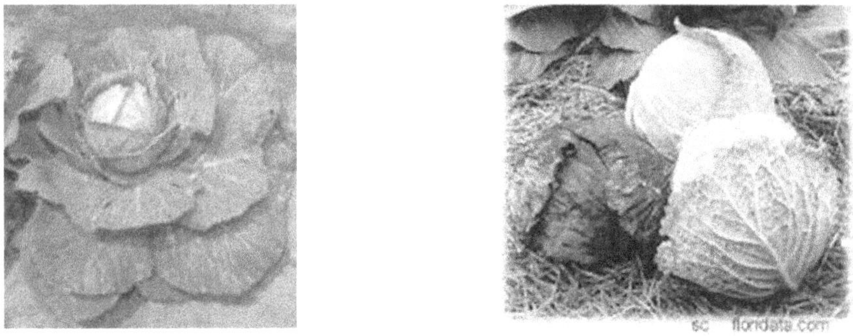

Source: Courtesy Floridata.

B. oleracea var. *botrytis*, broccoli and cauliflower

Cauliflower is derived from broccoli, being selected for short stout stems with a dense, terminal head or curd, made up of arrested inflorescence meristems, over topped by leaves (Figure 3.13). About 10% of the meristem mass will eventually develop into normal flowers and set seed (Sadik, 1962). Specific alleles of the *BoCAL*-a gene have been shown to be associated with discrete inflorescence morphologies (Smith and King, 2000; Purugganan, Boyles and Suddith, 2000). Smith and King (2000) present evidence suggesting that the cauliflower curd arose in southern Italy from a heading Calabrese broccoli via an intermediate Sicilian crop type.

Figure 3.13. **Head of cauliflower (left) and broccoli (right) *B. oleracea* var. *botrytis***

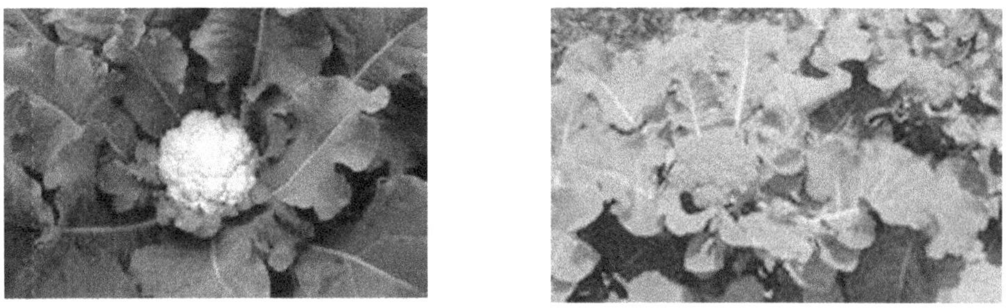

Source: Courtesy Cavaganaro, David/Sunset/Invision.

Broccoli differs from cauliflower in that broccoli flower heads tend to be smaller with more slender floret-stalks and are made up of arrested green (or purple) flower buds whereas the heads of cauliflower are formed by a condensed and thickened, malformed white (also purple or lime green) flower cluster. Both crops are biennial and, provided the plants have been vernalized, produce viable flowers and pods in the second year from the stump or parts of the head that remain. Vernalization requires a prolonged cold period of at least ten days with temperatures between 2°C and 10°C. The larger the plant when exposed to the cold treatment the greater the incidence of bolting. Plants of both crops are more susceptible to frost and less tolerant of heat and drought than cabbage. The cultural requirements of broccoli and cauliflower are similar but broccoli generally grows more rapidly. Most varieties are now F_1 hybrids.

The broccoli referred to above is more correctly known as "calabrese" broccoli. It produces a single head and is the form that is of greatest commercial importance. The "sprouting" broccoli, var. *italica*, produces a succession of small flowering heads over an extended period (Figure 3.14) while the "Romanesco" broccoli produces a head characterised by multiple cone shaped spirals consisting of masses of small flower buds (Figure 3.15).

Figure 3.14. **Sprouting purple broccoli**

Figure 3.15. **Romanesco broccoli**

Source: Courtesy Mr. Fothergill's Seeds Ltd. UK.

B. oleracea var. *viridis*, collards and kale

The kales and collards are biennials but are usually harvested in the first year for their edible leaves. They closely resemble their wild cabbage progenitors. Collards have large, smooth fleshy leaves with smooth margins (Figure 3.16). The leaves of kale are smaller and thinner than those of collards and many cultivars produce fringed, wavy-edged or feathery leaves (Figure 3.16). A thick flowering stem up to 1.5 metres tall emerges in the second year. One form called "Walking Stick" kale produces a tall straight stem which, when dried and polished, makes a fine walking stick.

Figure 3.16. ***B. oleracea* var. *viridis*, collard plant (left) and row of kale (right)**

Source: Courtesy Floridata.

B. oleracea var. *gemmifera*, Brussels sprouts

B. oleracea var. *gemmifera* plants are cool season biennials with simple erect stems up to 1 metre tall, bearing round to heart-shaped simple leaves with lengthy petioles. The leaves are glabrous with the colour varying from light to deep greyish-green. In the first year, auxiliary buds or sprouts are borne beneath the leaves on an elongated stem (Figure 3.17). The buds are modified leaves that form small heads up to 30 mm in diameter. Following vernalization, a seed head is produced from which a flower stalk emerges bearing perfect, self-incompatible flowers on terminal racemes. The seeds, weighing about 2.8 g/1 000, are borne in typical, two locule siliques.

Figure 3.17. ***B. oleracea* var. *gemmifera*, Brussels sprouts**

Source: Courtesy Limagrain.

B. oleracea var. *alboglabra*, Chinese kale

The var. *alboglabra* is widely grown throughout south-east Asia as a leaf and stem vegetable. The perennial plants are grown as annuals, producing dull or glossy thick green, glaucous, elliptic leaves about 25 cm long. The plants commonly called Chinese kale and kailan attain a height of up to 40 cm in the vegetative stage and 1-2 metres at the end of flowering. Upper stem leaves are oblong, petioled or non-clasping. The white flowered inflorescences develop siliques 5-8 cm long (Herklots, 1972).

Brassica napus

B. napus var. *napobrassica*, rutabaga or Swede

B. napus var. *napobrassica*, the common rutabaga or Swede, is a biennial with similar characteristics to the turnip. The bulbous root develops from the hypocotyl in the first year of growth (Figure 3.18). The surface of the root may be purple, white or yellow with the inner content solid yellow or white fleshed. The thick, smooth, dark green leaves emerge from the crown or neck of the root to form a ground covering rosette that shades out competing weeds. The presence of a root crown or neck distinguishes rutabagas from turnips. Early in the second year the flower stalk bolts from the root crown and the self-compatible flowers produce short beaked siliques on short pedicels containing two rows of round black seeds. Rutabagas are used for human consumption and for late fall cattle grazing.

Figure 3.18. ***B. napus* var. *napobrassica*, rutabaga or Swede**

Source: Courtesy Floridata.

B. napus var. *pabularia*, Siberian or rape kale

This sub-species has both annual and biennial forms with much branched erect stems up to 1.5 metres tall. Lower leaves are glaucous and lobed. Upper stem leaves are lanceolate, sessile and clasping (Figure 3.19). The much branched inflorescence is an elongated raceme producing siliques 5-11 cm long and 2.5-4 mm wide with a slender 0.5-3 mm long beak. The tap root produces many side branches. The crop is grown as a leafy vegetable and for fodder.

Figure 3.19. ***B. napus* var. *pabularia*, Siberian or rape kale**

Source: Courtesy Floridata.

Oilseed rape, *B. napus* var. *napus* f. *annua* and f. *biennis*

Oilseed *B. napus* has both an annual (spring) and a biennial (winter) form. The biennial form is less winter hardy than winter wheat which restricts its production to areas with mild winter conditions such as northern Europe and central China. The annual form is grown as a spring crop in western Canada and northern China but also as a winter crop in Australia and other countries with very mild winters.

Growth stages of annual *B. napus* plant development are illustrated in Figure 3.20. The glaucous lower leaves form a rosette from which the flowering stalk emerges bearing a dominant, indeterminate main raceme. The upper stem leaves are small, lanceolate, sessile and clasping. Plants of the species *B. napus*, *B. rapa* and *B. juncea* can be distinguished by their upper leaf attachment to the stem as illustrated in Figure 3.21. Flowering begins with the lowest bud on the main raceme and continues upward with three to five or more flowers opening per day. The buds, unlike those of *B. rapa*, are held above the uppermost open flowers. Flowers on the secondary branches begin to open about three days after the opening of the first flowers on the main raceme. The siliques are ascending on slender pedicles and about 7-10 cm long with a beak about 1.3 cm long. Seeds are dark brown to black, and weigh 2.5-5.5 g per 1 000 seeds.

Figure 3.20. **Growth stages of *B. napus* var. *napus* f. *annua***

Notes: a) Seedling cotyledons; b) cotyledons and first true leaf; c) rosette; d) flowering; e) pod set; f) mature plant.

Figure 3.21. **Upper leaves of *B. rapa*, *B. napus* and *B. juncea***

Notes: a) *B. rapa*, fully clasping stem; b) *B. napus* partially clasping; c) *B. juncea*, non-clasping.

Source: Downey (1983). Courtesy AAFC Reseach Station, Saskatoon (Photographer R.E. Underhood).

Brassica juncea

B. juncea vegetables

In China and south-east Asia many vegetable forms of *B. juncea* have been developed and classified as species or subspecies under numerous names, depending on the morphological features given the greatest importance. Kumazawa and Able (1955) examined some 200 East Asian cultivars of *B. juncea* vegetables grown in China, Japan, Nepal and Chinese Taipei on the basis of their plant size, root form, tillering and leaf characteristics. All accessions of *B. juncea* and its subspecies were described as annuals and placed in 25 different groups within 8 classes. These classes were further condensed into four subspecies. The authors state that the subspecies evolved from the leafy and oilseed forms of brown mustard, *B. juncea* (L.) Cross. From the collection, the authors illustrated a normal or "ordinal" root form, a turnip-like rooted form (also described by Dixon, 2007) and a little known form with a tuberous basal stem (Figure 3.22). The four subspecies were grouped and characterised as follows.

1. subsp. *napiformis* (Pailleux & Bois) Gladis, grown for its tuberous turnip-like root. This subspecies bolts late and has a high tolerance to cold.
2. var. *japonica* (Thunb.) L.H. Bailey, characterised by curled, narrow or dissected leaves.
3. subsp. *integrifolia* (H. West) Thell., characterised by entire or little lobed basal leaves. Herkots (1972) notes that some cultivars may form a tight head (Figure 3.23).
4. var. *rugosa* (Roxb.) N. Tsen & S.H. Lee, includes cabbage leafed forms with large entire or serrated radical leaves. The tuberous basal stem form (Figure 3.22) is included in this subspecies.

Herklots (1972) places var. *rugosa* within subsp. *integrifolia* but also puts forward the var. *sareptana* as characterised by lyrately-lobed basal leaves and var. *crispifolia* as having dissected, crisped lower leaves. More recently, a *B. juncea Biology Document* (Canadian Food Inspection Agency, 2007) quoted the grouping by Spect and Diederichsen (2001) into the following four sub-species:

1. subsp. *integrifolia*, used as a leaf vegetable in Asia.
2. subsp. *juncea*, cultivated mainly for its seeds, occasionally as fodder.
3. subsp. *napiformis*, used as a root-tuber vegetable. Dixon (2007) describes this subspecies as having a high tolerance to cold and an enlarged conical root.
4. subsp. *tsatsai* from which stalks and leaves are used as vegetables in China.

Figure 3.22. **Three forms of *B. juncea***

Bulbous root (a), normal or "ordinal" root (b) and tuberous basal stem (c)

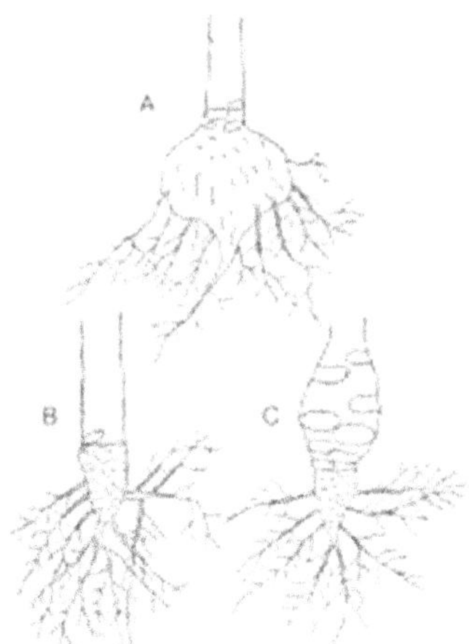

Source: Kumazawa and Able (1955).

Figure 3.23. ***B. juncea* subsp. *integrifolia*, heading mustard, BauSin**

Source: Courtesy AgroHaitai Ltd.

B. juncea, oilseed and condiment mustards

Plants of this species, grown for their seed oil or condiment production, are normally referred to simply as *B. juncea* without the attachment of a subspecies name. However, Spect and Diederichsen (2001) classify this plant group as *B. juncea* subsp. *juncea*. Plant for both oil and condiment are similar in their morphology but differ in seed oil percentage and the type and amount of glucosinolates present in the seed. These forms are annuals that grow to about 1.2 metres as spring-sown crops in western Canada and Europe. On the Indian sub-continent they are grown as a winter crop where, under short days, plants grow up to 2.1 metres tall. The plants are green and sometimes slightly glaucous. The lower leaves of the rosette are rather thin, elliptic to obovate and lyrately-lobed or divided. The upper stem leaves are small, narrow and not clasping (Figure 3.21). Depending on the day length and temperature the flowering stalk bolts and produces a raceme with no terminal flower. As with *B. napus*, the buds are borne above the open flowers. Apical dominance is present with the secondary racemes initiated about three days after flowers open on the main raceme. The silique is about 7 cm long containing seed weighing 2.5-3.0 g/1 000 seeds.

Geographic distribution, ecosystems and habitats, cultivation and management practices, centres of origin and diversity

Introduction

From an ecological and agronomic point of view, both the spring and winter forms of oilseed rape exhibit two undesirable characteristics. First, mature pods tend to shatter, leaving large but variable amounts of seed on the ground at harvest (see below on the contribution of *B. napus* harvest losses to persistence). Pod shatter not only results in lost yield but also sets the stage for large numbers of volunteer plants in subsequent crops. Fortunately *B. napus* seeds have no primary dormancy so if moisture and temperature are adequate, the vast majority of these seeds germinate and are killed by frost, herbicides, cultivation or predators (see below). The opportunity for *B. napus* to acquire primary dormancy is limited due to the vast majority of fields being sown each year with high germination certified seed.

The second undesirable characteristic is the tendency for a proportion of the shattered seed to acquire secondary dormancy. Such dormancy is induced by abiotic stresses (see section on persistence below). Although most of the shattered seed will quickly be reduced by fatal germination, predation, disease and abiotic stress, a small percentage can remain dormant and viable for ten years or more (Schlink, 1998; Lutman, Freeman and Pekrun, 2003). Thus, *B. napus* is able to establish seed banks within cultivated fields (see Lutman et al., 2005 and below). As a result, traits or genes that have been genetically silenced or augmented within improved varieties may be reintroduced. Examples would be the genetic blocking of the biosynthesis of erucic acid in rapeseed oil, the reduction in linolenic acid content and the augmentation of oleic fatty acids in the oil, or reduction of glucosinolates in the oilseed meal.

It should be noted, however, that there is considerable genetic variability within the species and its close relatives in both the degree of pod shatter and the percentage of induced dormancy. Until recently these characteristics have not been a priority for oilseed rape breeders but progress is possible. Wang, Ripley and Rakow (2007) have clearly demonstrated that selection for reduced pod shatter in *B. napus* can be achieved. In addition, Østergaard et al. (2006) have shown that expressing the *Arabidopsis*

FRUITFULL gene in *B. juncea*, using a CaMV 35S promoter, produces shatter-resistant plants. Although the shatter-resistant pods held their seed too tightly for combine harvesting, a weakened form of the FRUITFULL gene could result in an economically and environmentally valuable advance. It is unlikely that conventional breeding will lead to complete elimination of the shattering characteristic, but there appears to be considerable room for improvement. Further, Pekrun, Potter and Lutman (1997); Gruber, Pekrun and Claupein (2004); Gruber, Emrich and Claupein (2009); and Gulden, Thomas and Shirtliffe (2004) have all shown that among *B. napus* varieties, of both spring and winter forms, there is a wide range in the percentage of seed susceptible to induced dormancy. Thus, the application of conventional breeding techniques to select varieties producing seed resistant to secondary dormancy should greatly reduce the presence of volunteers in subsequent crops.

Ecologically, *B. napus* is described as a cultivated crop where escaped plants become colonisers of waste places. However, they are not invasive of natural habitats. Colonisers are defined as species that occupy disturbed sites or habitats but with populations that keep moving, founding new populations while losing old ones (Williamson, 1996). Feral populations of *B. napus* are most frequently found along road and rail verges, field margins and in disturbed soils. The reports on the abundance and persistence of such feral populations vary considerably from country to country and between the spring and winter forms. Williamson (1996) noted that colonising species are not the same as invaders, even if they have high intrinsic rates of increase, as exhibited by *B. napus*. He classifies *B. napus* in Britain as intermediate between naturalised and casual. On the other hand, recent intensive surveys of feral sites in mainland Europe have identified feral populations in higher frequencies than anticipated, with some sites able to sustain themselves in a semi-permanent state (Pivard et al., 2008). Such reports have given rise to concerns by some that a proportion of feral populations could become permanent and in time result in the invasion of natural habitats.

Although the species does have the weedy characteristics noted above, producing many propagules (seeds), plus the ability to cross with some weedy relatives, it is not competitive with perennial grasses that dominate the natural habitat. It should also be noted that oilseed rape has been part of the European landscape for a very long time as have the truly weedy, related species, *Sinapis arvensis* and *B. rapa*. However, none have become invasive of natural habitats. In recent years, the area of oilseed rape cultivation and intensity of production has increased worldwide. For example, since 1970 oilseed rape production in France and Germany has increased 4.5- and 8-fold, respectively. In the same period, the Canadian oilseed rape acreage has quadrupled, thus a wider and more frequent occurrence of feral populations is to be expected. The spring form of oilseed rape is much less likely to form feral populations or to be self-sustaining since fall germination is normally fatal while frosts will kill many seedlings that germinate in the spring. Although Knispel et al. (2008) has reported some transient feral population in the province of Manitoba, Canada, such roadside populations are rare over most of the Prairie Provinces, except near collection points and to a limited extent along railroad verges. This is because in western Canada most road verges are mowed in late August before feral populations set viable seed. Such roadside mowing is essential to prevent snowdrifts across roads that tall vegetation can cause. In contrast, in Europe, the winter form can avoid being killed by the fall road maintenance since mowing does not usually affect the established first year rosettes, leaving some plants to flower and set seed before the next fall mowing.

Different agronomic practices also influence the size and persistence of volunteer populations. In Europe, the large amount of straw remaining after harvest plus the short time between the July harvest and August sowing dates encourages ploughing down of residue, resulting in seed burial. In Canada on the other hand, ploughing is not practiced and most fields are spring- sown into undisturbed stubble (minimum or zero tillage) from the previous year's September harvest (Hall et al., 2005). Thus, seed burial is minimised and harvest seed losses are exposed to environmental hazards. The result is that in Europe, old or discontinued cultivars or genotypes will persist in the seed bank for a much longer time than in Canada. This is clearly illustrated in the changeover from high to low erucic acid *B. napus* varieties. In the German oilseed rape growing province of Schleswig-Holstein, it required ten years to reduce the commercial crop from the traditional high erucic varieties (50% erucic) to the desired level of 2% (Sauermann, 1987). In Canada, the same results were obtained in three years (Daun, 1983).

Geographic distribution

The genus *Brassica* and its wild relatives are part of the tribe Brassiceae that has its origin in the Mediterranean basin and in south-western Asia. However, the geographic centre is thought to be in the south-western Mediterranean region (Algeria, Morocco and Spain) where some 40 genera have been shown to be endemic or exhibit maximum diversity (Hedge, 1976; Gómez-Campo, 1999, 1980; Al-Shehbaz, 1984; Al-Shehbaz, Beilstein and Kellogg, 2006; Warwick and Hall, 2009). For the subtribe Brassicinae, Hedge (1976) leaves little doubt that it originated in the Mediterranean basin. The species distribution of the Brassicaceae family is concentrated in the northern temperate zone and south-western and central Asia (Holm et al., 1997). Few species are found in hot, humid tropics.

B. nigra

B. nigra or black mustard was widely grown for the sharp pungency of its seeds and as a leaf vegetable. Prakash and Hinata (1980) placed the species origin in central and south Europe. It is one of the oldest recorded spice crops, which undoubtedly resulted in its early and widespread distribution across Europe, Africa, Asia and the Indian sub-continent, and its dehiscing siliques with primary dormancy of the seed ensured its persistence. The GRIN describes the species distribution as widely naturalised in the following regions and countries. In Africa: countries along the south shore of the Mediterranean as well as Eritrea and Ethiopia. In temperate Asia: Afghanistan, Armenia, Islamic Republic of Iran, Iraq, Israel, Kazakhstan, Lebanon, Syrian Arab Republic, Turkey and northwest China. For the Indian sub-continent: India, Nepal, Pakistan. In Europe: all countries in western and eastern Europe as well as the Balkans and Greece. The crop was introduced to the Americas and Australia as a spice. However, in the 1950s it was displaced by the higher yielding, pungent *B. juncea* that was better suited to mechanical harvesting. Although in many regions black mustard is now a weed of waste places, it has never become established on the Canadian prairies, although it is present throughout much of the United States.

B. rapa

B. rapa is thought to have originated in the mountainous areas near the Mediterranean sea (Tsunoda, 1980). The time of domestication is unknown. Sinskaia (1928) proposed two main centres of origin, one being the Mediterranean and the other the Afghanistan-Pakistan region. The species appears to have attained a wide distribution throughout Europe, parts of Africa, Asia and the Indian sub-continent before recorded history. Excavations in China reported the presence of *B. rapa* seed at a 6 000-7 000-year-old archaeological site (Liu, 1985). Indian Sanskrit literature mentions the plant about 1599 B.C. (Prakash, 1961), and Renfrew (1973) indicated that *B. rapa* seed was consumed in Scandinavia as early as 350 B.C. *B. rapa* is grown as an oilseed crop in northern Europe, north-west China, the foothills of the Himalayas and northern India, while the vegetable forms were selected and modified in Asia, primarily in China. The oilseed form was introduced to Canada by a Polish immigrant about 1936 (Boulter, 1983) and Australia began its first investigations on the *B. rapa* crop in the early 1960s (Salisbury, 2002) but it has now been superseded by *B. napus* varieties. *B. rapa* also has a weedy form that differs from the cultivated plant in exhibiting primary dormancy and has a worldwide distribution (Figure 3.24).

Figure 3.24. **World distribution of *B. rapa* as a reported weed**

Source: Modified from Holm et al. (1997).

B. oleracea

The centre of origin for the *B. oleracea* species is along the European Atlantic coast while the wild related forms still grow on the islands and along the northern coast of the Mediterranean. The various forms of this species were developed in Europe and did not reach Asia until about the 16th century (Liu, 1985). The many cultivated forms of this species have been introduced and grown worldwide, with the exception of some tropical areas.

B. napus

B. napus is of relatively recent origin (<10 000 years; see the section on genetics at the end of this chpter and Figure 3.39) resulting from the interspecific cross between plants of *B. oleracea* and *B. rapa*. The cross must have occurred where the two species were growing in close proximity along the European Atlantic or Mediterranean coasts. Dispersal of the species is thought to have occurred throughout Europe in the 16th century with the introduction to the Americas in the 17th and 18th centuries and the Far East in the 19th century (Liu, 1985).

B. juncea

B. juncea is believed to have arisen about 10 000 years ago as the result of an interspecific cross or crosses between plants of *B. rapa* by *B. nigra*. Evidence suggests that one primary centre of origin is China, where the greatest divergence of forms evolved (Prain, 1898; Sinskaia, 1928; Vavilov, 1949). A second centre of origin is thought to be Afghanistan and adjoining regions (Olsson, 1960; Mizushima and Tsunoda, 1967; Tsunoda and Nishi, 1968) from where it spread to a secondary centre on the Indian subcontinent and became a major oilseed crop (Hemingway, 1995; Prakash and Hinata, 1980). GRIN (USDA-ARS, 2011) lists *B. juncea* as native to temperate Asia including China, Kazakhstan, Kyrgyzstan, Mongolia, eastern and western Siberia, Tajikistan and Turkmenistan. It has been introduced as a condiment crop to Europe, the Americas, Australia and New Zealand. It has been designated a weed of southern European Russian Federation, the Caucasus, central Asia and southern Siberia, and a casual or feral plant in southern and southeast Asia, Africa and America (Canadian Food Inspection Agency, 2007).

B. carinata

B. carinata, like *B. juncea*, is believed to have arisen about 10 000 years ago as a result of an interspecific cross between plants of *B. nigra* and *B. oleracea*. The cross is thought to have occurred in the Mediterranean region where both species were present. As the climate in North Africa became dryer, *B. carinata*, along with the flora of the moist Mediterranean region, moved south to the highlands (1 300-1 500 metres) of Ethiopia. The species distribution from its Ethiopian centre of origin has been limited to neighbouring east African countries. Recently it has been introduced as an oil crop to India and as a species of commercial interest in Canada and Spain.

Ecosystems and habitats where the species occurs natively and has naturalised

There are few areas of the world where members of the family Brassicaceae are totally absent. The exceptions are the Antarctic and some parts of the tropics. However, even in the tropics, the family is thinly represented by some introduced cosmopolitan weeds that have become established. The genera and species of the family occur in greatest number and diversity in the temperate zone of the northern hemisphere and in particular, the areas surrounding the Mediterranean basin and throughout the southwest and central regions of Asia (Figure 3.25) (Hedge, 1976). Although the generic and specific endemism in the family is highest in the Irano-Turanian region, the centre of the present-day subtribe, Brassicinae, lies in the Mediterranean basin (Hedge, 1976).

Feral populations in disturbed soils

Due to the large seed losses in commercial *B. napus* fields and the potential loss during transport and handling, the surviving seeds give rise to volunteers in subsequent crops and feral populations in non-cultivated areas (CETIOM, 2000; MacDonald and Kuntz, 2000; Orson, 2002; Pessel et al., 2001; Price et al., 1996). Volunteers are controlled by cultivation and herbicide application. In both Canadian and UK trials, the numbers of genetically modified (GM), herbicide resistant (HR) *B. napus* volunteers in the year following GM trials were comparable to, or less than, conventional *B. napus* (Crawley et al., 1993; Booth et al., 1996; Hails et al., 1997; Rasche and Gadsby, 1997; Sweet et al., 1999a, 1999b, 1997; Sweet and Shepperson, 1998; Norris et al., 1999). In their survey of Canadian commercial fields, MacDonald and Kuntz (2000) found the same trend, with similar numbers of volunteers in the year following cultivation of

GM HR canola compared to conventional varieties. Furthermore, prior to any field operations, they found an average over all fields of 200 volunteers/m^2. Initial soil disturbance was effective in controlling these emerged *B. napus* volunteers, but shallow cultivation resulted in the emergence of an even greater number of volunteers. A post-emergent weed control programme employed by the producer for the non-GM volunteers was also effective in controlling the GM volunteers (MacDonald and Kuntz, 2000). Downey and Buth (2003) reported that GM HR volunteers with single or stacked traits were readily controlled in western Canada by the same agronomic practices that are standard for controlling conventional canola volunteers. In Australia, post-harvest monitoring of GM HR (glufosinate or glyphosate) trial locations for six years indicated volunteer populations were adequately controlled by herbicide application or broadacre cultivation (either in-crop or by conservation tillage) (Salisbury, 2002).

Figure 3.25 **Approximate areas of the phytogeographic regions containing the world's greatest representation of Brassicaceae genera**

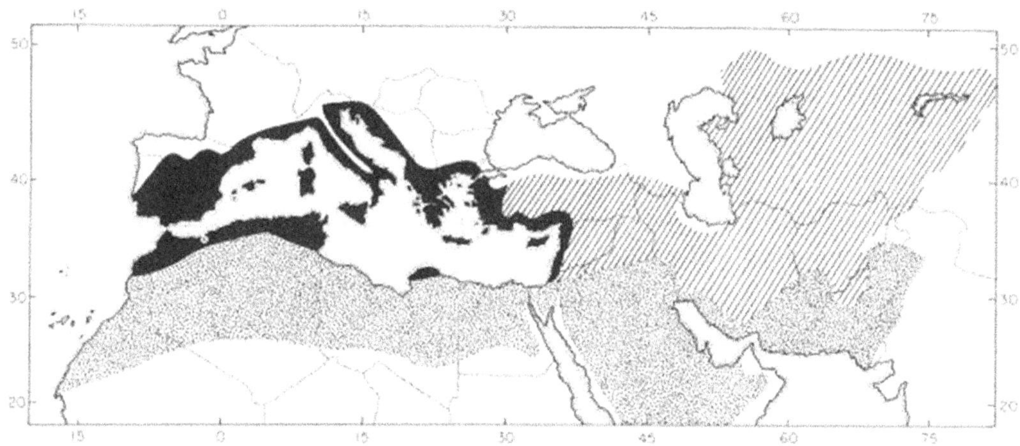

Note: They encompass the Mediterranean (black); the Irano-Turanian (striped) and the Saharo-Sindian (dotted) regions.

Source: After Hedge (1976).

Feral populations of *B. napus* can be found at various densities on road verges, along field margins and railway lines in all countries where it is grown (e.g. Crawley and Brown, 1995; Wilkinson et al., 1995; Squire et al., 1999; MacDonald and Kuntz, 2000; Agrisearch, 2001; Pessel et al., 2001; Orson, 2002; Salisbury, 2002). Populations may also become established in port areas where *B. napus* cargos are handled (Ramsay, Thompson and Squire, 2003; Saji et al., 2005; Aono et al., 2006). Annual recruitment to such sites is likely to be more from passing transport vehicles than from an established seed bank. *B. napus*, as with other *Brassica* species, is a coloniser of disturbed soils where it competes with other primary colonisers. However, *B. napus* is a poor competitor and is not regarded as an environmentally hazardous colonising species (European Commission, 2000, 1999, 1998a, 1998b; Beckie, Hall and Warwick, 2001; Dignam, 2001). Unless the habitats are disturbed on a regular basis, *B. napus* will be displaced (OECD, 1997).

In western Canada, roadside verges, field margins and railway lines were surveyed for canola plants (MacDonald and Kuntz, 2000). Only 13 and 27 volunteer *B. napus* plants were found in the mowed roadside over the respective 7 and 27 kilometres surveyed, and no plants were found in tall, unmowed grass. Surveys of rail beds leading from local grain elevators, approximately 3 and 5 kilometres long, identified 287 and 29 plants, respectively, growing at the interface of the rail bed gravel and the tall grass of the right of way. No plants were located on the rail tracks or in the tall grass of the right of way. Similarly in Australia, a survey, making 400 observations in 5 × 20 m areas along 4 000 kilometres of roads in oilseed rape growing areas, found *B. napus* plants in only 31%, 20%, 13% and 9% of the observation points in southern New South Wales, Western Australia, Victoria and South Australia, respectively. Nearly all the plants were growing within five metres of the roadside, with the vast majority close to or alongside the road edge, suggesting they originated from seed dropped from passing vehicles (Agrisearch, 2001).

In the United Kingdom, Crawley and Brown (1995) found that along undisturbed roadways, the persistence of *B. napus* is about three to four years and that the density of such feral populations is correlated with human activities, such as vehicle transport. In a three-year assessment of feral populations in Scotland, Wilkinson et al. (1995) found that the turnover of populations was high, with only 19% of the 1993 population persisting into 1994 and 12% of the 1994 population persisting into 1995. Crawley and Brown (1995) obtained similar results in southern England. In a study conducted in Germany from 2001 to 2004, Dietz-Pfeilstetter, Metge and Schönfeld (2006) found persistence rates for feral populations of 29% between 2001 and 2002, of 12% between 2002 and 2003 and 80% between 2002 and 2004. However, molecular profiling using ISSR-PCR (inter-simple sequence repeats-polymerase chain reaction) revealed that plants appearing in successive years largely belonged to different genotypes, suggesting new seed input and an even higher turnover of populations.

Reuter et al. (2008) investigated a 500 km^2 area in the region of Bremen, Germany and reported average densities of 1.19/km^2 and 1.68/km^2 of feral and volunteer oilseed rape populations in rural and urban areas, respectively. The investigation showed that population density varies between years and feral plants tend to be smaller in stature (by at least 40%) than plants growing on cultivated land.

Surveys by Agrisearch (2001) and MacDonald and Kuntz (2000) suggest that to survive spring, *B. napus* roadside populations need to be regularly replenished. However, in France, Pessel et al. (2001) found roadside feral populations contained plants of old varieties that had not been grown for eight to nine years, indicating that the seed source was not entirely from recent vehicle spillage. These results are in keeping with previous reports that seed of old rapeseed varieties can persist for at least five to ten years after they were last reported grown (Squire et al., 1999; Orson, 2002). Pessel et al. (2001) suggested that the analysed roadside feral populations arose from multiple spillages from different fields or germination of seed from a mixed seed bank or most likely, both.

In Austria, Pascher et al. (2006) genetically analysed plants from 9 selected feral populations consisting of 50-150 individuals. They found the feral populations were genetically more diverse than could be explained by the dominant varieties grown in the area in the previous five years. They concluded that even though the feral populations largely reflected the genetic makeup of the dominate varieties being grown, a significant portion of plants had originated from seed banks older than five years. They also found that feral populations disappeared more quickly under dense grass cover than at sites with little vegetation, but genetic diversity remained unchanged. Their results indicated that genetic migration from commercial varieties to feral populations was five times greater than the inverse.

Feral populations in natural habitats

In natural (undisturbed) ecosystems, *B. napus* is not considered to be invasive or even a significant component of any natural plant community (AAFC, 1994; Warwick, Beckie and Small, 1999; Beckie, Hall and Warwick, 2001; Dignam, 2001).

Production and agronomy of Brassica *oilseed crops*

The world demand for edible oils and more recently for biodiesel has led to a rapid growth in the production of most oilseeds, with total seed oil produced increasing by about 4% each year. The percentage growth in the world *Brassica* seed oil production increased some 60% between 1996-09 and 2006-10 (Table 3.1). The locations of the major rapeseed/mustard producing regions over the two decades 1995-2014 are shown in Table 3.4. The expanded *Brassica* oilseed production has resulted from both an increase in the area sown globally, as well as the yield per unit area that has increased in most regions (Table 3.4).

Table 3.4. **Area harvested, production and yield by major *Brassica* oilseed producing countries, averages 1995-99 to 2010-14**

Producing country	Area harvested ('000 ha)				Production ('000 tonnes)				Seed yield (kg/ha)			
	1995-99	*2000-04*	*2005-09*	*2010-14*[3]	*1995-99*	*2000-04*	*2005-09*	*2010-14*[3]	*1995-99*	*2000-04*	*2005-09*	*2010-14*[3]
E.U.[1]	3 944	4 255	5 407	6 693	11 038	12 221	18 108	20 865	2 786	2 862	3 082	3 120
Canada	4 917	4 366	5 322	7 757	6 866	6 237	10 510	15 171	1 398	1 415	1 790	1 954
China (People's Republic of)	6 708	7 244	6 740	7 244	9 391	11 573	12 070	13 315	1 399	1 597	1 842	1 836
India	6 541	5 110	7 280	6 303	5 756	5 045	7 239	7 417	884	982	1 079	1 178
Australia[2]	929	1 335	913	2 425	1 230	1 529	1 395	3 133	1 370	1 146	1 084	1 277
United States	292	497	452	568	443	762	665	1 009	1 505	1 555	1 635	1 780

Notes: 1. E.U. = Total production of the 27 (28 from 2013) member states of the European Union. 2. Extreme drought greatly reduced Australian production and seed yield in the 2005-09 period. 3. Columns added in January 2016.

Source: FAOSTAT.

Cultivation and management of oilseed crops

The small seeds of the *Brassica* oilseed crops require that the seed be sown at shallow depths, 2-3 cm below the soil surface, into a firm, moist seedbed. Under favourable growing conditions the seedlings emerge within four to five days of sowing. Cotyledon expansion is quickly followed by the formation of a rosette of seven to eight true leaves from which the flowering stalk bolts. The length of time the crop remains in the rosette stage can vary from less than 30 days to more than 210 days depending on climatic conditions and the species and form grown. The complete growth cycle may be as short as 70 days (*B. rapa*) or as long as 380 days for winter *B. napus* varieties in China (Sun et al., 1991).

Although the *Brassica* oilseed crops prefer a deep loam soil, it does well when sown in a wide range of soil types and conditions and can tolerate a pH range from 5.5 to 8. Compared to most other grain crops, *Brassica* oilseed crops require greater nutrient inputs to achieve high yields. Generally speaking, they need about 25% more nitrogen, phosphorus and potassium and up to 5 times more sulphur than a wheat crop. Harvested seed should be stored at no more than 9% moisture when cooled to 10°C to prevent deterioration due to fungal and/or insect activity. The usual rotation is as a break crop with cereals. Wheat yields following a *B. napus* crop invariably improve in Europe and Australia due the reduced level of cereal pathogens present and the control of grassy weeds (Almond, Dawkins and Askew, 1986).

North and South America

The oilseed rape/canola grown in North America is concentrated in the northern part of the Western Great Plains (Figure 3.26). The species and form grown is almost exclusively the spring or annual *B. napus*. In western Canada, less than 1% of the 5 million ha is sown to spring *B. rapa*. Production of the winter or biennial form of *B. napus* in North America is confined to a few thousand hectares in the Province of Ontario, Canada and a few west and central states in the United States. In South America, both spring and winter *B. napus* is produced on some 17 000 ha in central Chile.

Figure 3.26. **Areas of oilseed rape/canola production in North America**

Notes: Light grey indicates heavier production concentration.

Source: Courtesy Canola Council of Canada.

Cultural practices in the main oilseed rape production regions of western Canada and the United States have changed in recent years. Traditionally the crop was sown into summer fallow, land laid fallow the previous year. With the shift to continuous cropping and minimum tillage, *B. napus* is now sown into the undisturbed stubble of the previous year's cereal crop. Weed control, which would normally be a problem with this direct seeding system, can now be easily achieved with the new broad spectrum, post-emergence herbicides such as glyphosate, glufosinate and the imidazolinones. The adoption of these herbicides and their associated herbicide resistant varieties has been extremely rapid (Figure 3.27).

Figure 3.27. **Percentage of the total Canadian *B. napus* production area sown to herbicide resistant varieties, 1995-2008**

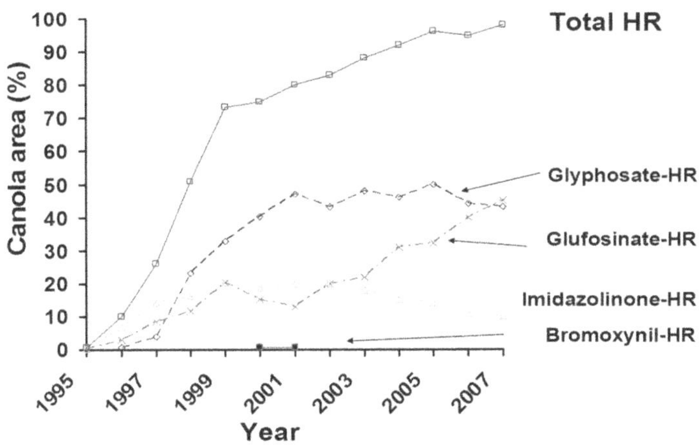

Note: HR = herbicide resistant.

Source: Adapted from Beckie (2011).

The double disc grain drill has now been largely replaced by large air seeders that place the seed and fertilizer some 2-3 cm below the soil surface, at a seeding rate of 5-8 kg per hectare. Seed is treated with an insecticide-fungicide coating. The herbicide glyphosate is usually spring applied prior to seeding to control early germinated or biennial weeds. In North America, seeding generally occurs in early May. The herbicide of choice is applied at the recommended rate when the weeds are small and the leaves of the *B. napus* plants have not fully covered the ground. *B. rapa* fields begin flowering in mid-June while *B. napus* fields begin to flower about two weeks later in late June or early July. Recommended fungicides and/or insecticides may be applied as a spray if the pest incidence warrants. At harvest, in late August through September, the crop is normally swathed into windrows to allow more uniform ripening and to protect against seed losses due to pod shatter. Combining the swaths is done when the seed is mature and dry. However, some straight combining of the standing crop is also practiced. Usually the seed is farm stored at less than 9% moisture until marketed.

Chile is the only country in South America that produces a significant quantity of oilseed rape with planting of winter and spring *B. napus* on some 17 000 hectares in the southern provinces of the central part of the country. The crop is predominantly winter *B. napus*. The winter crop is sown in March and April, flowers in October and November and is harvested by straight combining in January. Winter kill may occur in May or June due to the wet soil freezing and heaving, causing broken roots. The spring crop is sown in August-September, flowers in October and is harvested in late December or early January. The crop is normally sown on land broken out of grass pasture using a disk or mould board plough, disked twice with a double disk cultivator and packed. Seed is sown with a double disk seeder or the less satisfactory one-way disk at 7-8 kg per hectare. Fertilizer requirements vary widely due to the sharply different soil types encountered in rapeseed growing areas. Levels of macronutrients nitrogen, phosphorous and, in some soils, sulphur are very low. Also lacking in some soils are the micronutrients manganese, copper and boron. At harvest, a desiccant is applied and after the appropriate interval, the crop is straight combined. The seed is normally artificially dried to less than 9% moisture prior to storage or marketing.

European crop cultivation and management

Winter oilseed rape (*B. napus*) is the dominant species grown in both Western and Eastern Europe (Poland, western Russian Federation and Ukraine); however, the area sown to spring *B. napus* is rapidly expanding (Figure 3.28). Some spring and winter *B. rapa* is grown in Finland and Sweden. In Germany over the past 13 years, the area sown to spring oilseed rape decreased from 10% to 1% of the oilseed rape growing area. Spring *B. napus* is used primarily as a replacement crop on winter oilseed rape fields that have been winter-killed. The optimum date for sowing the winter form varies with the latitude and the onset of winter. In northern European countries, the optimum sowing date is the last half of August while more southerly regions in France and Germany can delay seeding until early September. The objective is to produce plants that are large enough and have stored sufficient food reserves to withstand the rigours of winter. It is recommended that plants entering the winter show a vigorous growth, a well-developed root system (taproot about 8-10 mm in diameter) and have at least 6-8 true leaves. Seed is sown into well-worked soil at 5.0-5.5 kg per hectare when drilled and 8-9 kg per hectare if broadcast, to obtain fall stands of 50-85 plants/m^2 to allow for some winter kill.

Figure 3.28. **Oilseed rape (*B. napus*) production regions in Europe showing millions of hectares of winter rape per country**

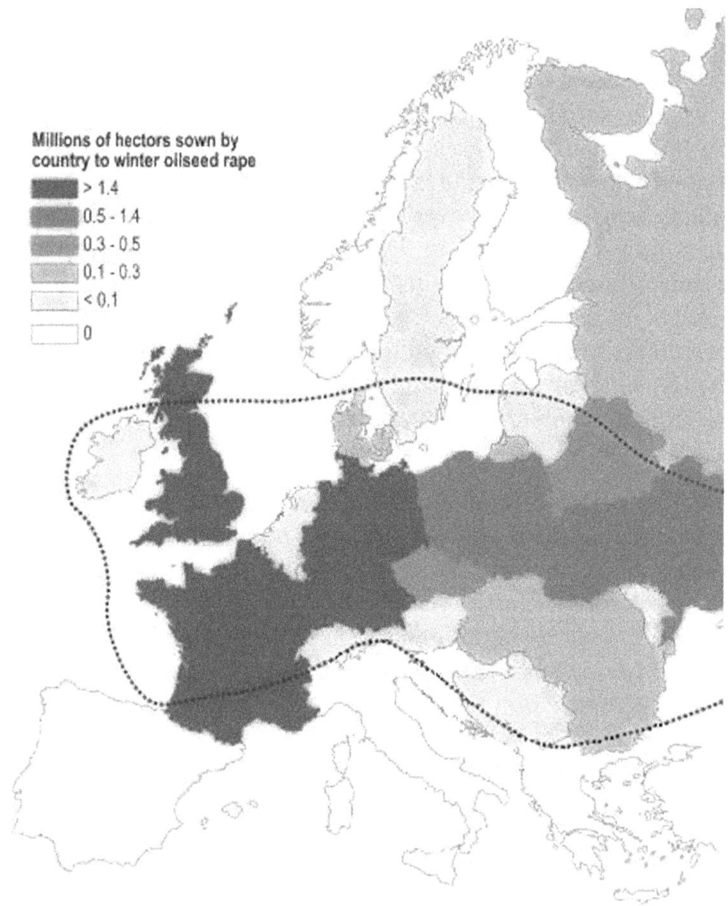

Notes: The dotted line encircles the primary growing region for winter oilseed rape. Spring rape production is concentrated in Eastern Europe, primarily the Russian Federation (>0.5 M ha) and Ukraine (0.1-0.5 M ha).

Source: Adapted from information supplied by Norddeutsche Pflanzenzucht.

The seeding rates recommended for precision drilled hybrid varieties with a high branching density and a 1 000 seed weight of 4 g/1 000 is 1.2-1.6 kg per hectare and for seed of 7 g/1 000, 2.1-2.8 kg per hectare. The seeding rates for drilled hybrids are lower than for open pollinated varieties since the hybrid seed is likely to produce a more vigorous plant that better withstands the winter. The optimum spring plant population is reported to be 80-100 plants/m^2. Winter varieties are heavy users of nitrogen so frequently some nitrogen is incorporated prior to planting, with the balance top-dressed in the spring. Excessive nitrogen promotes vigorous fall growth but tends to make the crop more susceptible to winter kill. Phosphorus and potassium are applied before planting at the recommended levels. Sulfur is used in early spring in combination with N-fertilization. Boron is often applied in late spring in combination with fungicides. Nearly all seed is treated with a fungicide-insecticide combination (often with more than two active ingredients) to control seedling pests.

Disease, insect and weed control in the emerged crop is achieved by spraying the recommended products when needed. Flowering in northern Europe begins the last days in April, and harvest starts with some swathing at the end of July with the vast majority of the crop straight combined a week or so later. Harvest can continue through to the end of August. In southern regions, harvest commences about one to four weeks earlier.

Australian crop cultivation and management

Oilseed rape production in Australia is relatively recent with the first commercial production undertaken in 1969. In the early years, both *B. rapa* and *B. napus* spring varieties from Canada were imported and grown in the winter season. Today production is almost exclusively from Australian-bred *B. napus* varieties. Canola is grown in most cropping areas of Southern Australia, including Western Australia (Figure 3.29). Most of the *B. napus* crop is sown in late autumn or early winter (April to June) during the rainy period. The seed is primarily sown with air seeders at seeding rates of 4-6 kg per hectare with hybrid varieties being sown at about 3 kg per hectare. All seed is treated to control blackleg disease (*Leptosphaeria maculans* [Desm.] Ces. et de Not.) and some seed is treated for control of the red-legged earth mite (*Halotydeus destructor* Tucker). Flowering occurs in August and September with harvest in late spring or early summer (November and December). The growing season ranges from about 150-210 days, depending on latitude, rainfall, temperature and sowing date. Growth and yield of the crop is almost always limited by the amount of water available to the crop, particularly during maturation.

Due to the age of Australian soils, macronutrients (particularly nitrogen, phosphorous and sulphur) and micronutrients are deficient. Deficiencies in boron, manganese, molybdenum and zinc have been reported for *B. napus* crops, as has toxicity on the more acid soils due to high levels of aluminium and manganese. Most soils are strongly acidic and liming is necessary to achieve high yields. Initially oilseed rape was sown into well-worked soil, but with the availability of glyphosate as a pre-planting herbicide and varieties resistant to triazine and imidazolinone herbicides, direct seeding has become standard practice.

Oilseed rape is most frequently preceded by a pulse crop or pasture while fallow and wheat are other alternatives. When the canola crop precedes wheat in the rotation, substantial wheat yield benefits occur.

Figure 3.29. **Areas and concentration of *B. napus* production by Australian government districts**

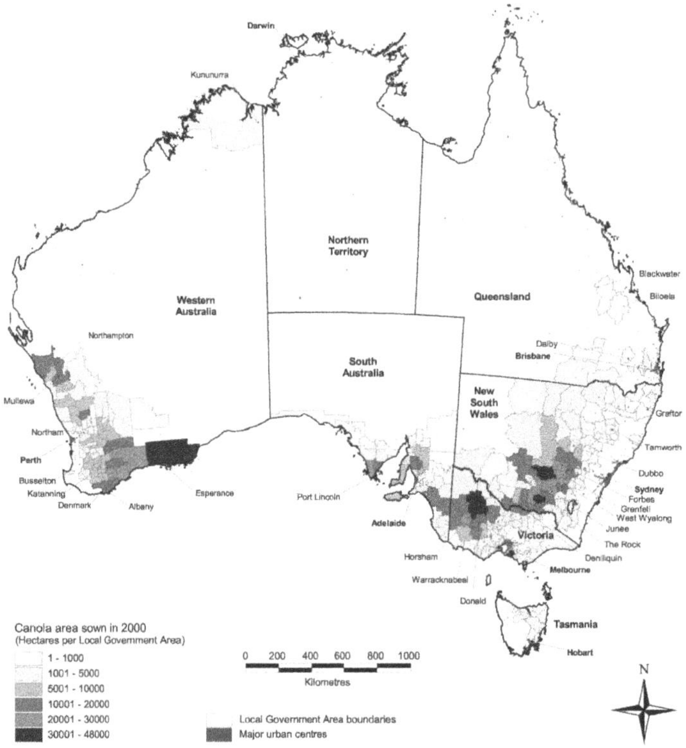

Data source: Australian Bureau of Statistics (2002).

Indian sub-continent cultivation and management

The dominant *Brassica* oilseed crop on the Indian sub-continent is *B. juncea*, although a limited hectarage is sown to the *B. rapa* form, toria, which is grown from September through December in northern areas. *B. napus* and *B. carinata* are grown to a limited extent in some irrigated and dry land areas of northern and central Indian states, respectively. The major crop of *B. juncea* as well as small pockets of yellow and brown sarson (forms of *B. rapa*) are sown in October or early November and harvested in late March or early April. Flowering occurs in early January. Production is centred in the northern half of the sub-continent, in what is called the mustard belt (Figure 3.30). The untreated seed is normally broadcast on the ploughed and levelled fields and the seed buried by drawing a heavy plank over the field. The traditional practice of sowing the *Brassica* species mixed with a cereal grain is no longer employed to any degree and the sowing of pure stands of each crop is now normal practice.

However, mixed cropping is still practiced in several areas by few farmers. Double cropping in the mustard belt is the standard practice with mustard sown on the same land each year following the summer crop, which may be pulses (mung and urd bean) or green manure. Other alternatives are rice, cotton or millets (such as sorghum or pearl millet).

The recommended seeding rate for *B. juncea* is 4-5 kg per hectare. Fertilization with nitrogen-phosphorous-potassium, in the ratio of 80-40-40 kg per hectare, together with 40 kg of zinc and 25 kg of sulphur, is recommended.

Figure 3.30. **Major production region (striped area) of oilseed mustard (*B. juncea*) and toria (*B. rapa*) on the Indian sub-continent**

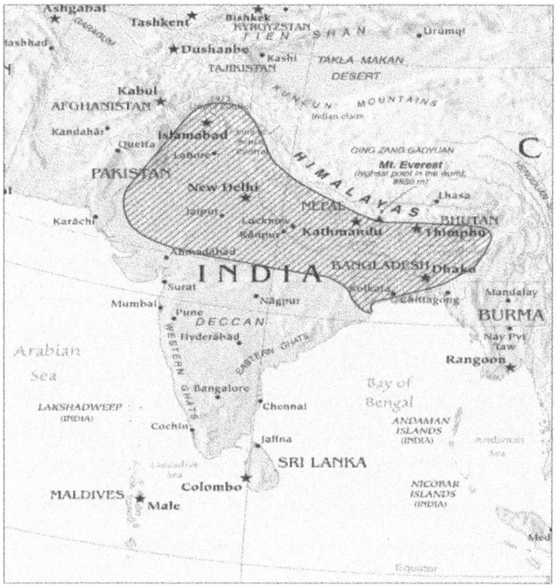

Source: Courtesy R.K. Downey

Chinese cultivation and management[3]

China is the world's largest producer of *Brassica* oilseed crops, annually producing some 11.5 million tonnes. Species contributing to this output include winter and spring *B. napus*, *B. juncea* and both winter and spring forms of *B. rapa*. Production is primarily from *B. napus* (representing 95% of the total), but both *B. juncea* (4%) and *B. rapa* (1%) oilseed crops are also grown at various concentration in the different provinces.

B. napus is grown throughout the country with the winter form dominating in the southern provinces and the spring form in the north.

The provinces along the Yangtze River provide the bulk of China's production. The level of winter hardiness required is not great. Indeed, Canadian and European varieties of the spring form have successfully survived the winters in the Chinese winter-growing region.

The spring-sown crops (*B. napus*, *B. juncea* and *B. rapa*) are sown in May, flower in June or early July and are harvested in September. The growing cycle for *B. napus* takes about 120 days. In the southern portion of the spring-growing area, half a season may be used to grow a forage or vegetable in conjunction with *B. rapa*. Because of the small field sizes, most are sown by hand or walking plough, although some large fields are mechanically sown. In the winter rape areas, the seed is sown into small seedling beds in September and the seedlings later transplanted into the production fields in mid- to late September. Flowering takes place in late March and harvest is in May. The total production cycle is about 220 days. The rotation in the triple cropping winter rape area is either rape-rice-rice or rape-maize-potato and in the double cropping regions rape-cotton or rape-rice.

Soil fertility is a limiting factor in production, with the area devoted to winter rape being particularly deficient in phosphorous. While all soils require nitorgen, phosphorous and potassium, significant areas are deficient in the micronutrients zinc and boron, while shortages of manganese, copper and iron also occur.

Herbicide resistant *B. napus*

B. napus is not considered a significant weed in managed ecosystems (AAFC, 1994). However, due to the high level of seed lost during harvest it can be an abundant weed in subsequent crops. Légère et al. (2001) ranked *B. napus* as 18th in relative abundance among Canadian weed species in western Canada, and Leeson et al. (2005) found *B. napus* plants in 10.5% of the fields surveyed. Studies in both Canada and Europe have shown that the incorporation of genes for resistance to specific herbicides imparts no altered weediness or invasive potential for glyphosate, including different events (AAFC, 1995b, 1996a; Norris et al., 1999: Crawley et al., 2001); glufosinate-ammonium, including its combination with the hybrid system (AFFC, 1995a, 1995d, 1996b; Rasche and Gadsby, 1997; Norris et al., 1999; MacDonald and Kuntz, 2000); bromoxynil (PBO, 1998) and non-GM imidazolinone (AAFC, 1995c). Experience in western Canada from 1995 through 2011, with all HR systems, have confirmed the validity of these earlier assessments (Beckie, 2011; Warwick, Beckie and Hall, 2009; Beckie et al., 2006).

However, GM-HR volunteers can occur in subsequent *B. napus* crops. The level will depend on the interval between oilseed rape crops in the rotation and how well the producer has controlled volunteer *B. napus* in the intervening years. The shorter the rotation and the less volunteer control, the greater the contamination level in the second planting. The presence of one GM-HR canola plant per square metre throughout a field of conventional oilseed rape calculates to a GM content of 2.5% in the harvested conventional crop (planted at 40 plants/m^2). This calculation assumes that the number of seeds produced by a volunteer plant is the same as that produced by the conventional plants (CETIOM, 2000). However, Gruber and Claupein (2007) report that volunteer winter *B. napus* plants, growing in a sown rapeseed crop only yield 45% of the seed produced by corresponding sown plants.

Off-type volunteer plants can come from multiple sources, including the seed bank from previous crops, movement of farm equipment and animals, pollen flow and contaminated seed stocks. In Australia, Stanton, Pratley and Hudson (2002) found sheep can excrete viable or germinable *B. napus* seed up to five days after ingestion. Similarly, Martens (2001) claimed that manure from oilseed rape-fed chickens resulted in volunteer plants when the manure was spread on a field 12 months later. In Canada, Downey and Beckie (2002) and Friesen, Nelson and Van Acker (2003) found certified pedigreed seed lots of conventional varieties contained unacceptable levels of GM seeds, apparently resulting from pollen flow in breeding nurseries. The seed industry quickly purified their breeding stocks but absolute exclusion cannot be guaranteed. Feral populations may disseminate genes to nearby oilseed rape crops but the incidence would be very small and far less than several of the sources noted above (CETIOM, 2000; Wilkinson et al., 1995).

In all oilseed rape growing regions, leaving the soil untilled for a period after harvest and using non-inversion tillage is an effective strategy for minimising the size of the seed bank (Gruber and Claupein, 2007; Gulden, Shirtliffe and Thomas, 2003a). Ploughing, as done in Europe, will bury the seeds below germination depth but when the field is again ploughed the dormant seeds will be brought to the surface. Pre-emergence and in-crop post-emergence herbicide applications are effective in controlling volunteers even if they contain one, two or three different herbicide-resistance genes (Table 3.5; Downey and Buth, 2003). In western Canada, where herbicide tolerant oilseed rape has been grown extensively for 15 years, there is no evidence that volunteer *B. napus* has increased or is more prevalent because of the herbicide resistance traits (Hall et al., 2000; Beckie et al., 2006, 2004).

Oilseed certified seed production

The production of oilseed *Brassica* sowing seed is normally undertaken within the areas where the *Brassica* crop is commercially grown. The rules under which pedigreed seed is produced and identified in the market place are stringent and extensive. Regulations vary from country to country but the minimum requirements for certified seed moving in international trade are governed by two international certification organisations. Both the OECD Seed Schemes and the Association of Official Seed Certifying Agencies (AOSCA) were developed to facilitate seed trade through mutual recognition of the official certification labels of member agencies. Member countries must meet OECD and AOSCA standards, but countries can – and most of them do – have domestic certification standards that exceed those minimums.

Table 3.5. **Number of herbicide products available for control of volunteer *B. napus* with nil, single or multiple herbicide tolerances in western Canada**

Herbicide system	Number of products
Susceptible	27
Liberty Link (LL)[1]	26
Roundup Ready (RR)[2]	25
Clearfield (CF)[3]	19
RR × LL	24
RR × CF	17
LL × CF	18
RR × LL × CF	16

Notes: 1. LL Glufosinate; 2. RR Glyphosate; 3. CF Imidazolinone.

Source: Downey and Buth (2003). Courtesy AAFC Reseach Station, Saskatoon.

AOSCA has a focus on the United States but its members include also Argentina, Australia, Canada, Chile, New Zealand and South Africa. AOSCA standards cover not only varietal certification of seed but also germination, physical purity, disease and other quality traits. Their varietal certification requirements include a maximum variety impurity "seed" standard that is used for post-control verification testing.

The OECD Seed Schemes, which largely reflect the requirements of the European Union seed certification system, are increasingly implemented at the global level. They comprise 58 member countries including most of the countries discussed above. China and Pakistan are currently not members of the OECD Seed Schemes (situation November 2012). However, China is developing standards for *Brassica* crops and Pakistan has regulations that are similar to those of India. OECD seed standards do not deal with germination or physical purity but focus on varietal certification, based mainly on morphological characteristics during inspections of seed-production crops. In addition, minimum requirements and standards for verification, using post-control field testing, are mandatory.

Seed classes allowed are normally designated by the breeder or maintainer of the variety. For *Brassica* oilseed, the seed multiplication factor for each generation is typically large (>1 000:1). Thus, the seed classes designated for its species are normally limited to three and identified under the OECD Seed Schemes as "basic", "certified 1st" and "certified 2nd" generations, with the equivalent generations designated under AOSCA as "breeder", "foundation" and "certified seed". Normally only one generation is allowed for the foundation and certified classes. The OECD seed regulations for Brassicaceae oilseeds require a five-year interval between crops of the same species. AOSCA standards for production of foundation seed of *B. napus*, *B. juncea* and *B. rapa* require four years between crops of these species and a two-year interval when producing certified seed. Under OECD regulations, basic and certified seed-production fields of *B. napus* must be isolated from any possible source of cross-pollinating pollen by a minimum of 200 m and 100 m, respectively. AOSCA regulations require foundation producing fields of *B. napus*, *B. juncea* and *B. rapa* to be isolated from any other crop of the same kind by 201 m, 402 m and 402 m respectively. For certified producing fields of these three species, the respective isolation distance required is 100 m, 402 m and 100 m. Both sets of regulations require all seed-production fields to be inspected by the designated authority at least three times for basic seed production and three times on each parental line for the production of certified seed of hybrid varieties, i.e. before the flowering stage, in the early flowering stage and before the end of the flowering stage. Fields must also meet stringent standards for varietal purity (visual characteristics) as well as freedom from cross-pollinating species and other crop kinds.

It must be emphasised that the above are minimum standards, with most countries having higher requirements as well as many seed companies exceeding the more stringent domestic regulations. Open-pollinated varieties of *B. napus* are rapidly being replaced by F_1 hybrid varieties, and a similar situation is likely to occur in *B. juncea* within the next few years. The requirement for nearly absolute purity of the female parent is mandatory if the hybrid is to produce the desired level of heterosis. The male restorer parent must also breed true for restoration of hybrid fertility. Thus, the hybrid regulations for isolation distances under AOSCA are much greater at 804 m while most seed companies use 1 000 m or more. Also, foundation and certified producing crops for hybrid seed production cannot be grown on land which has grown *B. napus*, *B. rapa*, *B. juncea* or oilseed *R. sativa* in the past five and three years, respectfully.

The studies by Downey and Beckie (2002) and Friesen, Nelson and Van Acker (2003) that identified some Canadian certified *B. napus* seed lots as containing undesirable levels of foreign herbicide resistance traits are often cited as sources of contamination. Regulators and the seed industry moved quickly to correct this situation. Today the Canadian Food Inspection Agency (CFIA) carries out seed testing of *Brassica* oilseed varieties for: 1) adventitious presence (AP) of approved events; and 2) herbicide trait purity of glyphosate and glufosinate ammonium resistant varieties. All official reference control samples for oilseed rape varieties submitted to the CFIA's Variety Registration Office at the time of registration of a new variety are subject to AP testing and if the variety is herbicide resistant, to herbicide purity trait testing. Furthermore, the CFIA also monitors AP and trait purity of foundation and certified seed. In instances where AP and/or trait purity issues are identified, the breeder of the variety is notified and appropriate action is taken (Canadian Food Inspection Agency, 2009).

The Canadian Seed Growers Association (CSGA) (2009) have also revised its "Regulations and Procedures for Breeder Seed Crop Production" so that seed certificates are only issued for breeder seed crops that are produced within a third-party audited quality management system (QMS) and verified to preserve varietal identity. Further, non-compliance with QMS requirements can lead to suspension or cancellation of the professional recognition of a plant breeder, which is required in both CFIA variety registration and CSGA seed crop certification.

Brassica *vegetable seed-production locations and management*

The market for *Brassica* vegetables has, in recent years, experienced a steady increase in demand. This expansion has been aided by widespread refrigerated transportation systems that can provide a year-round supply of such vegetables to most markets. The *Brassica* vegetable crop with the greatest demand for seed is cabbage, followed by the *B. rapa* Asian vegetables and broccoli. The world requirement for cauliflower seed is less while the demand for turnip, rutabagas and kohlrabi is relatively small. Accompanying the increased commercialisation of *Brassica* vegetable production has been the need to provide large quantities of seed of high quality and varietal purity. This requirement has resulted in the majority of the seed being produced in specific locations where climate and isolation from other *Brassica* crops are favourable for consistent high yield and quality. To aid the growing international trade in vegetable seed, the OECD has established a Scheme for the Certification or Control of Vegetable Seed which requires field and seed inspection by an accredited authority, within the country of origin, to ensure the seed meets varietal purity standards, including freedom from cross-pollinating species. The OECD Vegetable Seed Scheme provides for the production of "certified seed", and the designation of "standard seed", corresponding to two different control requirements. Other organisations that facilitate the seed trade include the International Seed Federation (ISF), which has defined trading terms and rules dealing with sales, can arbitrate settlements and assists with import and export licenses: ISF regional seed industry organisations, such as the Asia and Pacific Seed Association (APSA), which seeks to improve vegetable seed production and trade in the region (George, 2009). Many companies also use a QSM as described above for oilseed seed production.

Locations of concentrated vegetable seed production

In developing countries, vegetable seed is primarily supplied from farm-saved seed, and more rarely from the formally organised seed sector. In countries with strong agricultural and horticultural industries, nearly all the seed is from commercial pedigreed sources. For large-scale seed production of the biennial *Brassica* vegetables, seed companies have concentrated production in areas with relatively mild winters and moderate summer temperatures. In Europe, such areas are found in Belgium, Brittany (France), northern Italy and the Netherlands.

In North America, among the *Brassica* vegetables, broccoli has the greatest seed demand followed by cabbage and cauliflower. The market for the seeds of collard, Brussels sprouts and the Asian vegetables is much smaller. Seed production of these crops is concentrated in valleys of Oregon and Washington states (e.g. Oregon's Willamette Valley). Selected areas in California and Arizona are also important producers of broccoli and cauliflower seed. Essentially all broccoli and cabbage varieties produced in the United States are F_1 hybrids. In contrast, most cauliflower varieties are highly inbred and uniform, self-pollinating populations, but in recent years more and more F_1 hybrids have entered the market (Farnham, 2007). F_1 and inbred varieties of collards, Brussels sprouts and kale provide seed to the commercial market. In South America, Chile is a significant supplier of vegetable *Brassica* seed.

In Australia, seed production of *Brassica* vegetable crops is centred in Tasmania in the regions of the Coal River Valley, Derwent Valley, central East Coast,

Hagley/Westbury and Devonport. Tasmania is climatically suited for "counter-season" seed production for the northern hemisphere markets of Asia and Europe. Major seed crops produced in 2001 were hybrid cabbage (150 ha) and cauliflower (97 ha) (Government of Tasmania, 2003). Cabbage and cauliflower are high-value autumn planted crops while the lower value mustard and Chinese vegetable types are spring sown. Locations for hybrid seed production of cabbage and cauliflower are determined by the need for an isolation zone of 1.5-3 km from other crops of the same botanical family. Grower awareness and consultation between companies ensures adequate isolation distances.

In New Zealand, the Canterbury Plains and other smaller areas of the South Island (43° south) have become a major vegetable seed-production location, particularly for the Asian *Brassica* vegetables. In this region, the seed merchants and growers have put in place an isolation mapping system to avoid cross-pollination among different species and varieties. The system is operated by a government-owned company called AgriQuality that displays an Internet map of every farm field involved in seed production. When a seed contract is arranged and a field is selected, the seed merchant logs the details into the system and can see if there are any conflicts within the isolation distance required. Normal minimum isolation distance for the *Brassica* crops is 1 000 m, but that can be extended, particularly with hybrid seed production.

Not all seed-production regions are maritime based. In China, cauliflower and broccoli seed production is concentrated in semi-desert regions around the cities of Jiuquan and Jiayuguan in Gansu and in Yunnan provinces. In these high-elevation areas precipitation is minimal, but irrigation is available and the temperatures remain within the required range. Cabbage and Chinese cabbage seed production is located further south in Hebei, Henan, Shandong and Shanxi provinces (X.-W. Wang, personal communication). In these regions the normal isolation distances between production fields is 1 000 m.

On the Indian sub-continent, no concentrated areas for seed production were identified. However, small individual fields occur scattered in the foothill valleys of the Himalayas. For the production of certified cauliflower seed, the minimum isolation distance is 1 000 m (Indian Minimum Certification Standards).

In Japan, no concentrated area exists for large-scale seed production. However, various *Brassica* vegetables (*B. rapa* and *B. juncea*) are cultivated locally (Inomata, 2007) and seed production is practiced on a small scale. The minimum isolation distance required for seed production is 600 m.

George (2009) notes that most authorities recommend having a greater distance (up to 1 500 m) between different types of *B. oleracea* (cabbage vs. kohlrabi) than between varieties of the same type (two cabbage varieties, up to 1 000 m).

Vegetable seed cultivation and management

The optimum pH for cole crops is reported to be 6.0 to 6.5 with the generally recommended ratio of N-P-K nutrients being 1:2:2 at soil preparation, but it varies depending on the production region (George, 2009). The lower ratio of nitrogen is to avoid "soft plants" that are less winter hardy. Extra nitrogen is normally topdressed in the spring. It is important to ensure that adequate levels of sulphur as well as the micronutrients boron, manganese and molybdenum are available. The development of hybrids in Brussels sprouts has become very important (George, 2009) with self-incompatible and cytoplasmic male sterility (CMS) hybrids becoming more frequent in cabbage, cauliflower, kale and kohlrabi. The ratio of male to female in hybrid production fields is normally 1:1 or 1:2 (Takahashi, 1987).

Most cabbage and cole crops and some Asian vegetables are biennials and will not bolt until they have been exposed to temperatures of 4-7°C for 6-8 weeks. Day length has no affect on bolting or flower initiation (Nieuwhof, 1969). At the end of the first year, cabbage plants can withstand temperatures of -12°C to -14°C for extended periods, but lower temperatures can cause much damage, as can alternating periods of frost and thaw (Nieuwhof, 1969). The usual practice in producing cabbage seed is to sow in the summer with the plants over-wintering, bolting in the spring and to harvest the seed in summer. Cultivars differ in their winter hardiness with red cabbages the least hardy and savoy the hardiest. Summer temperatures are also important in determining seed yield. Temperatures above 25°C arrest growth and cause seed abortion. Because of these environmental constraints, commercial production tends to be concentrated in areas with mild winters, sufficiently cold to ensure vernalization without winter kill, combined with moderate summer temperatures. The availability of irrigation is also important to obtaining uniform high yields.

Seeding of the biennial crops in the northern hemisphere is normally done in mid-June to mid-August. If the seed is to be sown in beds for transplanting, rather than direct seeding into the field, seeding should be done about ten days earlier than the field sowing to allow for the plant setback brought on by transplanting (Nieuwhof, 1969). The recommended rate for field sowing is 3-5 kg per hectare, unless precision sowing is practiced, where only 1-2 kg per hectare is needed. Plants are thinned to 35-40 cm between plants within the row. To increase the over-wintering survival rate, plants may be earthed up covering the most sensitive plant portion just below the head. Weed control is critical, as in mild winters weeds may over grow the crop. Most of the cole crops are self-incompatible and depend on insects, primarily honey bees, to effect fertilisation. Harvesting is done once the pods have turned yellow and the seedsbrown. Depending on field size and seed value, harvesting may be done by various methods from hand cutting and threshing to straight combining. Kohlrabi, although a true biennial, can be vernalized by initiating germination through pre-soaking the seed for 8-9 hours at 20°C followed by a cold treatment of -1°C for 35-50 days. The treated seed can then be sown directly into the field in the spring and the seed crop harvested in the fall. Brussels sprouts and kale are grown for seed in the same manner as cabbages.

For cauliflower and broccoli crops, only a mild vernalization period is required so environmental limitations are less stringent. However, as a seed crop, these forms normally require an extended growing season. Selection of cauliflower varieties for a tighter curd has resulted in slow and incomplete bolting, thus further extending the required growing season. In Western Europe, cauliflower is sown in September and over-wintered under glass with transplanting to the field in early spring. Transplants are spaced on a 50 × 50 cm or smaller grid. Flowering occurs in July or August and the crop is harvested in September or early October. Seed production of tropical and subtropical cauliflower is discussed by Lal (1993).

Drying the harvested *Brassica* vegetable seed is frequently required. To maintain germination capacity, the maximum air drying temperature should not exceed 60°C. If seed is to be stored for a year, maximum moisture content should not exceed 9% with a storage temperature of 5-10°C.

The biennial turnips and Swedes (rutabaga) regenerate from growing points at or near ground level. This means they can benefit from a large underground source of nutrients for seed production. Thus, these crops are more winter hardy than cole crops and can be grown for seed over a much wider environmental range. However, the market for their seed is relatively small, so seed companies tend to contract their production with growers in areas already producing seed of other *Brassica* vegetable crops.

The *B. rapa* vegetables prefer a soil pH between 6.0 and 7.5 with an N-P-K fertilisation ratio at planting of 2:1:1. Additional nitrogen fertiliser is normally applied at anthesis (George, 2009). The seed is produced by either the head-to-seed or the seed-to-seed method described by Opeña, Kuo and Yoon (1988). As with the cole crops, the ratio of male to female in hybrid production fields is 1:1 or 1:2 (Takahashi, 1987).

Centres of origin and ancestors

Introduction

There are few areas of the world where members of the family Brassicaceae are totally absent. The exceptions are parts of the tropics, were the family is thinly represented, but where some introduced cosmopolitan weeds have become established. The genera and species of the family occur in greatest number and diversity in the temperate zone of the northern hemisphere and in particular, the areas surrounding the Mediterranean basin and throughout the southwest and central regions of Asia (Figure 3.31; Hedge, 1976). Although the generic and specific endemism in the family is highest in the Irano-Turanian region, the centre of origin of the current subtribe Brassicinae, lies in the Mediterranean basin (Hedge, 1976).

Using chloroplast DNA restriction sites together with cpDNA probes, Warwick and Black (1991) surveyed 33 diploid taxa of the Brassicinae. The phylogenetic results indicated there were clearly two ancient and distinct evolutionary lineages within the subtribe. They found the "Nigra" lineage to include *B. nigra*, *B. fruticulosa*, *B. tournefortii*, *Sinapis pubescens*, *S. alba*, *S. flexuosa*, *S. arvensis*, *Coincya cheiranthos*, *Erucastrum canariense* and *Hirschfeldia incana.* The other lineage, termed "Rapa/Oleracea", was made up of *Brassica rapa*, *B. oleracea* and subsp. *alboglabra*, the *B. rupestris-villosa* complex (*B. rupestris*, *B. drepanensis*, *B. macrocarpa*, *B. villosa*), *B. barrelieri*, *B. deflexa*, *B. oxyrrhina*, *B. gravinae*, *Diplotaxis erucoides*, *D. tenuifolia*, *Eruca sativa*, *Raphanus raphanistrum*, *R. sativus* and *Sinapis aucheri.* In the "Nigra" lineage, *B. nigra* was most closely related to the annual *Sinapis* species *S. arvensis* and *S. alba* (Figure 3.31). Only a single mutation difference was found between the crop and weedy accessions of *B. rapa* and between crop accessions of *B. oleracea* and wild accessions of *B. oleracea* subsp. *oleracea* and subsp. *alboglabra* (Warwick and Black, 1991). The weedy species *R. raphanistrum* and the crop species *R. sativus* differed by only four mutations.

Although the economically important *Brassica* species arose from ancestors in the Mediterranean region, wars and trade ensured their wide dispersal, resulting in islands of isolated environmental and selection pressure. The earliest widely distributed species were those that exhibited seed dormancy combined with useful traits. Seed dormancy allowed the introduced seed to survive long after its introduction. The fast-growing, weedy type of *B. rapa*, providing lamp oil and animal feed, and *B. nigra* as an oil and spice source, would be prime candidates. The Mission Trail in Southern California is a case in point: priests scattered *B. nigra* seed to mark the trail between the early Missions. Parts of those trails can still be seen each year as the black mustard blooms on the California hillsides.

Figure 3.31. **Phylogenetic tree for the subtribe Brassicinae, based on PAUP analyses of the chloroplast DNA restriction site/length mutations shared by two or more taxa/accessions**

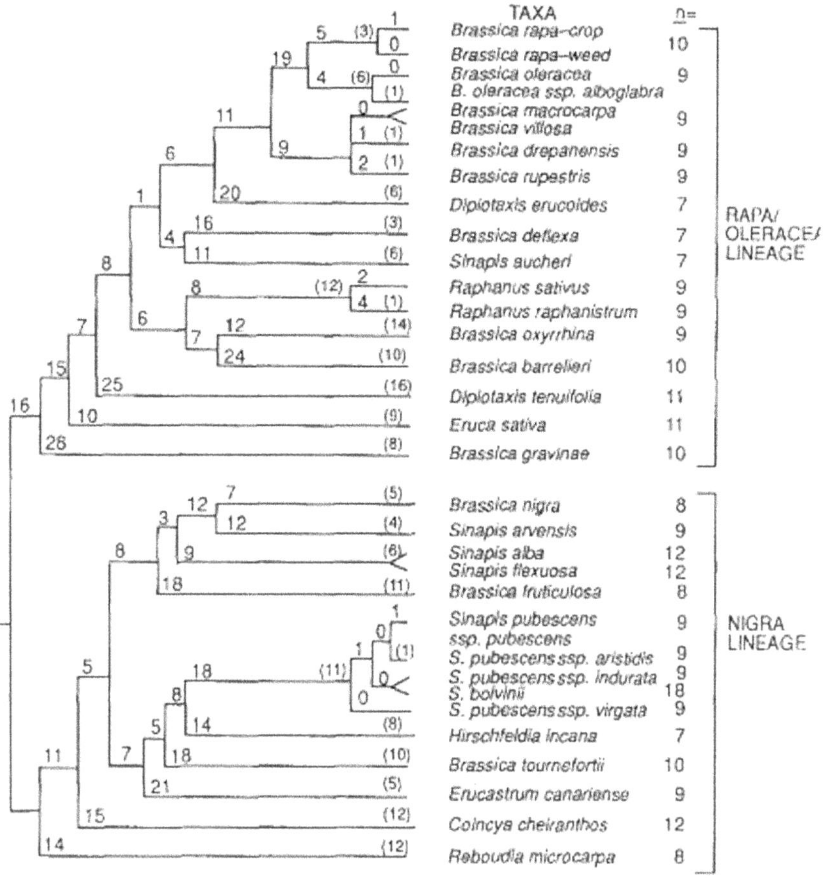

Notes: PAUP is a computational phylogenetics programme for Phylogenetic Analaysis Using Parsimony that infers evolutionary trees (phylogenies), Tree length in this tree is 489 steps, consistency index is 0.491. Tree topology indicates how accessions are related, and branch length (numbers above the branches) indicates the minimal number of mutational steps occurring during the evolution of a particular taxa. Mutations unique to a given species and to the genus *Raphanus* (number indicated in brackets at the end of a branch) should be added to determine terminal branch length. ANC indicates a common hypothetical common ancestor.

Source: Warwick and Black (1991). © Canadian Science Publishing or its licensors.

B. nigra

B. nigra is amongst the oldest recorded spices, being noted in the Sanskrit writings of about 3000 B.C. as "Sarshap" (Prakash, 1961). However, little is known about *B. nigra*'s true centre of origin. Hemingway (1995) placed it in Irano-Turanian, Saharo-Sindian region (Figure 3.32). However, Prakash and Hinata (1980) favoured an origin in Central and south Europe. Its use as a commercial spice ensured its very early, widespread distribution across Europe, Africa, Asia and India, and its dehiscing siliques ensured its persistence. The crop was grown for the sharply pungent chemical (allyl isothiocyanate) released when the crushed seed was mixed with a small amount of water, in the same way that *B. juncea* powdered mustard is used today. Until the 1950s *B. nigra* was the world's major source of pungent mustard, but because it shatters as soon as the pods are ripe it required hand harvesting. Thus, it was replaced in a single decade by highly pungent

B. juncea varieties well suited to mechanical harvesting. Today there is essentially no commercial production of *B. nigra* and it has become a weed of waste places in many regions. It is an introduced species to the Americas and Australia. It has never become established on the Canadian prairies although it is present throughout much of the United States.

B. rapa

B. rapa is generally believed to have originated in the mountainous areas near the Mediterranean sea rather than the coastal areas (Tsunoda, 1980). As with *B. nigra*, *B. rapa* had a wide distribution before recorded history. Indian Sanskrit literature first mentions the plant about 1599 B.C. as "Siddharth" (Prakash, 1961). Burkill (1930) proposed that the leafy vegetable forms were developed in China from the oilseed form about 2 000 years ago. Seeds of both *B. rapa* and *B. juncea* were found in excavations of the ancient village of Banpo, Xian, Shanxi Province, China that existed in Neolithic times 6 000-7 000 years ago (Liu, 1985). Turnip seeds were also found in pottery jars from the 5th century B.C. at the Yang-shao agrological site in Shensi Province (Chang, 1970). Cultivation of *B. rapa* is also mentioned in the oldest collection of Chinese poetry, Shi Jing (the book of Odes), written during the Chunqui period about 535 A.D. (Liu, 1985; Chapman and Wang, 2002). In Scandinavia, *B. rapa* seeds were being consumed as early as 350 B.C. as indicated by their presence in the stomach of the Tollund man (Renfrew, 1973).

Figure 3.32. **Evolutionary geography of *B. juncea*, *B. carinata* and *Sinapis alba***

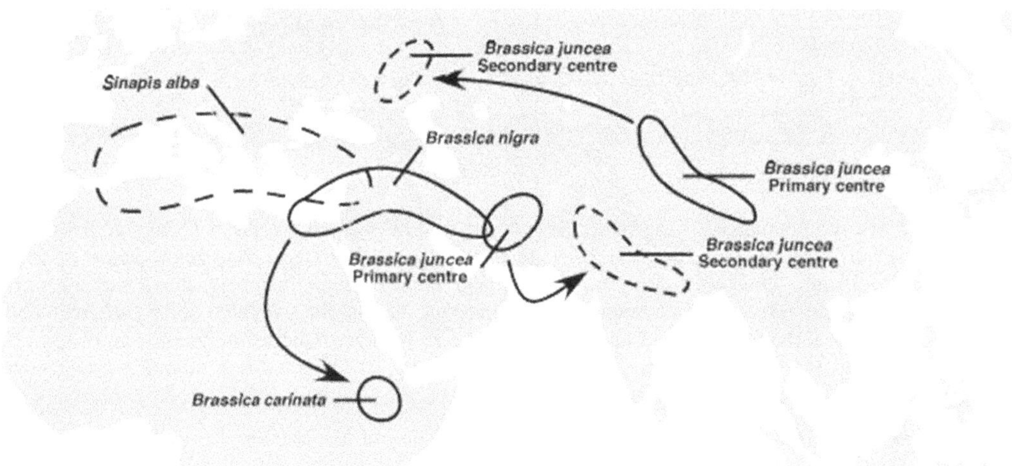

Source: Greatly modified from Hemingway (1995).

Sinskaia (1928) proposed two main centres of origin, with the Mediterranean area as the primary centre for the European form, and Afghanistan with the adjoining portion of Pakistan as the other primary centre. Asia Minor, the Transcaucasus and Iran were considered secondary centres. Alam (1945) concluded that the Sarson and Toria types of *B. rapa*, now grown as oil crops in India and Pakistan, evolved in the Afghan-Persian area and migrated south to India and further east. McNaughton (1995a) concluded that multiple domestication of the wild forms for oilseed occurred from the Mediterranean to India about 2000 B.C. with later selection for short stature and leafiness in the Far East (China) resulting in the numerous *B. rapa* vegetable forms. Tsunoda and Nishi (1968) proposed that, with selection for increased leaf number, subsp. *chinensis*, and *japonica* evolved and with increased leaf size and head forming, *pekinensis*, *narinosa* and *nipposinica* were selected. Cultivation of the oilseed form in Europe as a source of lamp oil is thought to have been under way by the 13th century, first as an annual form from which the biennial form was selected (Appelqvist and Ohlson, 1972). In northern Europe, turnip evolved from the biennial oilseed form through selection for bulbous roots (McNaughton, 1995a). Cartier in 1540 is credited with the first introduction of turnips into North America and more specifically to eastern Canada. They were also being grown in the Virginia colony by 1609 (Sauer, 1993). Canadian commercial production of the oilseed form began in 1943.

B. oleracea

B. oleracea has its centre of origin in the Mediterranean region (Snogerup, 1980). The wild forms of the *B. oleracea* complex still grow along the coast of the Mediterranean sea and Atlantic ocean from Greece to England (Figure 3.33). Snogerup, Gustafsson and Von Bothmer (1990) concluded from morphological and crossing studies among the wild *B. oleracea* forms, including *B. oleracea*, *B. cretica* Lam., *B. bilarionis* Post., *B. insularis* Moris., *B. villora* Biv., *B. incana* Ten., *B. macrocarpa* Guss. and *B. montana* Pourr., that these species should be considered subspecies of *B. oleracea* along with the cultivated forms. These conclusions were confirmed by Von Bothmer, Gustafsson and Snogerup (1995) through a crossing programme involving ten wild taxa and six major cultivated forms. Snogerup, Gustafsson and Von Bothmer (1990) reported that all wild forms of the *B. oleracea* complex were suffrutescent perennials, exhibiting no primary dormancy. They are also self-compatible and readily intercross within the group and with cultivated forms. They also identified some wild *B. oleracea* tetraploid plants and reported a higher fertility rate in F_1 hybrids between the wild *B. oleracea* and the cultivated forms than with the other wild subspecies.

Mutation, adaptation and selection within these populations yielded the present-day forms of cabbage, savoy, kales, collard, broccoli, Brussels sprouts, cauliflower and kohlrabi. The kales, several thousand years ago, were probably the first cultivated forms. They were grown as early as 600 B.C. by the Greeks while ancient Roman writers described heading cabbage and possibly kohlrabi (Thompson, 1976). De Candolle (1885) suggested cabbage was first domesticated somewhere in Western Europe by the Celts during the first millennium B.C. Support for this conclusion comes from the respective English, German and French common names "cabbage", "kopf or kohl" and "cabus", which are all probably derived from the Celtic word "cap" or "kap", meaning head (Prakash and Hinata, 1980). A number of authors have theorised, but lacked the research to support their views, as to which species in the *B. oleracea* complex gave rise to the various cultivated forms (Helm, 1963; Neutrofal, 1927; Schiemann, 1932; Schulz, 1936; Lizgunova, 1959). After considerable investigation, Snogerup (1980) concluded that: 1) headed cabbages originated from west European *B. oleracea* and savoy cabbage may have resulted from introgression with other cole crops; 2) branched bush kales originated from *B. cretica* in Greece; 3) stem kales probably arose from the *rupestris-incana* complex; 4) the origin of the inflorescence kales such as cauliflower and broccoli is uncertain although Schulz (1936) provided some evidence that *B. cretica* could be the ancestor; and 5) *B. alboglabra* originated from *B. cretica* in Greece and was carried east by traders. Today, *B. oleracea* var. *alboglabra*, or Chinese kale, is among the ten most important market vegetables in Southeast Asia, including Thailand and China (Rakow, 2004). Little is known as to when forms of *B. oleracea* arrived in Asia but Schafer (1977) noted that kohlrabi was being cultivated in Tang' times (600-900 A.D.).

Figure 3.33. **Distribution of wild "species" of *B. oleracea* in 1990**

Note: Introductions of *B. oleracea* outside its spontaneous area are not mapped.

Source: Modified from Snogerup, Gustafsson and Von Bothmer (1990).

B. napus

B. napus with its oilseed, forage and root forms is a relatively recent species. The Greeks and Romans knew of the Swede or rutabaga root crop, but reference to these forms does not appear in the ancient literature. Although Prakash and Hinata (1980) state that no wild *B. napus* populations have been found, Linné reported wild forms growing on the beaches of Gothland (Sweden), the Netherlands and Britain (cited by De Candole, 1885). Since the species is the result of an interspecific cross between a plant or plants of *B. rapa* and the *B. oleracea* complex, it could only have arisen in the Mediterranean or the European west coastal regions, where the two species were growing in close proximity (Figure 3.33). Olsson (1960) suggested that *B. napus* could have arisen several times by spontaneous hybridisation between different forms of *B. rapa* and *B. oleracea.* Evidence from chloroplast and mitochondrial DNA suggests that *B. montana* might be closely related to the maternal prototype that gave rise to *B. napus* (Song and Osborn, 1992). That *B. oleracea* was the maternal parent is supported by both Erickson, Straus and Beversdorf (1983) and Ohkawa (1986). However, Flannery et al. (2006), using SSR (simple sequence repeat) *Brassica* plastid markers, noted that *B. rapa* always grouped with *B. napus* and concluded that *B. rapa* is the more likely plastid genome donor. Further, Allender and King (2010), using chloroplast and nuclear markers, concluded that it is highly unlikely that *B. oleracea* or any of the C genome species are closely related to the maternal progenitor of most *B. napus* accessions. They suggest that a *B. rapa* strain from northern Italy called "spring broccoli raab" may be the closest extant relative of the *B. napus* maternal ancestor. However, the data also suggest that the interspecific cross may have occurred more than once, with *B. napus* having multiple origins. Thus, the

Swede or rutabaga could have originated in medieval gardens where turnips and kale grew side by side (McNaughton, 1995b). There is general agreement that the winter or biennial form of *B. napus* originated in northern Europe. On the other hand, forage rape almost certainly evolved from the oilseed form.

Cultivation of oilseed rape in Europe was under way by at least the Middle Ages (Appelqvist and Ohlson, 1972). It is only in relatively recent times that *B. napus* oilseed forms have been introduced to other parts of the world (Figure 3.34). *B. napus* did not arrive in China or Japan until about 1860-70, with the coming of European traders (Liu, 1985; Shiga, 1970). European immigrants introduced the forage and root crop forms into North and South America in the 17th and 18th centuries. In China, Japan and Korea *B. napus* proved to be more productive than the indigenous oilseed forms of *B. rapa*. Today most of the oilseed rape produced in China, Japan and Korea is harvested from *B. napus* cultivars that have been bred from interspecific crosses between introduced *B. napus* and the older indigenous *B. rapa* cultivars (Shiga, 1970). *B. napus* is less adapted to the Indian sub-continent due to the short days and warm growing conditions. Commercial production of the oilseed form did not occur until 1942 in Canada and 1969 in Australia.

Figure 3.34. **Dispersal of the *B. napus* species from a proposed centre of origin**

Notes: Distribution occurred throughout Europe in the 16th century, the Americas in the 17th and 18th centuries, and China and the Far East in the 19th century.

Source: Modified from Liu (1985).

B. juncea

B. juncea appears to have a much longer history than *B. napus*, even though it is also the result of an interspecific cross (*B. rapa* × *B. nigra*). Fraction 1 protein data (Uchimiya and Wildman, 1978) and chloroplast DNA analysis established that *B. rapa* functioned as the female parent in the formation of this species (Erickson, Straus and Beversdorf, 1983; Palmer, 1988; Palmer et al., 1983; Song, Osborn and Williams, 1988a, 1988b; Warwick and Black, 1991; Yang et al., 2002). However, Qi, Zhang and Yang (2007) reported that some Chinese phenotypes may have evolved with *B. nigra* as the maternal parent. They investigated the nuclear internal transcribed spacer (ITS) regions of ribosomal DNA from 15 different Chinese vegetable phenotypes and one oilseed form (pictures of the 16 phenotypes, including 2 root forms, are provided in the publication). They found that four of the accessions, including the oilseed form, apparently had *B. nigra* as the maternal parent, a finding at odds with the RFLP and chloroplast DNA investigations noted above. However, the difference may be related to the limited Chinese genotypes that were available to other researchers.

There has been much speculation in the literature as to the centre(s) of origin for *B. juncea.* However, Prain (1898), Sinskaia (1928) and Vavilov (1949) all agree that China, where the greatest divergence of forms occurs, is one centre of origin. In addition, Vavilov (1949) also identified Afghanistan and adjoining regions as a second primary centre. This observation was supported by Olsson (1960) and Mizushima and Tsunoda (1967) as well as Tsunoda and Nishi (1968), who found wild forms growing on the plateaus in Asia Minor and southern Iran. India and the Caucasus have also been put forward as secondary centres (Hemingway, 1995; Figure 3.32). There is strong evidence for China as a primary site. As noted in the *B. rapa* section above, *B. juncea* has a long history in China. Leafy, vegetable forms of *B. juncea* mustard are also consumed in great quantities in China and other Asian countries (Herklots, 1972; Nishi, 1980). The greatest range in leaf types occur in Sichuan Province within the varieties of *rugosa*, *japonica*, *integrifolia* and *cernua.* A root-forming type has also been selected and cultivated in northern China with the variety names of *napiformis* and *tumida* (Nishi, 1980; Chen et al., 2005).

The *B. juncea* from Afghanistan and Asia Minor is believed to have migrated south to Pakistan and India where a secondary centre of origin was established (Figure 3.32). The earliest direct reference to *B. juncea* is in the Indian Sanskrit literature about 1500 B.C., where it is mentioned as "Rajika" (Prakash and Hinata, 1980). The existence of two primary centres in China and the Middle East-India is supported by the fact that the Indian sub-continent and Chinese oilseed forms not only differ in morphological traits (Sinskaia, 1928), but also chemically and in day-length requirements. The seed from Indian *B. juncea* material contains mainly 3-butenyl glucosinolate and the crop is day neutral, while the Chinese spring-sown oilseed forms contain only 2-propenyl (allyl) glucosinolate and are long day requiring. The Chinese material also contains pure yellow seeded strains which are absent in the Indian material. The Russian material displays most of the same characteristics as the Chinese material and although it may also have resulted from an independent interspecific cross, more likely it was carried into the Russian Federation from China or Mongolia via the Northern Silk Road. Wu et al. (2009) investigated the relationships among 95 *B. juncea* accessions originating from China, France, India, Japan and Pakistan, using sequenced related amplified polymorphisms (SRAPs). They found the Chinese vegetable phenotypes formed a highly diverse group with the spring- and winter-sown oilseed forms split into two separate groupings. The winter-sown accessions exhibited more genetic diversity than the spring-sown accessions but less than the vegetable group. The SRAP markers did not provide a clear-cut separation between the Indian/Pakistan and Chinese winter-sown mustards. Srivastava et al. (2001), using AFLP markers, investigated the relatedness of oilseed *B. juncea* cultivars from Australia (2 cultivars), Canada (2), China (2), Europe (6), India (7) and Tibet (1). Their data separated the cultivars into an Indian/Chinese group and a second cluster of the remaining ones. Their findings and that of Wu et al. (2009) suggest a close relationship between the Chinese northern spring-sown oilseed cultivars and the European mustards, while the winter-sown cultivars are closely associated with the Indian form. The data from both Wu et al. (2009) and Qi, Zhang and Yang (2007) support the contention of Song, Osborn and Williams (1988b) that the vegetable and oilseed mustards had a polyphyletic origin and evolved separately.

B. carinata

B. carinata, commonly called Abyssinian or Ethiopian mustard or simply "carinata", is an amphidiploid species derived from and containing the full genomic complement of the putative parental species, *B. nigra* (black mustard) as the female and *B. oleracea* as the male (Uchimiya and Wildman, 1978; Palmer et al., 1983; Song, Osborn and Williams, 1988b; Erickson, Straus and Beversdorf, 1983). The plant is cultivated on a small scale on the Ethiopian plateau. *B. carinata* may have originated from a hybrid between kale, which is grown on the plateau, and wild *B. nigra*, which is also present. However, this species, as with others in this group, almost certainly originated in the Mediterranean basin where the two putative parental species were growing in close proximity. It is believed that the cross occurred many eons ago when the climate on the African side of the Mediterranean was moist and lush. However, as the climate of this region became dryer and hotter, *B. carinata*, together with the plant community of the region that included castor oil plant and coffee, moved to the south and became isolated in the Ethiopian highlands. Thus, Ethiopia in effect preserved the environment of the centre of origin of *B. carinata* (Figure 3.32). Farmers of northeast Africa grow the plant both for its leaves, which are plucked, boiled and eaten, and for the edible oil in the seed. The local common name for the crop is *gomenzer*. The interspecific cross that created this species does not appear to have occurred elsewhere in nature or, if it did, the progeny did not survive. There is no commercial production of this species, other than in Ethiopia and neighbouring countries where the crop is grown on small holdings or in kitchen gardens. However, the species is being investigated and bred for potential commercial production in Australia, Canada, India and Spain.

Sinapis alba

Sinapis alba has its centre of origin in the eastern Mediterranean region (Figure 3.32) and wild forms are present around most of the Mediterranean littoral (Hemingway, 1995). In China, *S. alba* appears to have been cultivated by the middle of the first millennium A.D. (Hemingway, 1995).

Reproductive biology

Generation time and duration under natural and managed conditions

Generation and flowering times are discussed in the above sections dealing with cultivation and management.

Reproduction

Floral biology[4]

The basic floral characteristics of all the *Brassica* species included in this chapter are essentially the same, differing only in flower size. The floral arrangement in *Brassica* species is typically a corymbiform raceme. Flowering is indeterminate beginning at the lowest part of the main raceme and auxiliary branches, and continuing upward. The inflorescence may attain a length of 1-2 m. The buds begin opening under the pressure of the rapidly growing petals. The process of flower opening begins in the afternoon and is all but complete very early the following morning. The stigma is receptive from three days before to three days after the flower opens (Mohammad, 1935). Day length can play a critical role in initiating bolting of the flowing stem. Species such as *S. alba* are very day-length sensitive while some cultivars of *B. napus* and *B. juncea* are day neutral.

Both the onset of flowering and duration of the flowering period are variable and quite dependent on weather, particularly temperature. Low temperatures decrease the rate of plant development and hence the onset and rate of flowering is delayed. Low plant density results in secondary branching, thus extending the flowering period. If plants are pruned back when still green, regrowth and a second flush of flowers can be obtained. Flowers produced on regrowth are typically smaller and less productive than the first formed flowers (Downey, Klaasen and Stringham, 1980).

The flowers of the *Brassica* species are regular, bisexual and hypogenous. The differentiation of the flower proceeds through the successive development of four free sepals in two whorls, medium and transverse, six stamens, two carpels and four free diagonally placed petals (Figures 3.2 and 3.35). The flowers have one pair of lateral stamens with shorter filaments and four median stamens with longer filaments. When the anthers are a few millimetres in length, the pollen mother cells, after meiosis, give rise to the tetrads. The pollen grains are 30-40 µm in diameter and have three germination pores. The sutures of the anthers are introse in the bud stage, but the four long anthers become extrose as the flower opens (except in the *B. rapa* Yellow Sarson form where they remain introse).

Figure 3.35. **Typical flower of *B. napus***

Note: This photo shows the typical four petals with the stigma in the centre surrounded by four median stamens and a pair of shorter lateral stamens.

Source: Downey, Klaasen and Stringham (1980).

Two functioning nectaries are located at the base of the short stamens and two non-functional nectaries at the base of the pairs of the long stamens. The anthers dehisce when the petals completely unfold. The pollen is shed through two longitudinal slits on the upper side of the anthers. If the weather is warm and dry, nearly all the pollen is shed the day the flower opens. In the evening the flowers tend to close, approaching a funnel shape but open again the following morning. On the third day the flower remains almost closed and the petals and sepals begin to wilt.

Studies on pollen-tube growth indicate that fertilisation is effected within about 24 hours of pollination (Khanna and Chowdhury, 1974). The two carpels (although flowers on some plants may produce three or four carpels) form a superior ovary with a "false" septum and two rows of campylotropous ovules. After fertilisation, the ovary develops into a bivalve silique with a longitudinal septum (Figure 3.2). When the buds are about 5 mm long, the megaspore in each ovule divides twice, producing four cells, one of which becomes the embryo sac, while the others abort. The nuclear tissue is largely displaced by the remaining embryo sac and at flower opening, the ovules mainly consist of two integuments and the ripe embryo sac.

Pollination, pollen dispersal and viability

Brassica pollen, although heavy and slightly sticky, can still become airborne and float on the wind due to its minute size (30-40 μm). In addition to wind, pollen can be transferred by insects, primarily honey bees (Williams, Martin and White, 1987, 1986; Scheffler, Parkinson and Dale, 1993; Paul, Thompson and Dunwell, 1995; Timmons et al., 1995; Thompson et al., 1999). Physical contact between flowers of neighbouring plants also results in pollen dispersal while animals, including humans, passing through flowering *Brassica* fields can act as pollen vectors.

Pollen movement can be detected using pollen traps for airborne pollen or by using bait plants (either male sterile or emasculated) to detect outcrossing, usually through the use of marker genes such as herbicide resistance. An effective pollen trap, developed in Germany, combines a sampler that determines pollen deposition rate (Sigma-2 sampler) and a pollen mass filter apparatus that collects sufficient pollen for polymerase chain reaction (PCR) analysis (VDI Richtlinien, 2007). Pollen from the Sigma-2 sampler is analysed as to species and amount under a light microscope and/or by automated imaging analysis. Strategically located bee hives can also be used to monitor pollen flow whereby pollen in honey and bee bread samples is concentrated and analysed under a light microscope or subjected to PCR analysis (VDI Richtlinien, 2006).

Under natural conditions, Ranito-Lehtimäki (1995) reported a gradual decrease in pollen viability over four to five days. In the laboratory, Mesquida and Renard (1982) found pollen remained viable between 24 hours to 1 week. However, Chiang (1974) reported that *B. oleracea* pollen stored at 4°C germinated above 20% for the first 10 days, and even after 6-7 weeks an average 4.5% of the test pollen remained viable.

The greatest pollen outflow from flowering *Brassica* fields is undoubtedly wind borne. Studies have shown that the vast majority of the pollen cloud travels less than 10 m and approximately half the pollen produced by an individual plant falls to the ground within 3 m (Lavigne et al., 1998). In a two-year study, Bilsborrow et al. (1998) reported that the pollen concentrations at 10 m was reduced by 48% and 67% compared to that recorded 2 m from the field border. McCartney and Lacey (1991) found that the amount of pollen detected at 20 m from the field border was 90% less than that recorded at the field edge. Over longer distances of 360 m and 400 m, relative to the field margin, Timmons et al. (1995) and Thompson et al. (1999) reported reductions of 90% and 95%, respectively. These findings, combined with outcrossing data, established that *Brassica* pollen follows a leptokurtic distribution i.e. the presence of pollen shows a steep decline with distance, but with a long tail containing long-distance events (Figure 3.36; Thompson et al., 1999; Staniland et al., 2000). These data indicate that at a distance of 50 m from the pollen source, the level of outcrossing is less than 0.5%, even when male sterile bait plants are used as pollen recipients (Figure 3.36).

Figure 3.36. **Outcrossing percentages as affected by distance from the pollen source**

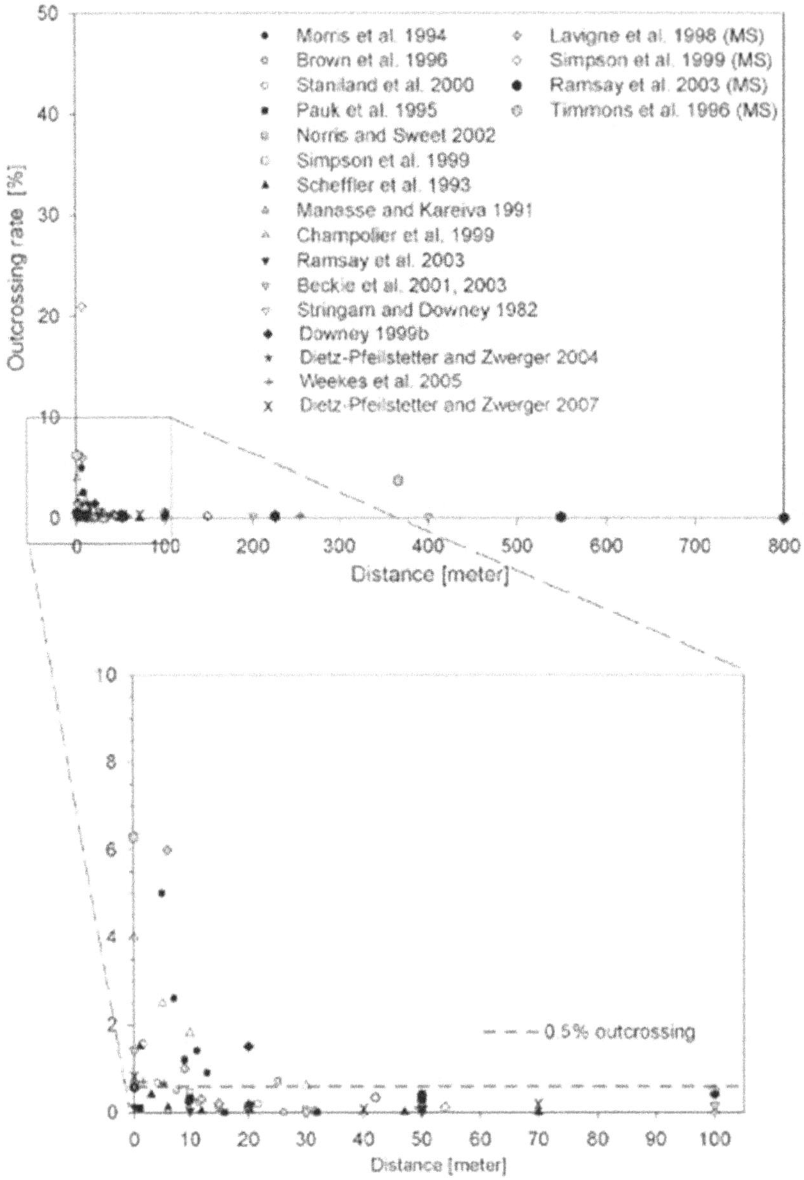

Note: (MS) indicates male sterile bait plants used.

Source: Modified from Andersson and de Vicente (2010).

Where fields are large (>60 hectares) and/or production regions are extensive, as in Australia, Canada and India, wind is considered to be the primary pollen vector since bee populations cannot service the vast number of exposed flowers. However, in the United Kingdom and other parts of Europe where field size is small and bees and pollen beetles are abundant, insects play an important role in pollen dispersal, especially over long distances (Ramsay, Thompson and Squire, 2003; Ramsay et al., 1999; Thompson et al., 1999). Pollen distribution by insect can vary greatly depending on the production region, the environment and the experimental design (Barber, 1999; Thompson et al., 1999; Ramsay, Thompson and Squire, 2003). Honey bees visiting a new field are covered with pollen from that field after visiting about four flowers, thus reducing the chances of cross fertilisation between plants of the new field and fields previously visited (Cresswell, 1994). Honey bees are also more efficient pollinators than wind-borne pollen over longer distances. This is to be expected since to effect fertilisation, wind-borne pollen must fall from the sky and land on an unfertilised stigma. Using published measurements of pollen dispersal, Hayter and Cresswell (2006) estimated that when bees are scarce, wind can contribute to pollination of fields 1 km distant at a level of up to 0.3%, but only up to 0.007% when bees are abundant. However, with a non-GM pollen source 500 m from a beehive and a GM field 800 m from the same hive, Ramsay et al. (1999) detected some pollen grains from the GM field in largely non-GM pollen loads. They concluded that there was either switching between fields or a long persistence of pollen grains on the bees, or there was pollen mixing within the hive. Ramsay et al. (1999) also found that honey bee colonies can forage up to 2 km from their hive, indicating a potential for pollen transfer around the hive covering an area 4 km in diameter. The maximum 4 km distance for pollen dispersal by bees corresponds closely with the 4 km maximum for the wind-borne pollen model reported by Timmons et al. (1996).

A number of models have been developed to predict the level of gene flow that might be expected among *B. napus* fields and feral populations as well as interspecific crosses with *B. rapa* (among others, Bateman, 1947a, 1947b; Lavigne et al., 1998; Colbach et al., 2005; Klein et al., 2006; Devaux et al., 2007; Graziano Ceddia, Bartlett and Perrings, 2007). However, as many biotic and abiotic factors affect gene flow, the models currently only provide an approximation. Further, the models have tended to focus on pollen dispersal and its arrival on the stigma, and have paid little attention to hybridisation and introgression.

Outcrossing in the field

Although *B. napus* is self-compatible (autogamous), pollen from neighbouring and distant *B. napus* plants compete with the plant's own pollen to effect fertilisation. There are no genetic or morphological barriers to cross-pollination among *B. napus* plants, so crossing between fields does occur (Becker, Damgaard and Karlsson, 1992; Becker et al., 1991; Rakow and Woods, 1987). The outcrossing rate within fields varies considerably, averaging between 20% and 40%, mainly depending on the environmental conditions during flowering (see Becker, Damgaard and Karlsson, 1992 and references therein). It is estimated that one hectare of spring oilseed rape produces 9.3 ± 0.5 kg of pollen each 24 hours during a 17-day flowering period with *B. rapa* fields producing 20.2 kg/ha/day, more than twice that of *B. napus* (Szabo, 1985). Most of the crossing occurs between neighbouring plants (Rakow and Woods, 1987), but long-distance pollen transfer can occur by both wind and insects (primarily bees). The measurement of pollen flow via wind or insects, or estimating the amount of outcrossing using male sterile or emasculated bait plants, provides information on the potential for outcrossing; however, it is not an accurate indicator of the actual outcrossing level that can occur between fully fertile oilseed rape crops. In reality, male sterile plants would normally be growing in association with fully fertile plants, so data from male sterile bait plants significantly overestimate the level of outcrossing that would normally be expected. Ramsay, Thompson and Squire (2003) concluded that bait plants over-estimate the outcrossing level by at least one order of magnitude.

Numerous experiments have been undertaken in recent years to determine the frequency of outcrossing that occurs between two populations of *B. napus*, with increasing distance between the pollen donor and recipient populations. The availability of HR genes and other markers have facilitated the detection of such genes in non-HR *B. napus* plots and fields and multiple HR types in single HR crops. However, measurement of the rate of outcrossing is complex as it can vary with the experimental design, environmental conditions, cultivars grown, synchrony of flowering, insect pollinator activity, local topography, and the relative size and arrangement of the donor and recipient populations. Two types of designs have been used in these studies. In the continuous design, the recipient population surrounds the donor, while in the discontinuous designs the recipient populations are distributed in locations at increasing distances from the pollen source (Hüsken and Dietz-Pfeilstetter, 2007). Using continuous designs, over short isolation distances (0-30 m), researchers observed a rapid decline in outcrossing rates as they sampled from the field edge into the recipient population (Scheffler, Parkinson and Dale, 1993; Morris et al., 1994; Brown et al., 1996; Staniland et al., 2000; Reboud, 2003; Dietz-Pfeilstetter and Zwerger, 2009, 2004). Examples from such studies, conducted in Canada, the United Kingdom and the United States are given in Table 3.6. These results underline the importance of determining outcrossing data across the whole field and not just the level at a particular spot or distance into the field. However, using commercially sized fields in a discontinuous design, Rieger et al. (2002) found that fields situated within 100 m of the pollen source showed very little edge effect while fields far from donor sources displayed a low and variable edge effect.

Table 3.6. **Short distance pollen mediated gene flow from *B. napus* pollen donor to recipient field/plots in Canada, the United Kingdom and the United States**

Metres into recipient field	Outcrossing %	Reference, location and trial year
0.5	4.8	Scheffler, Parkinson and Dale (1993) United Kingdom, 1991
1.0	1.5	
3.0	0.4	
6.0	0.11	
12.0	0.016	
24.0	0.004	
36.0	0.001	
0.0	2.0/3.5	Morris et al. (1994) United States, 1992 east/west wind direction
0.3	1.0/1.5	
0.6	0.75/1.2	
3.0	0.65/0.6	
4.6	0.50/0.6	
0.0	0.70	Staniland et al. (2000) Canada 1994-95 Data averaged over wind directions and years
2.5	0.30	
5.0	0.10	
10.0	0.07	
15.0	0.08	
20.0	0.07	
25.0	0.04	
30.0	0.03	

The mean rate of outcrossing at various isolation distances is a valuable statistic, but the more important question might be "what is the maximum outcrossing that might be expected at various distances?" There is now considerable evidence that the highest rate of outcrossing that might be expected at 50-100 m is <0.5% and at 200 m the maximum would be <0.1% (Tables 3.7 and 3.8 and Figure 3.36).

In most of the small plot trials, arranged in a continuous design, the area occupied by the donor population is small in relation to the recipient populations, the ratio being about 1:4. This unequal availability of pollen tends to dilute the amount of donor pollen accessible to both wind and bee vectors. As a result, outcrossing rates reported for small plot trials with isolation distances of over 30 m tend to be lower than those recorded in larger scale investigations where the area devoted to the pollen donor are substantially greater (Table 3.7).

Table 3.7. ***B. napus* to *B. napus* outcrossing rates, by isolation distances, reported from small plot trials and/or large fields**

Isolation distance	Small plot trials		Large field trials (0.05 ha or more)	
	% outcross	Reference	% outcross	Reference
30-60 m	0-0.0003	Scheffler, Parkinson and Dale (1993)	<0.01	Champolivier et al. (1999)
	0.022	Manasse and Kareiva (1991)	0.1	Simpson[2]
	0.11-0.16	Sweet et al. (1999a)	0.2	Beckie, Hall and Warwick (2001)
	0.02-0.24	Monsanto[1]	0-0.4	Downey (1999a, 1999b)
	0.05-0.33	Simpson et al. (1999) (MS)	0.1-0.65	Norris[2]
	2.1	Stringam and Downey (1982)	0.02	Ramsay, Thompson and Squire (2003)
	0.02	Staniland et al. (2000)	<1	CETIOM (2000)
	0.05	Von Ernst et al. (1998)	0.05	Wilkinson et al. (1995)
	0	Lavigne et al. (1998) (MS)	0.1-0.08	Dietz-Pfeilstetter and Zwerger (2004)
	0.02-0.05	Wilkinson et al. (1995) (MS)	0.2-0.4	Weekes et al. (2005)
	0.33	Ramsay, Thompson and Squire (2003)	0.00-0.09	Rieger et al. (2002)[3]
90-150 m	0.01-0.02	Manasse and Kareiva (1991)	0.05	Simpson[2]
	0.00-0.07	Kamler (2000)	0.1	Downey (1999a, 1999b)
	0.11-0.22	Simpson et al. (1999)	0.15	Beckie, Hall and Warwick (2001)
	0.01-0.13	Simpson[2] (FB)	0.25-0.5	Norris[2]
	0.01-0.21	Monsanto[1]	<0.5	CETIOM (2000)
	0.5	Timmons et al. (1996) (MS)	0.01-0.02	Weekes et al. (2005)
175-225 m	0.02-0.03	Simpson[2] (FB)	<0.1-0.2	Norris[2]
	0.017-0.6	Dietz-Pfeilstetter et al. (1998)	0.2	Beckie, Hall and Warwick (2001)
	0-0.9	Monsanto[1]	0.02	Ramsay, Thompson and Squire (2003)
	0.15	Scheffler, Parkinson and Dale (1995)	0.00-0.005	Rieger et al. (2002)[3]
	0.21	Ramsey, Thompson and Squire (2003) (MS)		
360-400 m	0.0038	Scheffler, Parkinson and Dale (1995)	0.1	CETIOM (2000)
	0.06	Simpson[2] (FB)	0.14	Beckie, Hall and Warwick (2001)
	0.6	Stringam and Downey (1982)	0.00-0.025	Rieger et al. (2002)[3]
	0.0	Monsanto[1]		
	3.7	Timmons et al. (1996) (MS)		
500-800 m	0.02-0.1	Ramsey, Thompson and Squire (2003) (MS)	0.00-0.053	Ramsay, Thompson and Squire (2003)
			0.001-0.03	Rieger et al. (2002)[3]

Notes: 1. Cited by Salisbury (2002). 2. (FB) indicates use of fertile bait plants, cited by Easthan and Sweet (2002). 3. Ranges estimated from published graph. (MS) indicates the use of male sterile bait plants.

Crawford, Squire and Burn (1999) estimated that a square donor plot of at least 400 m^2 would be needed if a sharp decline in the effectiveness of donor pollen is to be avoided. Positioning of the donor and recipient fields can also affect the outcrossing measurements. Ingram (2000) noted that the rate of outcrossing would be higher when the long sides of donor and recipient fields faced each other. Hüsken and Dietz-Pfeilstetter (2007) statistically analysed published outcrossing results for both continuous and discontinuous designed studies. Their data indicate that with the discontinuous design, the mean outcrossing rate between *B. napus* fields at 50 m and 100 m would be 0.11% and at 200 m, 0.05% with lower rates for the continuous design studies (Table 3.8).

Under short isolation distances, surrounding the pollen source with a synchronous flowering recipient border may be effective in reducing pollen outflow (Staniland et al., 2000; Reboud, 2003). Staniland et al. (2000) found that surrounding a spring *B. napus* pollen donor with a 15 m and 30 m wide *B. napus* border/pollen trap, separated from the pollen donor by a cultivated 1.5 m strip, reduced the outcrossing level to 0.02% at 30 m, a level they equated to the outcrossing rates observed at 200 m by Scheffler, Parkinson and Dale (1995) (Table 3.7). They concluded that under western Canadian conditions, the current regulations, which require a 10 m wide continuous border surrounding the pollen donor, would effectively contain the majority of pollen-mediated gene flow, but would not completely eliminate gene escape.

Table 3.8. **Mean outcross percentages of pollen donor to *B. napus* recipient populations, for various isolation distances and two design classes**

Distance from pollen source (m)	Continuous design			Discontinuous design		
	Mean	Standard deviation	Number of data points	Mean	Standard deviation	Number of data points
0-10	1.78	2.48	26	0.94	0.51	10
10-20	0.33	0.45	7	0.40	0.47	8
20-50	0.05	0.05	10	0.14	0.11	11
50-100	0.04	0.04	3	0.11	0.11	11
>200	n.d.	n.d.	n.d.	0.05	0.05	6

Note: n.d. = insufficient data.

Source: Hüsken and Dietz-Pfeilstetter (2007).

Field size experiments by Reboud (2003), using 24 m borders, indicated that for short isolation distances gaps of bare ground between the donor and recipient plots/fields should be avoided. Outcrossing declined more rapidly when there were intervening plants, e.g. when the pollen donor was separated from the recipient field by a 3-4 m gap the level of outcrossing was similar to that found 1 m into the crop where the gap was zero. The same effect was noted by Dietz-Pfeilstetter and Zwerger (2004) when a bare gap between donor and recipient fields was increased from 0.5 m to 10 m.

In the large field studies, not all the factors contributing to gene flow have been controlled. Weekes et al. (2005) found the level of outcrossing to be considerably higher in winter than in spring oilseed rape (Table 3.9) while Ramsay, Thompson and Squire (2003) found the opposite to be true. They attributed the low value in the winter rape trial to poor pollinating weather in May.

However, Reboud (2003) and Dietz-Pfeilstetter and Zwerger (2009) observed that varieties used as pollen donors differed significantly in their outcrossing potential. The outcrossing values in some fields in the Rieger et al. (2002) study may have been overestimated since seed sown in the recipient fields was not tested as to the possible presence of imidazolinone-tolerant seeds (Salisbury, 2002). Such contaminant HR seed could have been present in seed sown in the recipient fields as a result of outcrossing or admixture during the breeding and multiplication of the donor and recipient varieties, as was observed in Canada by Downey and Beckie (2002) and Friesen, Nelson and Van Acker (2003).

Also, it has been suggested that outcrossing levels were underestimated due to the segregation of the two genes required to provide full tolerance to the selective herbicide. Hall et al. (2000) identified some herbicide-resistant seedlings from recipient plants situated some 650 m from an HR field. However, Downey (1999b) suggested the seed may have been transported by the farmer's swathing and harvesting equipment as observed in the Dietz-Pfeilstetter and Zwerger (2009) study.

The outcrossing percentages reported by Stringam and Downey (1982) are substantially higher than recorded for other studies listed in Table 3.7. However, it should be noted that in the Stringam-Downey trials the pollen donors were fields of >60 hectares which resulted in the overloading of the small 42 m^2 recipient plots with donor pollen. Similar high outcrossing rates were recorded by Ramsey, Thompson and Squire (2003) where blocks of ten male-sterile plants were placed at increasing distances from a large commercial field. These results have implications for feral populations situated near commercial fields. Other observations suggest that field-to-field crossing is likely to be highest in fields just commencing or finishing flowering when a nearby field is in full bloom.

Table 3.9. **Predicted outcrossing rates for spring and winter oilseed rape at three isolation distances (with 95% confidence limits), based on 2000-03 multilocation UK field trials**

Oilseed rape type	Percent outcrossing		
	2 m	50 m	150 m
Spring	0.46 (9.97)[1]	0.02 (0.39)	0.01 (0.14)
Winter	0.76 (12.25)	0.04 (0.84)	0.02 (0.40)

Note: 1. 0.46 is the average percent outcrossing with a 5% chance that outcrossing could be as high as 9.97%.

Source: Weekes et al. (2005).

Downey and Beckie (2002) and Friesen, Nelson and Van Acker (2003) illustrated how easily pedigree seed can be contaminated in breeding nurseries. Admixture during seeding, harvesting or cleaning was also identified as a contaminant source (Downey and Beckie, 2002). These studies alerted seed companies to the problem of contamination in breeders seed stocks, leading to tighter controls (see the section "Oilseed certified seed production").

However, the present rapid development and acceptance of *B. napus* hybrid varieties dictates that certified seed-production fields will contain at least 66-75% male sterile plants. This increases the risk of outcrossing. In Canada, all hybrid producing seed fields are regulated and inspected to ensure that they are isolated from other rapeseed plants and fields by at least 800 m and free of certain *Brassica* weeds within the production field and the regulated isolation area. The isolation distance used by most seed companies for hybrid seed production of *B. napus* in Canada is at least 1.6 km (Wescott and Nelson, 2001). To further reduce the possibility of fertilisation by foreign pollen, the fields are heavily stocked with honey bees. Such fields are also saturated with leaf cutter bees (*Megachile rotundata* [Fabricius]), which have a short foraging range, to ensure the desired rapid and complete fertilisation of the male sterile female parent.

Seed development, production and natural dispersal

After fertilisation the endosperm develops rapidly, while embryo growth does not start for some days. The embryo is generally still small two weeks after pollination but by three to five weeks has almost completely absorbed the endosperm and filled most of the seed coat. Nutrient reserves for germination are stored in the cotyledons which are folded one over the other so that there is a smaller inner and a larger outer cotyledon (Figure 3.37).

Figure 3.37. **Development stages of the *Brassica napus* zygotic embryo**

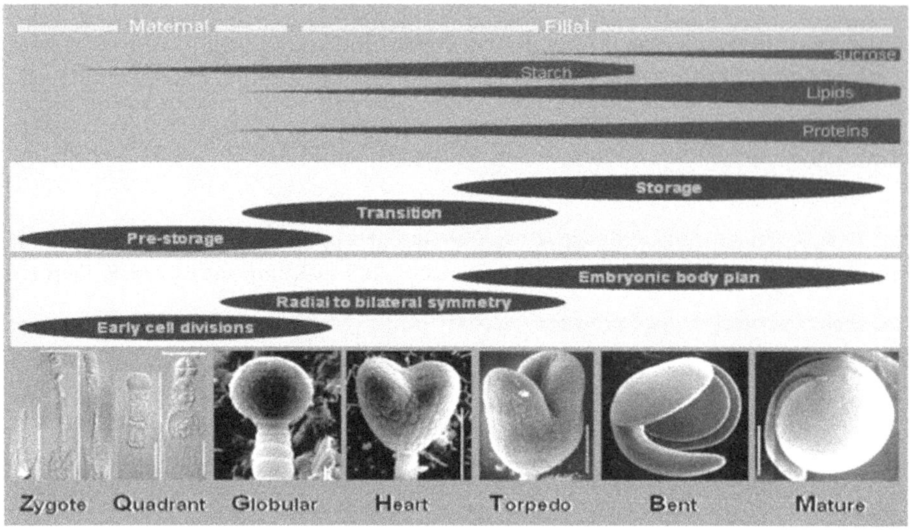

Source: Courtesy Plant Biotechnology Institute, Saskatoon, Canada.

The size of seeds can be defined by both their physical dimensions and weight. The range in seed weight among the *Brassica* crop species is given in Table 3.10. Typical seeds of *Brassica* species and subspecies are illustrated in Figure 3.38. These drawings, produced by the USDA many years ago, are still valid and can be used as a starting point to distinguish many of the species and subspecies according to the reticulation patterns on the seed surface. The different patterns are the result of variation in the size of the palisade cells that form the outer cell layer of the seed coat.

Table 3.10. **Typical seed weight ranges (or averages) of *Brassica* crop plants by species and form**

Species	Form	g/1 000 seeds	Source
B. napus	Winter oilseed rape	4.5-5.5	Bengtsson et al. (1972)
	Spring oilseed rape	2.5-4.6	Elliott, Franke and Rakow (2008)
B. rapa	Winter turnip rape	3.0-4.0	Bengtsson et al. (1972)
	Spring turnip rape	2.0-3.0	Bengtsson et al. (1972)
B. juncea	Condiment and oilseed mustard*	2.5-3.0	Rakow and Rode (2009); Rakow et al. (2009)
B. oleracea	Cabbage	3.6	Ohio State University (2009)
	Broccoli	2.7-5.8	Heather and Sieczka (1991)
	Brussels sprouts	2.8	George (2009)
	Kohlrabi	3.2	George (2009)

Note: * The Indian cultivar Pusa Bold has larger than normal seed at about 5.3 g/1 000.

Sources: Bengtsson et al. (1972; Elliott et al. (2008); Rakow and Rode (2009); Rakow et al. (2009); Ohio State University (2009); Heather and Sieczka (1991); George (2009).

Vaughan and Whitehouse (1971) investigated and described the seed surface and general features of some 200 Brassicaceous species including shape, colour, mucilage production and hilum characteristics. Koul, Nagpal and Raina (2000) also examined the seed surface architecture of 78 accessions from the 3 subtribes – Brassicinae, Raphaninae and Moricandiinae – at both low magnification (x80) as well as the fine structure using a scanning electron microscope (x640, x1260). They noted that the seed coat patterns at high magnification were generally species-specific. However, significant seed coat pattern variations were found at the intraspecific level among the *Brassica* diploids, *B. rapa*, (two types), *B. nigra* (one type) and *B. oleracea* (two types), with the patterns of *B. rapa* and *B. oleracea* resembling each other. The seed coat patterns in most of the amphidiploids were intermediate to their putative parents, although one *B. carinata* and one *B. napus* accession exhibited patterns of their respective *B. nigra* and *B. rapa* parents. Thus, employing seed coat reticulations for species identification is not foolproof, but it provides a good starting point to identify the adventitious presence of foreign species in commercial seed lots.

Figure 3.38. **Distinguishing *Brassica* species by their seed coat characteristics**

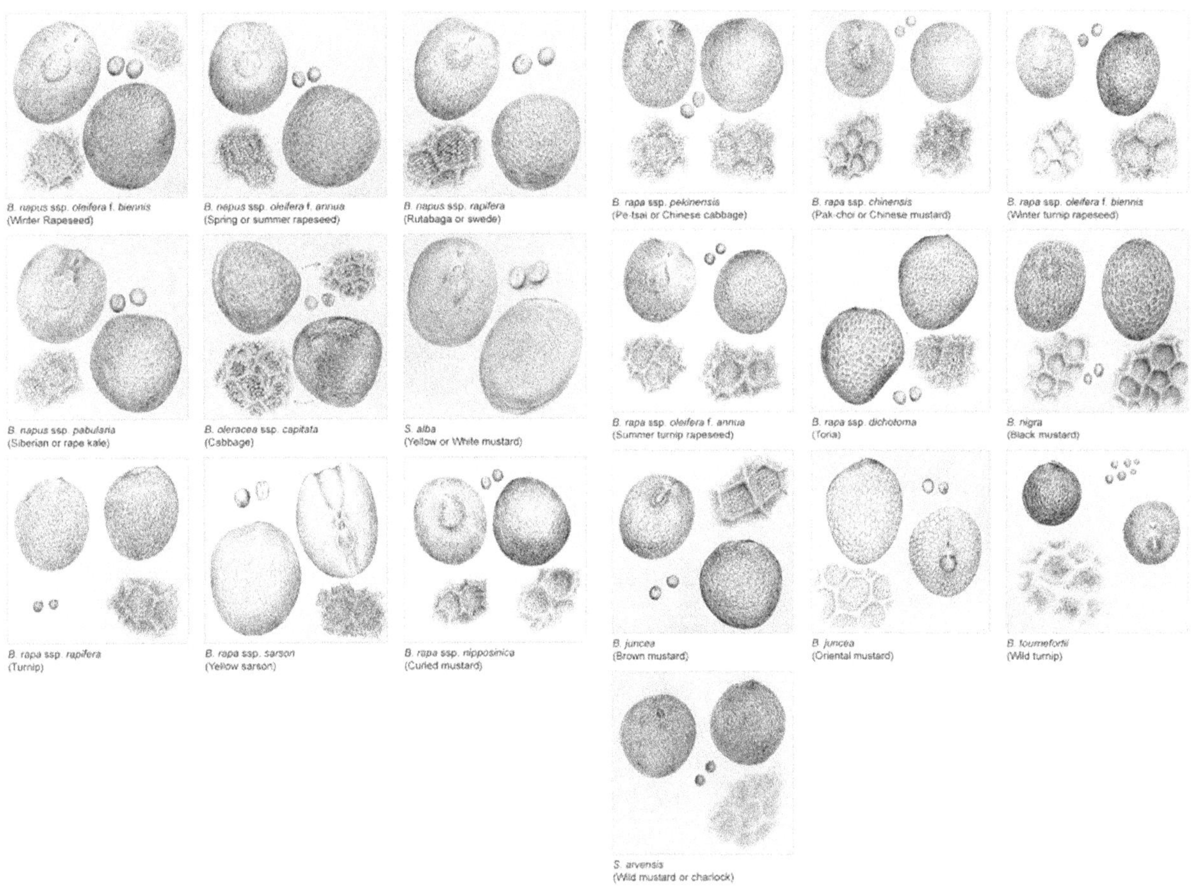

Notes: The small seeds shown in each compartment are about three times their natural size. The greatly enlarged surface detail is not drawn to scale but a relative proportion is maintained throughout.

Source: USDA.

The fruit of major *Brassica* crops is a glabrous silique, which is 4-5 mm wide and can be over 10 cm long, with 2 rows of seeds lying along the edges of the replum (false septum, an outgrowth of the placenta). A silique normally contains 10-30 seeds. Three to four weeks after the flower opens, the silique attains its full diameter and length. When ripe, the silique has a tendency to dehisce and shatter, dispersing its seed. Species and varieties differ in their susceptibility to shattering. The physical forces of silique hitting silique or other plant parts causes a separation of the valve walls from the placenta, starting at the pedicel end and working toward the unattached end. The exposed seeds attached to the placenta are soon dislodged by wind action. Threshing operations easily separate the seed from the intact siliques.

All commercially grown *Brassica* crops, as well as weedy species, tend to shatter their seed when ripe. However, the ease or degree of shattering varies among species. Within the oilseed crops, *B. napus* has the greatest tendency to shatter its seed, with *B. rapa* intermediate and *B. juncea* the least. Breeding work is developed to transfer the shatter-resistant characteristic from *B. juncea* to *B. napus* (Wang, Ripley and Rakow, 2007). The vegetable *Brassica* species follow a similar pattern. However, with the high-value F_1 seed of *B. oleracea* hybrids and the relatively small fields used for seed production, every precaution, sometimes including hand harvesting, is taken to ensure little or no seed is lost. Pod shatter is rare in the closely related *S. alba* (yellow or white mustard) species, but some loss of intact ripe pods, due to wind or mechanical action, does occur at harvest.

Seed that falls to the ground can be dispersed by wind and water as well as by birds and other animals. Because the seed is small and round it is difficult to prevent some loss during transportation of farm equipment from field to field, or from field to bin and from bin to its ultimate destination. Significant losses can occur from truck containers of uncovered oilseed rape due to the wind vortex caused by the movement of the truck. The faster the truck goes, the greater the loss. The distribution of seed from the truck vortex will depend on seed size and the direction and velocity of the wind prevailing at the time of loss. For spring *B. napus*, the distance such seed will travel at various wind speeds has been calculated (Table 3.11), although for average spring and winter *B. napus* seed, which is larger and heavier than that used to calculate the table, the wind-borne dispersal distance would be reduced.

Table 3.11. **Estimated dispersal distances of spring *B. napus* seed released from transport vehicles at various heights above adjacent fields**

	Wind speed in km/h					
Height (metres)	10	20	30	40	50	60
	Horizontal dispersal in m					
1.0	1.4	2.7	4.1	5.5	6.8	8.2
2.0	2.1	4.1	6.2	8.3	10.3	12.4
3.0	2.6	5.1	7.7	10.2	12.8	15.3
4.0	3.1	6.2	9.3	12.3	15.4	18.5
5.0	3.6	7.2	10.8	14.4	18.0	21.6
6.0	4.1	8.2	12.4	16.5	20.6	24.7
7.0	4.6	9.3	13.9	18.5	23.2	27.8
8.0	5.1	10.3	15.4	20.6	25.7	30.9

Note: 1. Estimates based on small seeds of spring *B. napus*, calculated to weigh 2.2 mg with a diameter of 1.8 mm, that are the most likely to become air borne and travel the farthest.

Source: Hertz (1999).

Seed viability, longevity and dormancy, germination, seedling establishment

Well-developed, fully mature *Brassica* oilseeds may remain viable for at least 25 years if dry seed is refrigerated in sealed containers (Ellis et al., 1994). As of 2009, seed of oilseed *Brassica*, harvested in 1977 and stored in manila envelopes at -20°C in the Saskatoon AAFC Seed Bank, had retained its high germination (Downey, personal communication). Viability of seed lost during harvest is an important factor in determining the presence and amount of volunteer plants and populations in subsequent crops. Harvest losses can be substantial and the survival and persistence of this seed is greatly influenced by environment, seed dormancy as well as crop and field management.

Contribution of *B. napus* harvest losses to persistence

Harvest losses in the United Kingdom, when the winter *B. napus* crop is straight combined under ideal conditions, ranged from 2% to 5%, but under unfavourable harvest environments could amount to 50% (Price et al., 1996). Pekrun et al. (1998) placed these losses between 200-300 kg/ha or about 5 000-7 000 seeds/m^2. Lutman et al. (2005) in the United Kingdom and Gruber, Pekrun and Claupein (2004) in Germany recorded average harvest losses of 3 000-3 500 seeds/m^2. Similarly, French studies estimated harvest losses to be between 1.5% and 8.5% of the average yield. This calculates to 50-300 kg/ha of seed remaining on the field after harvest or 1 100-6 700 seeds/m^2 (CETIOM, 2000; Messéan et al., 2007). In Canada, Gulden, Shirtliffe and Thomas (2003a) reported that spring *B. napus*, harvest losses averaged 5.5%, or about 3 590 seeds/m^2, while Légère et al. (2001) estimated the losses at 2 000/m^2. Similarly, Warwick et al. (2003) reported spring *B. napus* harvest losses averaging 5.5%, or about 3 590 seeds/m^2. Salisbury (2002) estimated Australian losses would be similar to those found in Canada. However, a vast majority of the seed remaining in the field after harvest will not survive the first year. The *Brassica* oilseed density of the seed bank in western Canada is reported to drop ten fold in the first year and to decline slowly thereafter, due to replenishment of the seed bank by uncontrolled volunteer plants. However, where post-harvest tillage is shallow and delayed and volunteers in subsequent crops are controlled, very few plants are found four years after a spring *B. napus* crop (Gulden, Shirtliffe and Thomas, 2003b).

Seed dormancy

Seed dormancy can play an important part in determining the amount and persistence of volunteer *Brassica* plants in subsequent crops. There are two main types of seed dormancy: primary and secondary. Primary dormancy is when seed germination is prevented during the seed maturation process and for some time after the seed has been removed from its parent (Karssen, 1980/81; Hilhorst and Toorop, 1997). To overcome primary dormancy, a period of after-ripening is usually required. Secondary dormancy is a reduction in seed germinability that develops after the seed is separated from the parent plant and may, in some cases, be induced prior to the complete alleviation of primary dormancy. Primary dormancy does not occur in ripe seeds of any of the cultivated *Brassica* oilseed, vegetable or condiment crops. For seed certification status, these crops require a minimum germination of at least 90%. However, during seed maturation, germination percentages may be low in spring and winter *B. napus* but increase with maturity (Finkelstein et al., 1985) to where at harvest no primary dormancy occurs (Schlink, 1995). However, secondary dormancy can be induced in *B. napus* and cultivated *B. rapa* under certain conditions (Hails et al., 1997; Pekrun, Lutman and Baeumer, 1998; Adler et al., 1993). An exception to the rule occurs in the weedy forms of *B. rapa*, where primary dormancy is present as a recessive trait in weedy *B. rapa*. Thus, crossing between weedy and cultivated *B. rapa*, as well as between weedy *B. rapa* and *B. napus*, will produce seed that does not exhibit primary dormancy (Linder, 1998; Landbo and Jorgensen, 1997; Adler et al., 1993).

The main factors contributing to secondary dormancy of *B. napus* seed are elevated temperatures, darkness, osmotic stress and limited oxygen (Gulden, Thomas and Shirtliffe, 2004; Pekrun et al., 1997). Studies in Europe (Pekrun, Potter and Lutman,

1997; Gruber, Pekrun and Claupein, 2004) and China (Momoh et al., 2002) suggested that genotypes differ in their predisposition to undergo secondary dormancy. Indeed, it has been clearly shown that genotype is the principal factor controlling its potential in *B. napus* (Gulden, Thomas and Shirtliffe, 2004; Pekrun et al., 1997; Gruber, Emrich and Claupein, 2009; Gruber, Pekrun and Claupein 2004). Gulden, Thomas and Shirtliffe (2004) found seed size was of secondary importance, with large seed more likely to undergo secondary dormancy, while maturity and pre- and post-harvest environment had little influence. The occurrence of secondary dormancy is reduced by alternating temperatures (Pekrun, Potter and Lutman, 1997; Momoh et al., 2002), while cold stratification readily releases secondary dormancy as does exposure to continuous light (Schlink, 1995). Exogenous applications of gibberellic acid (0.2 mg l^{-1}) will also reverse secondary dormancy (Pekrun, Lutman and Baeumer, 1998).

In Germany, Gruber, Pekrun and Claupein (2004) evaluated the persistence and secondary dormancy in the seed of four winter oilseed rape varieties. They found that of the 3 000-3 500 seeds/m^2 lost during harvest, 60-75% of that seed either died or was scavenged within a few months. Similar levels of seed disappearance were observed by Gruber, Pekrun and Claupein (2003) when investigating the effect of different tillage treatments on seed persistence. Six months after harvest, no seed of the variety Artus could be detected in the soil seed bank while the other three varieties – Bristol, Liberator and Capital – respectively contributed 4.3%, 9.3% and 11% of their lost seed to the seed bank. Laboratory tests for the presence of secondary dormancy closely corresponded to that observed in the field. Gruber, Emrich and Claupein (2009) also laboratory tested seed from over 40 varieties for their tendency to undergo secondary dormancy. The seed was harvested from one site for three years and a second site for two years. They found that, over several years, varieties consistently ranked high, medium or low in percentage of seed exhibiting secondary dormancy. However, the rate of secondary dormancy varied significantly with harvest years, dry years having the lowest incidence. They concluded variety rank, rather than the actual percentage of secondary dormancy, should be used to characterise a variety. Thus, selection for varieties without secondary dormancy could be easily achieved and would greatly reduce the incidence of *B. napus* volunteers in subsequent crops. It should probably be made mandatory for all new *B. napus* varieties to be free of the secondary dormancy trait.

At shallow burial depths, *B. napus* and closely related species exhibit low seed bank persistence (Schlink, 1995; Pekrun and Lutman, 1998; Sparrow, Knight and Conn, 1990; Gulden, Shirtliffe and Thomas, 2003a). At 10 cm depth Gulden, Thomas and Shirtliffe (2004) found seed-bank populations shifted from a germinable to an ungerminable state and no seedling recruitment was observed. Masden (1962) reported that 1% of buried *B. napus* seed germinated after five years, and that trace amounts of *B. rapa* seed emerged after ten years. Schlink (1998) and Lutman, Freeman and Pekrun (2003) found that approximately 1% of *B. napus* seed in undisturbed soil could survive for ten years. Jørgensen, Pavlo Hauser and Bagger Jørgensen (2007), sampling a deep soil layer, identified viable seeds of a variety sown in the field 17 years earlier. In Canada, Beckie and Warwick (2010) reported a small population of volunteers resistant to the herbicide bromoxynil in a field that had not grown oilseed rape since the sowing of a bromoxynil-resistant variety seven years previously. The volunteers persisted in low-lying areas of the field which were too wet to plant or spray with herbicides between 2001 and 2007. No volunteers were detected in either 2008 or 2009. There is general agreement that secondary dormancy will be induced in a significant percentage of deeply buried *B. napus* seed.

Persistence

Very few seeds of oilseed rape survive in the seed bank compared with their wild relatives (Chadoeuf, Darmency and Maillet, 1998). Most seeds of the cultivated *Brassica* crops, if left on or near the soil surface, will germinate and be killed by frost or cultivation or be eaten by rodents, birds and insects. Nevertheless, a small proportion may not germinate and secondary dormancy may be induced, particularly if the seed is buried. Studies in Europe with winter *B. napus* found that when seeds were buried immediately after seed shed, 30% of the seed bank survived one winter compared to only 0.1% when seeds were left on the undisturbed soil surface (Pekrun and Lutman, 1998). Similarly for spring *B. napus* in western Canada, Gulden, Shirtliffe and Thomas (2003a) found spring seedlings, from fall-sown seeds buried at a 1 cm depth, to be only 0.1-1.5% of the original seed bank. In Canada, oilseed rape is typically grown on the same land once in four years with most of the volunteers occurring in the year following oilseed rape production. However, volunteers can occur four to five years after production (Légère et al., 2001; Simard et al., 2002; Beckie and Owen, 2007). Harker et al. (2006) found that if first-year volunteers were prevented from producing seed, the densities of volunteers in subsequent years were reduced to levels that would not require herbicidal intervention. Surveys in southern Australia by Baker and Preston (2008), where zero and minimum till are practiced, found zero germination of seed sampled from fields 3.5 years after the last *B. napus* harvest. But in Germany, Förster and Diepenbrock (2002) reported more than 0.5 plants/m^2 of winter *B. napus* three years after the last oilseed rape harvest. However, no information on timing or type of post-harvest cultivation was provided. In France, two conventional oilseed rape varieties, one of which was dwarf, were planted on fields that had grown three different HR varieties three to eight years before (Messéan et al., 2007). The percentage of GM HR seed occurring in the harvest of the conventional varieties was determined. HR seed from two of the GM varieties never exceeded 0.9% of the conventional harvested seed. However, one GM variety that was grown five years previous made up 4-18% of the conventional harvest, with the highest values occurring in the seed harvested from the dwarf variety. Since all oilseed rape volunteers were removed from the rotation crops in the intervening years, the volunteers must have arisen from dormant seed in the seed bank. The results illustrate the importance of breeding varieties without the secondary dormancy trait, not only for GM varieties, but more generally for the production of pure seed stocks and segregation of specialty oil types.

In the United Kingdom, Lutman et al. (2005) recorded a large average harvest seed loss (3 575 seeds/m^2) from four *B. napus* winter varieties grown in multiple-site, multi-year trials. Within six months, the number of seeds present declined by an average of 63%, with a slower decline recorded at 18 and 30 months. Appreciably more seeds were found on sites that were ploughed immediately after harvest compared to sites where cultivation was delayed by about four weeks. These data support the recommendations of Pekrun et al. (1998) and Gulden, Shirtliffe and Thomas (2003a) that cultivation of *B. napus* stubble should be avoided for several weeks after harvest. Regression models applied to the Lutman et al. (2005) data predicted that it would take an average of nine years to reduce the seed in the soil bank by 95%. However, other studies (Lutman, Freeman and Pekrun, 2003) indicate that the 95% reduction would occur in three to four years. Indeed, Beismann and Roller (2003) in Germany reported that no viable *B. napus* seeds could be found in soil sample cores taken from sites where transgenic plots were sown five and six years before.

Studies in the United Kingdom and Canada with winter and spring forms of *B. napus* indicate that seed bank persistence is less in lighter than heavier, clay containing, soils

(López-Granados and Lutman, 1998; Gulden, Thomas and Shirtliffe, 2004). In general, if post-harvest tillage is delayed and volunteers are controlled in the intervening years, the evidence indicates the presence of volunteers from the seed bank decline by at least 90% by the fourth year (Lutman, Freeman and Pekrun, 2003: Gulden, Shirtliffe and Thomas, 2003b; Baker and Preston, 2008). Failure to follow the above-recommended practices can extend the presence of seed bank volunteers by several years (Lutman et al., 2005).

Linder and Schmitt (1995) assessed the persistence, in field and greenhouse trials, of GM *B. napus* lines with elevated levels of stearate and laurate fatty acids in their seed oils. They concluded the risk of persistence of the high stearate and high laurate genotypes, compared with their parental non-GM types, was low. No interspecific hybrid seed could be obtained from hand-crossing GM high stearate *B. napus* × wild *B. rapa*. Greenhouse trials using seed from the high laurate *B. napus* × *B. rapa* cross indicated that such hybrids "will not possess seed bank dynamics promoting reproduction".

Genetics

Relevant detailed genetic information

Cytology

Mitotic metaphase chromosomes of the Brassicaceae are very small. Conventional cytological protocols condense *Brassica* meiotic chromosomes to tiny rods or dot-like shapes. Their small size, lack of distinctive cytological features and the difficulties of pachytene investigations make cytological identification of individual chromosomes almost impossible. Although the small chromosome size of the Brassicaceae family has limited the direct cytology approach, the sequencing of the *Arabidopsis thaliana* (The *Arabidopsis* Genome Initiative 2000), *B. rapa* (Wang, 2010; The *B. rapa* Genome Sequencing Project Consortium, 2011), *B. oleracea* and *B. napus* genomes (Bayer CropScience, 2009) are providing a much clearer picture of species interrelationships. 2014 saw the culmination of a major effort worldwide to generate "reference" annotated *Brassica* genome sequences, and some are available online for *B. napus*, *B. oleracea* and *B. rapa*. From 2015, the focus is on a range of "re-sequencing" efforts (The Multinational Brassica Genome projet, 2015).[5]

Comparative mapping, using more than 20 linkage maps for *B. oleracea*, *B. rapa*, *B. nigra*, *B. napus* and *B. juncea*, has contributed greatly to the understanding of chromosome homology and colinearity (Lysak and Lexer, 2006). In addition, great strides have been made in determining the extent of genome colinearity, and rates and modes of evolution in the Brassicaceae family. Comparative cytogenetic studies now employ a wide array of techniques including, among others, rDNA probes, nucleolus organizer regions (NORs), variation in centromeric satellite repeats, genome *in situ* hybridisation (GISH), fluorescence *in situ* hybridisation (FISH), combined with bacterial artificial chromosomes (BAC FISH) and large-scale comparative chromosome painting (CCP). Such techniques have helped to unravel the genomic evolution of *A. thaliana*, *B. oleracea*, *B. rapa*, *B. juncea* and *B. napus* as well as the time frame in which the species arose.

Research into the genome microstructure of the Brassicaceae species indicates the family originated from an ancestral karyotype that evolved after the monocot/dicot split. The ancestral karyotype had a basic chromosome number of x=4 and underwent a genome duplication some 65 million years ago (Mya) followed by diploidisation (Song,

Osborn and Williams, 1988a; Rana et al., 2004; see Figure 3.39). From this progenitor, the ancestral Brassicaceae form evolved with x=8 chromosomes (Lysak et al. 2006). This was followed by the divergence about 20 Mya of the ancestral genera of *Arabidopsis* and tribe Brassiceae. Genome triplication via allohexaploidy occurred about 14-16 Mya (Lysak and Lexer, 2006), followed by diploidisation and chromosome number reduction resulting in the evolution of the ancestral Brassiceae karyotype with x=6 chromosomes (Lysak et al., 2005; Yang et al., 1999). It is estimated that the separation of the Nigra and Rapa/Oleracea lineages took place about 7.9 Mya (Lysak et al., 2005). The *B. oleracea* and *B. rapa* divergence is estimated to have occurred about 4 Mya (Inaba and Nishio, 2002), with the interspecific crosses, forming *B. napus*, *B. juncea* and *B. carinata*, taking place less than 10 000 years ago (Song, Osborn and Williams, 1988a; Rana et al., 2004; Lysak et al., 2005).

Figure 3.39. **Illustration of major events in the evolution of selected *Brassica* species and *Arabidopsis thaliana***

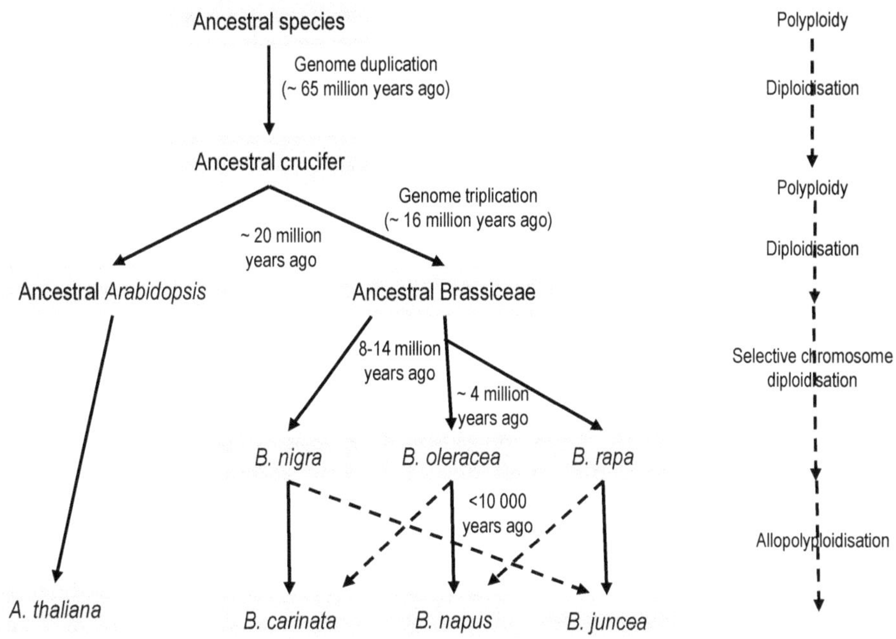

Note: The dotted lines indicate the species believed to be the maternal parent in the interspecific cross.

Source: Modified from Song, Osborn and Williams (1988a; 1988b); Rana et al. (2004).

A slightly different scenario of the polyploidy events in the evolution of the Brassiceae genomes has been put forward by Mun et al. (2009), following a *B. rapa* and *A. thaliana* genome-wide comparative analysis. They suggest that a whole genome duplication (WGD) occurred twice, once about 55-63 Mya and again at 23-30 Mya, between the existence of an ancient ancestral species and the evolution of the ancestral Brassicaceous karyotype. They suggest that the second WGD resulted in the divergence of *Arabidopsis* from the *Brassiceae* linage about 13-17 Mya. This was followed by a whole genome triplication in the Brassiceae about 11-12 Mya with the divergence of *B. rapa* from *B. oleracea* taking place about 8 Mya. Their data also suggest that the allopolyploidisation that resulted in the species *B. napus* occurred only 0.7-1 Mya.

Genome mapping of *B. rapa* and *B. oleracea* has shown the gross organisation of their genomes to be highly collinear (Lagercrantz and Lydiate, 1996) but their genome size and complexity differ. The genome size of *B. rapa* is *ca.* 500 Mb compared to the much larger and more complex genome of *B. oleracea* at *ca.* 600 Mb (Arumuganathan and Earle, 1991). Comparative studies have shown that within the amphidiploids species, *B. napus*, *B. juncea* and *B. carinata*, the chromosomes within the respective putative diploid genomes have remained more or less intact (Parkin et al., 1995; Sharpe et al., 1995; Axelsson et al., 2000). DNA sequence data indicate that the A genome of *B. rapa* and the C genome of *B. oleracea* are very closely related while *B. nigra*, with its B genome, is from an earlier divergent lineage (Mizushima, 1972; Song, Osborn and Williams, 1988b; Prakash and Chopra, 1991). Song et al. (1995) reported there was rapid genome change after polyploidisation in *B. napus* and *B. juncea*, which suggests that the micro-structural changes observed in the *Brassica* lineage happened shortly after genome duplication, followed by a slow but ongoing rate of change (Rana et al., 2004).

The techniques of fluorescence *in situ* hybridisation (FISH) facilitates the integration of genetic and physical chromosome maps as it allows chromosomal location of labelled DNA probes to be directly determined (Snowdon et al., 2007). Since molecular markers can now be ordered and physical distances measured, it is possible to construct molecular karyotypes and distinguish individual chromosomes of the A and B genomes that make up *B. napus* (Fukui et al., 1998; Armstrong et al., 1998; Snowdon et al., 2002). Snowdon, Lühs and Friedt (2007) provides a consensus genetic linkage map of molecular markers for *B. napus* where linkage groups (LGs) N1-N10 correspond to the *B. rapa* A genome LGs of A1-A10, and LGs N11-N19 correspond to *B. oleracea* C genome LGs of C1-C9.

Nuclear genome size

The genome size of the *Brassica* diploids (approximately 500-700 Mbp) are more than four times that of the related Brassicaceous species *A. thaliana* (approximately 157 Mbp; see Table 3.12). The gene content of *A. thaliana* is believed to be very similar to *Brassica* diploids with more than 87% sequence identity in the coding regions (Parkin et al., 2005). Although it is believed that the diploid *Brassica* evolved through a common hexaploid ancestor (Parkin et al., 2005), the necessary genome triplication would be insufficient to explain the differences in genome size. Therefore, this important difference in genome size is likely to reflect a different rate of non-coding DNA accumulation.

Possible extent of repetitive or non-coding DNA sequences

Transposable elements (TEs) constitute a major fraction of non-coding DNA in plant species. Good estimates of TE distribution and density are presently only available for the *B. oleracea* genome, based on a partial draft genome sequence (Zhang and Wessler, 2004). Class 1 (retro) elements were the most abundant TE class with long terminal repeat (LTR) and non-LTR elements comprising the largest fraction of the genome. However, several families of class 2 (DNA) elements have amplified to very high copy numbers in *B. oleracea* compared to *A. thaliana* and have contributed significantly to genome expansion. Approximately 20% of the *B. oleracea* genome was estimated to be composed of class 1 and class 2 TEs.

Table 3.12. **Ploidy level, chromosome number, genome size and map length of *A. thaliana* and *Brassica* species of "Triangle of U 1"**

Species	Ploidy level	Chromosome number	1C nuclear DNA[2] content (Mb)[3]	Observed map length (cM)[4]
A. thaliana	2	10	157	437 and 501
B. nigra	2	16	634-765	855
B. oleracea	2	18	696-765	820-1 738
B. rapa	2	20	528-784	1 455
B. carinata	4	34	1 280-1 548	–
B. juncea	4	36	1 070-1 500	2 073
B. napus	4	38	1 127	1 441-1 765

Notes: 1. The Triangle of U is a theory about the evolution and relationships between members of the plant genus *Brassica* (*Source:* Wikipedia, the Free Encyclopedia). 2. The C value refers to the haploid DNA content of the species. 3. Data adopted by Lysak and Lexer (2006) compiled from Bennett and Leitch (2004) and Johnston et al. (2005). 1 pg = 980 Mb. 4. From Lysak and Lexer (2006), choice based on map marker coverage.

Lim et al. (2005) describe the morphology and molecular organisation of heterochromatin domains in the interphase nuclei and mitotic and meiotic chromosomes of the ten chromosomes of *B. rapa,* using DAPI staining and FISH of rDNA and pericentromere tandem repeats. They characterised the centromeric repeat sequences, which fell into two classes, CentBr1 and CentBr2, occupying the centromeres of eight and two chromosomes, respectively. The centromere satellites encompassed about 30% of the total chromosomes, particularly in the core centromere blocks of all the chromosomes. Interestingly, centromere length was inversely correlated with chromosome length.

Main genetic diversity or variability

Considerable genetic diversity has been found within the six cultivated *Brassica* species using nuclear restriction fragment length polymorphisms (RFLP) markers (Song, Osborn and Williams, 1988b). These results suggested that: 1) *B. rapa* and *B. oleracea* have multiple centres of origin; 2) *B. nigra* originated from one evolutionary pathway whereas *B. rapa* and *B. oleracea* came from another pathway; and 3) amphidiploid *B. napus* and *B. juncea* arose from different combinations of diploid morphotypes, indicating polyphyletic origins may be a common mechanism for the natural occurrence of amphidiploids in *Brassica*.

The genetic diversity within *B. napus* is considerably less than that found within either of the diploid ancestral species. This is probably a result of *B. napus* being a relatively modern species, fixed as a product of human civilisation and with no truly wild populations. Most of the diversity within *B. napus* has been introduced from its diploid progenitors. Variation in the A genome has been increased by natural *B. napus* × *B. rapa* crosses whereas variation in the C genome is more limited. Recent molecular marker analysis has identified more extreme genetic variation in exotic vegetable and fodder genotypes as well as newly resynthesised *B. napus* lines (Snowdon and Friedt, 2004 for a review). In *B. juncea* the A genome is mostly conserved and the C genome is significantly changed, more so than the considerably altered C genome in *B. carinata*. Similar genetic information, with much duplication, is contained in all three genomes (Slocum, 1989; Slocum et al., 1990; Chyi, Hoenecke and Sernyk, 1992; Jackson et al., 2000; Parkin, Sharpe and Lydiate, 2003). However, the chromosomal organisation and the genetic distribution within the genome is different (Truco et al., 1996). New high throughput and very informative simple sequence repeat (SSR) and single nucleotide polymorphism (SNP) molecular markers are now being used routinely to expedite the introduction of novel genetic variation in *Brassica* breeding programmes.

Maternal and/or paternal inheritance of organelle genomes

Analysis of the chloroplast DNA of the cultivated diploid *Brassica* species, and their close relatives, divided the subtribe Brassicinae into two ancient evolutionary lineages (Warwick and Black, 1997), the "Nigra" lineage, which contained the diploid *B. nigra* and the related wild mustard *Sinapis arvensis*, and the "Rapa/Oleracea" lineage, which contained the diploid progenitors of *B. napus* (Figure 3.31). There has been little work studying the origins of the cultivated amphidiploids *B. carinata* or *B. juncea*. However, studies of organellar and nuclear DNA of *B. napus* and related species suggested that a species closely related to *B. montana* gave rise to the cytoplasm of both *B. rapa* and *B. oleracea* (Song and Osborne, 1992). The same study and an earlier study on chloroplast evolution in amphidiploid *Brassica* species (Palmer et al., 1983) suggested that oilseed rape (*B. napus*) evolved from multiple hybridisations between *B. oleracea* and the closely related n=9 species, *B. montana* and *B. rapa*. Some of these lineages may have been subject to introgression from post-hybridisation with their diploid progenitor.

Self- incompatibility, "S" alleles

Self-incompatibility (SI) occurs in many flowering plants and is one of the most important systems to prevent inbreeding (Takayama and Isogai, 2005). SI is defined as the inability of plants to produce functional gametes to effect fertilisation upon self-pollination or when crossed with certain relatives (De Nettancourt, 1971). Although the amphidiploid *Brassica* species, *B. napus*, *B. juncea* and *B. carinata* are largely self-pollinating (autogamous), the diploid species, with some exceptions, are self-incompatible and are obligatory out crossers. Among *Brassica* species and their close relatives, 50 out of 57 species are self-incompatible (Hinata, Isogai and Isuzugawa, 1994). The self-/non-self recognition in most species is controlled by a single locus, termed the "S locus" that inhibits the self pollen from penetrating the style when the same S-allele specificity is expressed by both the pollen and pistil. In the *Brassica* incompatibility system, over 30 *B. rapa* alleles and 50 *B. oleracea* alleles have been identified: S_1, S_2, S_3....S_{50+} (Nou et al., 1993; Ockendon, 2000). Self-compatible (S_f) alleles are also known.

Among angiosperms there are two major types of physiological SI systems: gametophytic (GSI) and sporophytic (SSI) (Briggs and Knowles, 1967). In a GSI system, the pollen reaction is controlled by the genotype of the individual pollen grain, i.e. a plant heterozygous at the S-locus would produce two possible types of pollen with each microspore receiving one of the two possible S-alleles. However, in the SSI system that is present in the Brassicinae, all pollen released by a plant has the same phenotype with respect to the compatibility reaction, regardless of the genotype of the individual pollen grain. The S-locus consists of at least three tightly linked transcriptional units arranged in pairs, with one functioning as the female determinant and the other the male. This multi-gene complex at the S-locus is inherited as one segregating unit so the gene complexes are called "*S*-haplotypes". Self-/non-self recognition operates at the level of protein-protein interaction of the two determinants (Takayama and Isogai, 2005). When the SI system is activated in a Brassicinae species, a recognition reaction occurs between the papilla cells of the stigma and the pollen (Hinata and Nishio, 1980).

There are three highly polymorphic genes involved in the SI response. The two female determinants consist of the S-locus glycoproteins (SLGs) and the S-locus receptor kinase (SRK). SRK consists of an SLG-like extracellular domain, a transmembrane domain and an intercellular serine/threonine domain. SLG and SCR expression occurs just before the flower opens, primarily in the stigma papilla cells. They also exhibit allelic sequence diversity (Takayama and Isogai, 2005). The male determinant genes named *SP11* (*S-locus protein 11*) or *SCR* (*S-locus cysteine rich*) code for the secretion of small, cystine-rich proteins, *SP11/SCR*, in anther tapetum cells and gametophytically in the microspores (Takayama et al., 2000). These genes are tightly linked and behave as a single Mendelian locus, displaying multiple allelic versions (Takayama and Isogai, 2003). The SI-response occurs when stigma and pollen share at least one allele. Upon pollination, SP11, carried in the pollen coat, penetrates the papilla cell wall and binds with SRK. The binding induces autophosphorylation of SRK starting a signalling cascade that causes the rejection of self-pollen by preventing hydration and further development of the pollen tube (Takayama and Isogai, 2005). SLGs are not present or active in all members of the mustard family (Kusaba et al., 2001). If there is a compatible reaction, the papilla cells provide moisture for pollen germination; however, with self-pollination, the absorption of water and germination are disrupted (Dickinson, 1995) and a callus deposition may occur at the attachment site (see Hinata, Isogai and Isuzugawa, 1994 for a review). If the incompatible pollen is able to germinate, the pollen tube growth is slowed or inhibited due to the inability of the pollen tube to grow through the papilla cell wall.

For vegetable crops, the National Vegetable Research Station at Wellesbourne, England, maintains a collection of all known S alleles together with their internationally accepted nomenclature (Dickson and Wallace, 1986). The genotypes of most self-incompatible *Brassica* plants will be heterozygous at the S locus, since cross-fertilisation is mandated by the self-incompatibility specificities of the S alleles present. Dominant and recessive interactions occur between *S*-haplotypes (Thompson and Taylor, 1966). The interaction is complex with the *S*-haplotypes classified as class I or class II, based on the nucleotide sequences of *SGL* and *SRK* alleles (Nasrallah, Nishio and Nasrallah, 1991). The class I *S*-haplotypes are normally dominant over class II *S*-haplotypes in the pollen. The S allele specificities of the pollen and the stigma can be co-dominant, which occurs more frequently than the dominance/recessive. Dominance/recessive relationships occur more frequently in the pollen than the stigma and are not identical for *S* alleles between the stigma and pollen (Watanabe and Hinata, 1999). Among the *SP11/*SCR alleles in Class I S-haplotypes, the dominance relationship is non-linear whereas Class II *S*-haplotypes exhibit linear dominance (Takayama and Isogai, 2003; Hatakeyama et al., 1998). The molecular mechanism of the dominance relationship in the stigma is an active area of investigation and is not fully understood (Takayama and Isogai, 2003; Fujimoto et al., 2006). Selfed seed of most incompatible plants can be obtained through bud pollination i.e. applying pollen to the stigma one to four days before the flower opens since the SGLs and SRK are not expressed until just prior to the flower opening (Takayama and Isogai, 2005). Various other methods have been utilised to overcome the SI system including stigma mutilation, stigma treatment with various organic acids, solvents, oils and ionic solutions, thermally aided pollination as well as elevated carbon dioxide treatment and momentary high temperature application (Hinata, Isogai and Isuzugawa,1994).

The SI system of *S*-haplotypes has been used by vegetable breeders to capture heterosis by producing top cross, double or three-way F_1 hybrids. However, from the perspective of intra- and interspecific outcrossing in the field, it has been noted that the incidence of interspecific crossing in mixed species populations is likely to increase as the number of plants in the self-incompatible species decreases, due to scarcity of pollen of the same species and increasing pollen competition from other nearby species.

Although nearly all the mono-genomic *Brassica* species are self-incompatible, the natural amphidiploids species – *B. napus*, *B. juncea* and *B. carinata* – are all self-compatible (Takahata and Hinata, 1980). Okamoto et al. (2007) note that interspecific crosses between *B. rapa* and *B. oleracea* are difficult to make and, when the chromosome complement is doubled, produce self-incompatible amphidiploids plants (Beschorner, Plümper and Odenbach, 1995; Nishi, 1968). They suggest that a single mutation in a dominant *S*-haplotype could result in a self-compatible *B. napus* plant that could reproduce itself through the production of self seed. Amphidiploid plants without such a mutation would be forced to cross with one or the other diploid parent and rapidly be assimilated into one or the other parent species. Fujimoto et al. (2006) provide evidence for such mutations in *B. rapa* and *B. oleracea.*

Interspecific hybridisation and introgression

Introduction

With the introduction of genetically modified (GM) *B. napus*, the potential for inserted genes to transfer and introgress into related Brassicaceae species has been the subject of much speculation and research. There are many conditions which have to be met for such an event to occur. First, the cross of interest must occur. However, crossing success depends on a series of preconditions that include physical proximity of the parents, pollen movement and longevity, synchrony of flowering, breeding system of the parents, flower characteristics, pollen-style compatibility and competitiveness of foreign pollen. If all these pre-fertilisation conditions are met, the next series of hurdles include sexual compatibility, embryo-endosperm imbalance as well as hybrid fertility and viability in nature. In addition, the hybrid must have sufficient fitness to backcross with the recipient parent producing fertile progeny through several generations. For example, Wei and Darmency (2008) found crosses between male sterile *B. napus* and *B. juncea, B. nigra, H. incana* and *R. raphanistrum* produced only small seed, resulting in poor seedling establishment of the hybrids under field conditions. Even if all the conditions are met, introgression will not occur unless there is pairing between a chromosome of the recipient parent and a donor parent chromosome segment that carries the inserted gene. Gene transfer cannot occur in nature if any one of these requirements is not met. However, it has been speculated that strong selection pressure over many backcross generations could result in the transgene existing in a stable strain carrying an extra chromosome pair (Chèvre et al., 2001).

Modern researchers have overcome many of the natural barriers to interspecific and intergeneric crosses within the tribe Brassiceae. Techniques such as ovule, ovary and embryo culture, as well as protoplast fusion have produced hybrids that would otherwise fail due to sexual barriers. Success has also been achieved by crossing induced polyploids from one or both parents. Such techniques have been used to try to integrate important agronomic or quality traits from a foreign species into a cultivated crop. However, success using such techniques is no indication that the same result could occur through sexual crossing in nature.

The development of male sterile *B. napus* parental lines, for the production of commercial varieties, has also provided a means to investigate intraspecific, interspecific and intergeneric crossing on a field scale, without pollen competition. The results have shown that where male sterile plants were used, the frequency of interspecific crosses was significantly higher as indicated in the following species cross reports below. Thus, the presence of male sterile *B. napus* plants in commercial fields was seen as increasing the incidence and/or risk of unwanted species hybrids.

Some of the first developed hybrid *B. napus* varieties used a seed-production system termed "synthetic hybrids". Commercial production fields growing such hybrids consisted of about 80-90% male sterile hybrid plants with the remaining fully fertile plants (10-20%) providing the pollen cloud necessary to fertilise the male sterile plants in the rest of the field. Fortunately, this "synthetic hybrid" system has been replaced with new systems that reverse the ratio of fully fertile to male sterile plants in commercial hybrid fields. Today only a small percentage (15-20%) of male sterile plants may occur as off-types in these hybrid varieties. Such plants would be saturated with pollen from the surrounding *B. napus* plants, thus greatly reducing the risk of pollination by a foreign pollen source.

Chèvre et al. (2004) identified 14 species related to *B. napus* to which gene introgression from *B. napus* could be of concern to oilseed rape growing countries in Europe and North America. The reports of interspecific and intergeneric sexual crossing attempts between these species and *B. napus* are summarised in Table 3.13. Each species cross is discussed in the following paragraphs.

Warwick, Francis and Gugel (2009) have compiled a complete list of reports on interspecific and intergeneric hybridisation within the Brassicaceae that includes studies that use sexual as well as special techniques to effect a cross.

Table 3.13. **Interspecific and intergeneric sexual crossing attempts, degree of success and potential for gene introgression**[1]

Interspecific cross	Sexual cross	Field cross	Seeds/ cross	BC ♂	BC ♀	Potential		References
						Natural cross	Introgression	
Brassica napus								
B. napus × *B. carinata*	Y	NR	8		Y	L	L	U (1935); Roy (1980, 1977); Alam et al. (1992); Gupta (1997); Rashid, Rakow and Downey (1994); Fernandez-Escobar et al. (1988); Sacristan and Gerdemann (1986); Navabi et al. (2010)
B. carinata × *B. napus*	F, Y,	NR	<1		Y	L	L	
B. napus × *B. juncea*	Y	Y	4	Y	Y	H	H	Bing, Downey and Rakow (1991); Bing et al. (1996); Alam et al. (1992); Frello et al. (1995); Jørgensen (1999); Jørgensen et al. (1998); GoshDastidar and Varma (1999); Choudhary and Joshi (1999); Kirti et al. (1995); Davey (1959); Sharma and Singh (1992); Heyn (1977); Roy (1984, 1980); Dhillon et al. (1985); Shpota and Podkolzina (1986); Sacristan and Gerdemann (1986); Wei and Darmency (2008)
B. juncea × *B. napus*	Y	Y	0.54	Y	Y	H	H	
B. napus × *B. fruticulosa*	Y	NR	0.008			VL	VL	Heyn (1977); Plümper (1995); Siemens (2002); Salisbury (2002)
B. fruticulosa × *B. napus*	Y	NR	F			VL	EL	
B. napus × *B. maurorum*	Y					EL	EL	Bijral et al. (1995)
B. maurorum × *B. napus*								
B. napus × *B. nigra*	Y	Y	0-0.09	Y	F	L	L	Bing, Downey and Rakow (1991);This et al. (1990); Brown and Brown (1996); Struss, Bellin and Röbbelen (1991); Daniels et al. (2005); Wei and Darmency (2008)
B. nigra × *B. napus*		F	0.01	F	F	VL	L	
B. napus × *B. oleracea*	Y	NR		Y	Y	VL	VL	Gupta (1997); Ford et al. (2006)
B. oleracea × *B. napus*	Y	Y				VL	VL	
B. napus × *B. rapa*	Y	Y	Many	Y	Y	H	H	Bing, Downey and Rakow (1991); Bing et al. (1996); Brown and Brown (1996); Gupta (1997); Jørgensen and Andersen (1994); Landbo and Jørgensen (1997); Mikkelsen, Jensen and Jørgensen (1996); Vijayakumar et al. (1994); Metz et al. (1997); Choudhary and Joshi (1999); Daniels et al. (2005)
B. rapa × *B. napus*	Y	Y	Many	Y	Y	H	H	
B. napus × *B. tournefortii*	Y	NR	0.69			L	L	Nagpal et al. (1996); Gupta (1997); Lokanadha and Saria (1994); Liu, Landgren and Glimelius (1996); Salisbury (2002)
B. tournefortii × *B. napus*	F					VL	VL	
B. napus × *D. catholica*	Y	NR	NR	NR	NR	VL	VL	Bijral and Sharma (1998)
D. catholica × *B. napus*	NR	NR	NR	NR	NR	EL	EL	
B. napus × *D. muralis*	Y	NR	0.28			L	VL	Bijral and Sharma (1996a)
D. muralis × *B. napus*	NR	NR						
B. napus × *Eruca sativa*	Y	NR				L	VL	Bijral and Sharma (1996b)
Eruca sativa × *B. napus*	NR	NR						
B. napus × *Erucastrum gallicum*	Y	F	0.1	Y	Y	VL	VL	Lefol, Seguin-Swartz and Downey (1997); Batra, Shivanna and Prakash (1989); Warwick et al. (2003)
Erucastrum gallicum × *B. napus*	F	F	0			VL	VL	
B. napus × *H. incana*	Y	Y	2	Y	Y	H	L	Chadoeuf, Darmency and Maillet (1998); Lefol, Danielou and Darmency (1996) ; Lefol et al. (1995, 1991); Eber et al. (1994); Chèvre et al. (1996); Kerlan, Chèvre and Eber (1993) ; Kerlan et al. (1992); Wei and Darmency (2008)
H. incana × *B. napus*	Y	Y	2×10^{-5}		Y	H	L	

Table 3.13. **Interspecific and intergeneric sexual crossing attempts, degree of success and potential for gene introgression**[1] (*cont.*)

Interspecific cross	Sexual cross	Field cross	Seeds/ cross	BC ♂	BC ♀	Potential		References
						Natural cross	Introgression	
B. napus × *R. raphanistrum*	Y	Y	Y	Y	Y	H	L	Baranger et al. (1995); Chadoeuf, Darmency and Maillet (1998); Darmency, Lefol and Fleury (1998); Eber et al. (1994); Lefol, Seguin-Swartz and Downey (1997); Rieger et al. (1999); Chèvre et al. (1998, 1997a); Wei and Darmency (2008)
R. raphanistrum × *B. napus*	Y	F	$10^{-4,-8}$	Y	Y	VL	VL	
B. napus × *R. sativus*	Y	NR	0.6	NR	NR	VL	EL	Gupta (1997); Ammitzbøll and Jørgensen (2006)
R. sativus × *B. napus*	NR	F	0	NR	NR	VL	EL	
B. napus × *S. alba*	Y	NR	Y			VL	EL	Bijral, Sharma and Kanwal (1993); Ripley and Arnison (1990); Mathias (1991); Lelivelt et al. (1993); Chèvre et al. (1994); Brown et al. (1997); Sridevi and Saria (1996)
S. alba × *B. napus*	F	NR				EL	EL	
B. napus × *S. arvensis*	Y	F	0.18			L	VL	Bing, Downey and Rakow (1991);Moyes et al. (2002); Inomata (1988); Brown et al. (1996); Sweet et al. (1997); Eastham and Sweet (2002); Daniels et al. (2005); Lefol, Danielou and Darmency (1996)
S. arvensis × *B. napus*	Y	F	F	F		EL	EL	
B. napus × *D. erucoides*	NR	NR		Y		VL	VL	Ringdahl, McVetty and Sernyk. (1987)
D. erucoides × *B. napus*	Y	NR						

Notes: Y = successful cross by hand pollination or in the field; F = cross attempted but failed; NR= not reported. Probability of crossing in nature and/or gene introgression H = high, L = low, VL = very low, EL = extremely low.

B. napus – Raphanus raphanistrum

R. raphanistrum is an economically damaging weed with a worldwide distribution but its range is limited to areas with acid soils. Hand crosses between *B. napus* and *R. raphanistrum* have produced reciprocal hybrids with a higher number of hybrids obtained with *B. napus* as the female (Kerlan et al., 1992: Chèvre et al., 1996). In France, when *R. raphanistrum* served as the female, only three hybrids have been identified, even though tens of thousands of seeds were examined (Eber et al., 1994; Baranger et al., 1995; Chèvre et al., 2000, 1998, 1997b; Darmency, Lefol and Fleury, 1998; Darmency and Fleury, 2000). Chèvre et al. (2000) estimated the hybridisation frequency to be 10^{-7} to 10^{-5} while Australian and Canadian studies reported respective frequencies of 4×10^{-8} (Rieger et al., 2001) and 3×10^{-5} (Warwick et al., 2003).

Guéritaine, Bazot and Darmency (2003) found that under field conditions the F_1 hybrid emergence was lower and slower and seedling survival significantly less than both parents. A six-year UK monitoring programme of natural populations of *R. raphanistrum* growing near fields of HR *B. napus* showed no evidence of intergeneric crossing (Eastham and Sweet, 2002). Similarly in the United Kingdom, Daniels et al. (2005) found no *R. raphanistrum* × *B. napus* plants or progeny when they sampled *R. raphanistrum* plants growing in or near four fields sown to glufosinate resistant *B. napus*. Further, no hybrids were found in a Swiss survey (Thalmann, Guadagnuolo and Felber, 2001). When *R. raphanistrum* was the female, no hybrids were found in any of these studies. The frequency of hybridisation can vary depending on the *B. napus* parental variety and the population source of *R. raphanistrum*. When *B. napus* male sterile plants were used as females, the frequency of hybrids was greatly increased, ranging from <0.2% (Chèvre et al., 2000; 1996) to as high as 90% in Danish and French field trials (Eber et al., 1994; Baranger et al., 1995; Ammitzbøll and Jørgensen, 2006). These findings would be of concern if the use of synthetic hybrids became standard, as the vast majority of plants in commercial oilseed rape fields would be male sterile. However, as indicated earlier, this hybrid system has now been phased out.

In the *B. napus* by *R. raphanistrum* cross, the majority of the F_1 hybrids had half the chromosomes of each species (ACRr, 2n=28) while one hybrid had all the chromosomes of *R. raphanistrum* and half the *B. napus* chromosomes (RrRrAC, 2n=37) (Chèvre et al., 2000). Thus, the fertility of the hybrids is very low (Baranger et al., 1995; Chèvre et al., 1998, 1996; Darmency, Lefol and Fleury, 1998; Pinder et al., 1999; Thalmann, Guadagnuolo and Felber, 2001; Warwick et al., 2003). However, Rieger et al. (2001) reported two fertile amphidiploids hybrids with a genome complement of AACCRrRr, 2n=56. Chèvre et al. (2000) also reported four fertile amphidiploids but questioned their genetic stability due to the presence of univalents and multi/quadrivalents at meiosis. The fitness of F_1 hybrids produced on *B. napus* male sterile plants was assessed in the field by Guéritaine, Bazot and Darmency (2003). They found that the hybrids were slower to emerge and less likely to survive than either parent, particularly when subjected to crop competition. The hybrids also flowered later than either parent, which limited the opportunities for backcrossing to *R. raphanistrum*. It should also be noted that if crossing between these species were to occur, it would most likely take place in a field of oilseed rape. Thus, most of the crossed seed would be harvested and only a very small proportion of the original hybrid seed would remain (Rieger et al., 2001). The few surviving hybrids would germinate among *B. napus* volunteers with backcrosses to *B. napus* much more likely than with wild radish.

When *B. napus* herbicide resistant (HR) hybrids were surrounded by *R. raphanistrum* plants in the field, the seed set was less than one seed per hybrid plant (Darmency, Fleury and Lefol, 1995). Despite the low fertility and poor fitness of the hybrids, the fertility and fitness of the backcross progeny improved with each backcross generation but the percentage of HR plants decreased (Chèvre et al., 1999, 1998, 1997b; Darmency, Lefol and Fleury, 1998; Benabdelmouna et al., 2003; Guéritaine, Bazot and Darmency, 2003). In each generation the progenies were selected for herbicide tolerance and only HR plants advanced to the next backcross (BC). None of the HR plants in the BC_3 to BC_5 had the chromosome number of *R. raphanistrum* (2n=18) indicating that no genomic introgression had occurred (Chèvre et al., 1998; Guéritaine et al., 2002). Backcrossing to *R. raphanistrum* was continued up to BC_7 followed by random mating and selection pressure in generations (G) G8 through G11 (Al Mouemar and Darmency, 2004). Root tip cytology of HR G9 plants established that all 32 plants were either carrying extra chromosomes or, as indicated by the non-Mendelian segregation of the progeny, did not have the HR gene stably introgressed into the *R. raphanistrum* genome. The authors concluded that "the prospect of stable introgression of herbicide tolerance to wild radish in nature seems remote".

B. napus – B. rapa

B. rapa, a widespread weed of cultivated and disturbed lands, is also grown as a vegetable and oilseed crop. The weedy type differs from the cultivated oilseed form only in the primary seed dormancy trait. Plant breeders of *B. rapa* and *B. napus* have known for many years that these two species readily cross in nature and they were not surprised that natural interspecies gene flow was demonstrated in several countries, including Denmark (Landbo, Andersen and Jørgensen, 1996; Hansen et al., 2001), Canada (Warwick et al., 2003; Beckie et al., 2003; Yoshimura et al., 2006), the United Kingdom (Daniels et al., 2005; Allainguillaume et al., 2006), the United States (Halfhill et al., 2002) and the Czech Republic (Bielikova and Rakousky, 2001).

Normally the highest hybrid frequencies occur when individual, self-incompatible plants of *B. rapa* are present in *B. napus* fields (Jørgensen et al., 1996). In the field, more hybrids are produced on *B. rapa* plants than on *B. napus* plants (Jørgensen and Andersen, 1994; Hauser, Jorgensen and Ostergard, 1997; Jørgensen et al., 1998), primarily due to their respective self-incompatible and self-compatible breeding systems. However, in reciprocal hand crosses, more hybrids per cross are found when *B. napus* is the female (Downey, Klaasen and Stringham, 1980). Natural interspecific hybridisation between *B. rapa* and *B. napus* varies widely, depending on the environment under which the plants develop and the design of the experiment, particularly the ratio of *B. rapa* to *B. napus* plants. In Danish trials, up to 95% hybrids were found in *B. rapa* progeny (Mikkelsen, Jensen and Jørgensen, 1996), while in New Zealand Palmer (1962) reported a range of 10-88%. In contrast, others in Canada (Bing, Downey and Rakow, 1991) and England (Wilkinson et al., 2000) found less than 1% hybridisation. In Canadian field experiments (two in the east and one in the west), *B. rapa* plants were grown at various positions within and alongside HR *B. napus* plots. Approximately 7% of the harvested *B. rapa* seed was found to be triploid hybrids (AAC, 2n=29) (Warwick et al., 2003). Similarly, in commercial *B. napus* fields containing sparse populations of weedy *B. rapa*, the hybrid frequency was approximately 13.6%. However, the frequency of hybrids from weedy *B. rapa* growing in a harvested corn field with HR *B. napus* volunteers was only 0.023% (Warwick et al., 2003). In New Zealand field studies with ratios of *B. rapa* to *B. napus* plants of 1:400 and 1:1, the hybrid frequencies ranged from 2.1% to 0.06% with the total for the experiment of 0.46% (Jenkins, Conner and Frampton, 2001). A study of *B. rapa* populations growing outside *B. napus* fields in the United Kingdom found few hybrids (0.4-4.5%) in 7% of the populations, and no hybrids in the remaining 93% (Scott and Wilkinson, 1998).

Hybridisation also occurs with *B. napus* as the female; however, most of the hybrid seed that is formed will be removed from the field at harvest. Any hybrids that volunteer the following year are almost certain to be surrounded by *B. napus* volunteers. Thus, any backcrosses will quickly revert to *B. napus* form and chromosome number.

Compared to the parent species, natural interspecific hybrids have reduced fertility and poor seed set, averaging two to five seeds per pod (Jørgensen and Andersen, 1994). The survival rate of hybrid seedlings is also low, with <2% survival (Scott and Wilkinson, 1998), reducing the rate of introgression (Jørgensen et al., 1996; Sweet et al., 1999b). Interspecific vegetative and reproductive competition strongly impacts the relative and absolute fitness of the hybrids (Hauser et al., 2001). When Mikkelsen, Jensen and Jørgensen (1996) sowed interspecific hybrids within a *B. napus* population, no *B. rapa* × hybrid BC progeny were found among 2 000 offspring raised from 30 *B. rapa* plants. Further, the hybrids lacked primary seed dormancy (Linder, 1998). This may explain why Landbo and Jørgensen (1997) found interspecific hybrids in feral *B. rapa* populations, but no hybrid seed in the seed banks at those sites. Introgression of HR transgenes from *B. napus* to *B. rapa* has occurred in Europe (Jørgensen, 1999; Hansen et al., 2001; Norris and Sweet, 2002). However, no evidence of introgression was found in seed samples taken from *B. rapa* plants in the field, indicating there may be selection pressure against backcross individuals (Norris and Sweet, 2002).

The rate of introgression of a *B. napus* trait into the *B. rapa* genome will greatly depend on the selection pressure exerted on the gene (Scott and Wilkinson, 1998; Sweet et al., 1999a; Snow and Jørgensen, 1999). The introgression of a gene into the *B. rapa* genome might be slowed by positioning it in the C genome of *B. napus* but the findings of Stewart, Halfhill and Warwick (2002), where 12 independent *B. napus* transformations distributed across both the A and C genomes all generated backcrosses at similar rates, suggests this theory may not be valid. Leflon et al. (2006) found that the transmission rate of the C chromosomes depended on which C chromosome was involved, and that a gene carried on a C chromosome is less likely to be transferred in a *B. rapa* background than if it was on an A chromosome. The presence of an introgressed HR gene in *B. rapa* did not increase its fitness or weediness relative to conventional non-GM *B. rapa* including glufosinate resistant BC_3 hybrids (Snow, Andersen and Jørgensen, 1999) or BC_2F_2 glyphosate hybrids (Warwick, 2007). It should be kept in mind that if introgression of an R gene does occur, the resulting HR *B. rapa* plant(s) can be controlled with other herbicides or cultivation. In Canada, with 16 years of experience growing millions of hectares of HR *B. napus* each year, no significant agronomic problems with HR *B. rapa* have been encountered (Beckie et al., 2006).

B. napus – Hirschfeldia incana

H. incana is an important weed in some European countries and eastern Australia, but not in Canada or the Indian sub-continent. Hand crosses between *B. napus* and *H. incana* produced 1.3 and 3.1 hybrids per 100 pollinations when *H. incana* and *B. napus*, respectively, were used as the female (Kerlan et al., 1992). In the field, when male sterile *B. napus* was used as the female, 1.9 hybrids were recorded per pollinated flower (Eber et al., 1994). However, in three years of field trials, isolated *H. incana* plants growing in *B. napus* plots only produced 0.6 hybrid seeds per plant. Most F_1 plants had reduced fitness with seedling emergence over three years being <1% (Chadoeuf, Darmency and Maillet, 1998). However, some hybrids were at least as competitive as the wild parent (Eber et al., 1994; Lefol, Fleury and Darmency, 1996).

When F_1 plants were backcrossed to *H. incana* and only HR progeny were selected for further backcrossing, fewer seeds were produced in each generation. BC_3 produced only one seed with no viable seeds obtained in BC_4 (Darmency and Fleury, 2000). It is suggested that a *H. incana* gene inhibits homeologous pairing, resulting in an expulsion of *B. napus* chromosomes (Kerlan et al., 1992; Lefol, Fleury and Darmency, 1996). Thus, although interspecific F_1 hybrids will frequently occur in areas where *H. incana* is prevalent, their persistence will be short and the possibility of gene introgression from *B. napus* remote.

B. napus – B. juncea

B. juncea is primarily a crop plant grown in China, the Russian Federation and on the Indian sub-continent as a major source of edible oil, and in Canada and a few other countries as a condiment crop. However, it is present as a weed in parts of Europe and Australia. Since *B. juncea* (AABB) and *B. napus* (AACC) have a common genome, the chance of interspecific crossing is enhanced. In Canadian co-cultivation experiments, Bing et al. (1996) identified five interspecific hybrids in seed harvested from 469 *B. napus* plants and 3 out of 990 plants when *B. juncea* was the female. Jørgensen et al. (1998) noted that as the ratio of *B. juncea* to *B. napus* plants increased from 1:3 to 1:15, the hybridization frequency on *B. juncea* plants decreased from 2.3% to 0.3%. Warwick (2007) reported gene flow from HR *B. napus* to neighbouring fields of *B. juncea* at a rate of 0.245% at the adjacent *B. juncea* field border and 0.030%, 0.021% and 0.005% at 50 m, 100 m, and 200 m, respectively.

The viability of F_1 pollen is reported to be low (18-26%) (Frello et al., 1995; Choudhary and Joshi, 1999; GoshDastidar and Varma, 1999), but spontaneous backcrossing with improved fertility has been reported (Alam et al., 1992; Bing, Downey and Rakow, 1991; Bing et al., 1996; Jørgensen, 1999). Given this background of results, the introgression of genes from *B. napus* could be expected to occur where these two species are widely grown.

B. napus – Sinapis arvensis

S. arvensis is a serious weed in all oilseed rape growing countries. In a five-year study of *S. arvensis* growing in and around GM *B. napus* crops in the United Kingdom, Sweet et al. (1997) and Norris et al. (unpublished, cited in Eastham and Sweet, 2002) failed to detect any hybridisation with *S. arvensis*. Also in the United Kingdom, Daniels et al. (2005) tested 60 768 progeny from 818 *S. arvensis* plants, growing in or close to 23 glufosinate resistant *B. napus* fields. No resistant plants were found in the parents or their progeny. Similarly, Warwick et al. (2003) found no interspecific hybrids among 43 828 *S. arvensis* progeny from plants growing in HR *B. napus* fields in western Canada. Bing et al. (1996) also found no hybrids in Canadian co-cultivation experiments involving the assessment of 7 500 *S. arvensis* seeds. Similar results were reported from UK trials where 9 688 *S. arvensis* seedlings were screened (Moyes et al., 2002) and in France, Lefol, Danielou and Darmency (1996) found no hybrids among the 2.9 million *S. arvensis* seeds tested. However, when male sterile or emasculated *B. napus* plants were pollinated with *S. arvensis* pollen, either naturally or artificially, a small number of hybrids were obtained. Chèvre et al. (1996) found 0.18 hybrids per 100 pollinations while

Lefol, Fleury and Darmency (1996) detected 6 hybrids in 50 000 seeds analysed. In hand crosses using *S. arvensis* females from different UK and French populations, Moyes et al. (2002) detected one completely sterile hybrid. No such hybrid had previously been reported without embryo rescue or ovule culture (Inomata, 1988; Kerlan et al., 1992; Bing et al., 1996, 1991; Chèvre et al., 1996; Lefol, Fleury and Darmency, 1996). All hybrids produced were weak, largely or completely sterile, and unlikely to survive in nature (Moyes et al., 2002). None of the hybrids were able to backcross to *S. arvensis*.

Daniels et al. (2005) identified a single plant in the United Kingdom that they believed to be a *S. arvensis* × *B. napus* hybrid. It was growing in a patch of *S. arvensis* plants adjacent to a field that had grown a crop of glufosinate resistant *B. napus* the previous year. The hybrid classification was based on a null reaction to the application of glufosinate to a single leaf followed by a positive DNA test for the glufosinate resistance gene. However, only morphological characteristics were used to classify the plant as a *S. arvensis* × *B. napus* hybrid. The lack of any information on chromosome number and/or markers, and in the light of previous studies, the question remains as to whether the plant was indeed a *S. arvensis* × *B. napus* hybrid rather than another interspecific cross such as *B. rapa* × *B. napus*. In the words of the report's reviewer "such a finding needs to be interpreted with caution."

Despite the one hybrid produced by Moyes et al. (2002) on an emasculated *S. arvensis* plant, there is general agreement among researchers that the possibility of gene flow between *B. napus* and *S. arvensis* is extremely low (Moyes et al., 2002) to non-existent (Downey, 1999a; 1999b).

B. napus – Raphanus sativus

R. sativus is a vegetable crop in many parts of the world, but when grown for seed it can escape from cultivation and colonise disturbed sites such as roadsides, fields and coastal sand dunes (Snow, Uthus and Culley, 2001). Daniels et al. (2005) reported flowering of *R. sativus* plants could coincide with either winter or spring *B. napus*. In *R. sativus* plants growing in or near a field of glufosinate resistant *B. napus* in the United Kingdom, Daniels et al. (2005) found no *R. sativus* × *B. napus* hybrids. Further, progeny from the sampled *R. sativus* plants were all susceptible to glufosinate. Hybrids between *B. napus* and *R. sativus* have been obtained in several studies with the aid of ovule culture or embryo rescue (Lelivelt et al., 1993; Paulmann and Röbbelen, 1988; Sundberg and Glimelius, 1991; Metz, Nap and Stiekema, 1995; Takeshita, Kato and Tokumasu, 1980) and also by hand pollination (Gupta, 1997). All artificially produced hybrids were male sterile. However, in natural crosses Ammitzbøll and Jørgensen (2006) obtained an average of 0.6 seeds per pod when male sterile *B. napus* plants were used as the female and a radish cultivar as the pollen parent. Huang et al. (2002) in hand crosses also produced many hybrids on Ogura male sterile plants. All seeds produced proved to be F_1 triploid hybrids with low pollen fertility (0-15%). It is highly probable that the presence of radish cytoplasm in the male sterile *B. napus* parent greatly facilitated *R. sativa* pollen penetration of the stigma. Further studies need to be carried out with this cross since *R. sativa* crosses easily with *R. raphanistrum* (Snow, Uthus and Culley, 2001).

B. napus – Erucastrum gallicum

E. gallium is a self-compatible, annual or winter annual with very small seeds. It is a minor weed of cultivated fields and waste places in many oilseed rape growing countries. Batra, Shivanna and Prakash (1989) obtained three hybrids from the cross *E. gallicum* ×

B. napus using embryo rescue. Lefol, Seguin-Swartz and Downey (1997), using reciprocal hand crosses, obtained one slow-growing *B. napus* × *E. gallicum* F_1 hybrid with pollen viability of 28%. Indications were that the F_1 would not survive in competition with a *B. napus* crop. No seed was produced when *E. gallicum* served as the female parent. The F_1 hybrid was backcrossed in all combinations and many seeds were obtained when *E. gallicum* was the male and a few when *B. napus* was the female. Backcross seed from the hybrid produced plants identical to *E. gallicum*, suggesting that the *B. napus* chromosomes were lost. A survey of 22 000 seedlings of *E. gallicum* from western Canadian *B. napus* fields yielded no hybrids, indicating that the possibility of hybridisation between *B. napus* and *E. gallicum* is very low ($<5 \times 10^{-5}$) (Warwick et al., 2003).

B. napus – B. nigra

B. nigra is a minor weed and an occasional crop in warmer, shorter day-length locations of oilseed rape growing regions. Interspecific hand crosses between *B. napus* and *B. nigra* have been difficult to obtain, with some success using oilseed rape as the female (Davey, 1959; Heyn, 1977; Diederichsen and Sacristan, 1988; Nishiyama, Sarashima and Matsuzawa, 1991; Bing, Downey and Rakow, 1991; Bing et al., 1996; Kerlan et al., 1992; Struss, Quiros and Röbbelen, 1992; Zhu, Struss and Röbbelen, 1993). The F_1 hybrids were moderately to highly sterile but a few F_2 and BC seeds were obtained (Bing, Downey and Rakow, 1991; Zhu, Struss and Röbbelen, 1993). Using controlled crosses hybridization levels were extremely low (Raybould and Gray, 1993; Scheffler and Dale, 1994). In the cross *B. napus* × *B. nigra*, Brown and Brown (1996) observed the pollen tubes of *B nigra* were short and twisted with only a few penetrating the style. No hybrids were found in natural crosses when *B. nigra* was the female (Bing, Downey and Rakow, 1991; Leckie, Smithson and Crute, 1993; Daniels et al., 2005).

B. napus – B. oleracea and *Brassica* vegetables

Gene flow from oilseed rape to *B. napus* vegetables (Swedes, rutabaga, Siberian kale) is possible since they are all within the same species. Similarly, gene flow to *B. rapa* vegetables (e.g. turnip, Chinese cabbage, etc.) is possible since they have the A genome in common. However, *B. napus* and *B. rapa* vegetables are not considered weedy. In addition, they are generally harvested prior to flowering.

Hand crosses between *B. napus* and *B. oleracea* have been successful but at a very low frequency (Chiang, Chiang and Grant, 1977) and natural crosses have only been successful with the assistance of embryo rescue (Ayotte, Harney and Souza Machado, 1987; Takeshita, Kato and Tokumasu, 1980; Quazi, 1988; Habman et al., 2010). However, amphidiploid F_1 hybrids were fertile and readily backcrossed to either parent (Sundberg and Glimelius, 1991; Kerlan et al., 1992; Chèvre et al., 1996).

No spontaneous hybrids between *B. napus* and *B. oleracea* were found in two UK surveys of wild *B. oleracea* populations (Scheffler and Dale, 1994; Wilkinson et al., 2000). However, a later UK survey of two wild *B. oleracea* populations, growing within 25 m of *B. napus* fields, identified one triploid F_1 hybrid and nine introgressants based on flow cytometry and crop-specific microsatellite markers (Ford et al., 2006). The fertility of these plants has not been reported.

B. napus – Sinapis alba

S. alba is commercially grown as a condiment crop but weedy forms occur in the Mediterranean region and in some countries where *S. alba* is used as a green manure crop. The cross *B. napus* × *S. alba* is difficult to make even with hand pollination, usually requiring embryo or ovule culture (Ripley and Arnison, 1990; Mathias, 1991; Bijral, Sharma and Kanwal, 1993; Lelivelt et al., 1993; Chèvre et al., 1994; Brown et al., 1997; Sridevi and Saria, 1996). No field crosses have been reported (Daniels et al., 2005) and the possibility of such an occurrence is very low.

B. napus – Other weedy species

Hand crosses have been made in enclosed environments between *B. napus* and a number of weedy species within the tribe Brassiceae (e.g. *B. fruticulosa*, *B. tournefortii*, *B. maurorum*, *Diplotaxis muralis*, *D. tenuifolia*, *Rapistrum rugosum*, *Eruca sativa*) while protoplast fusion and embryo or ovule rescue have produced F_1 plants in *B. napus* crosses with *B. oxyrrhina*, *B. barrelieri*, *B. elongata*, *B. gravinae*, *B. souliei* and *Diplotaxis tenuisiliqua*. No field interspecific or intergeneric hybrids have been reported between *B. napus* and the above species (Salisbury, 2002).

Ecology

Interactions in natural and agricultural ecosystems

Glucosinolates and their ecological interaction

Virtually all plants of the Brassicaceae produce sulphur compounds called glucosinolates (Kjaer, 1960). Although there are some 250 of these allelochemicals that occur in 16 botanical families of the order Brassicales (Verkerk et al., 2009), only about 20 are commonly found in *Brassica* species (Sarwar and Kirkegaard, 1998). A single species will usually contain significant amounts of 4 different glucosinolates but a single plant may contain as many as 15 different glucosinolates. They are present in varying amounts in all tissues of the plant and directly or indirectly impact their biological environment (Brown and Morra, 1997). They are the source of the flavour and odour of the *Brassica* vegetables and the hot component in mustards. The kind and quantity of glucosinolate varies within and among species and even between stages of plant development as well as between plant parts e.g. cotyledon, leaf, root, flower buds and seed. The highest concentration of glucosinolates is normally found in flower buds and seeds.

All glucosinolates have the same basic structure consisting of a β–D-thioglucose group, a sulphonated oxime group and a side chain "R", derived from one of the amino acids, methionine, phenylalamine, tryptophane or a branched-chain amino acid (Figure 3.40). Glucosinolates accumulate in plant cell vacuoles. They can be broken down (hydrolyzed) by the enzyme myrosinase which is located separately in the idoblast cells. When plant cells are crushed or broken, and moisture is present, the glucosinolates and myrosinase are released and the enzyme catalyses the hydrolysis of the glucosinolates into glucose, sulphate and thiocyanates, isothiocynates and nitriles plus sulphur (Figure 3.40). The intact glucosinolates have little biological activity but their thiocyanate and isothiocynate breakdown products have broad biocidal activity (Brown and Morra, 1997).

The glucosinolates serve as an advance-prepared system of protection that is activated only when plant tissue is damaged by a disease or insect attack. The destruction of the plant cells results in the hydrolysis of the glucosinolates by the myrosinase enzyme, thus releasing the volatile isothiocyanates that have a wide spectrum of anti-microbial effects and act as attractants or repellents to some insects and herbivores (Vašák, 2002; Brown and Morra, 1997; Fenwick et al., 1983).

Brassica crops are also used as biofumigants based on the release of the bioactive isothiocyanates in the soil when seed meal amendments or green manure are incorporated, or *Brassica* crops are used in the rotation (Brown and Morra, 1997). It is also suspected that the volatile isothiocyanates, from residue of *Brassica* crops, result in inhibitory effects on some subsequent crops (see the section on "Allelopathy").

The glucosinolates also impact on the health and nutrition of animals and humans as well as the quality and usefulness of products from *Brassica* crops. These aspects are discussed in the following section.

Figure 3.40. **Glucosinolate chemical structure and enzymatic breakdown products formed in broken *Brassica* plant cells with moisture**

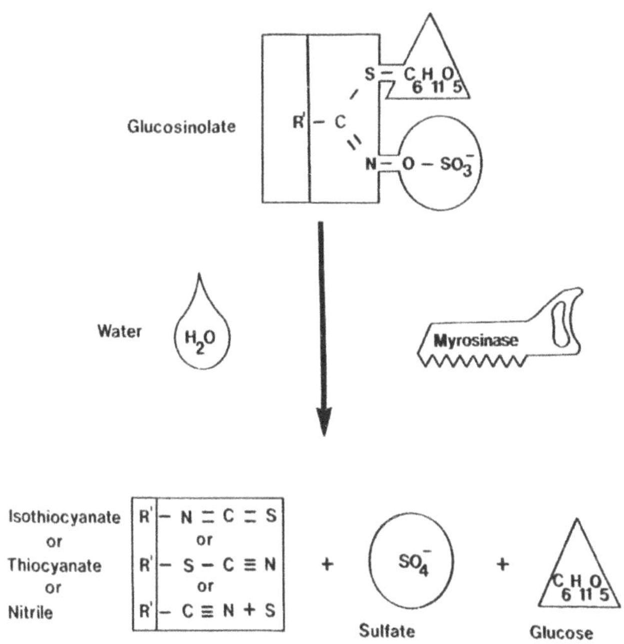

Damaging insects

Brassica species are important components of temperate climate ecosystems. They provide forage for many insects as well as wild life. The complex of insects that feed upon the Brassicas is one of the important factors limiting the production of commercial *Brassica* crops (Ekbom, 1995; Lamb, 1989). Brassicaceous plants produce a family of sulphur compounds called glucosinolates, whose breakdown products are attractants and stimuli for feeding and oviposition but, on the other hand, act as deterrents or toxins for herbivores not adapted to plants of the Brassicaceae. A list of insects important to *Brassica* plants is given in Table 3.A1.1 of the annex. Some of the more important insects are discussed below.

Phyllotreta spp. – Flea beetles

Flea beetles feed on spring-sown seedlings and in some years the second generation may attack green foliage and pods in the fall. Several species of flea beetles occur in different *Brassica*-growing areas of the world. Damage by these small beetles is characterised by feeding holes in cotyledons and first true leaves and is most severe under warm, dry conditions. Some Brassicaceous species (e.g. *Sinapis alba*, *B. villosa*) avoid damage due to the presence of hairs (trichomes) on cotyledons, leaves and stems. Attempts at biological control have not been successful, but research is underway to develop *B. napus* plants expressing large numbers of trichomes as a means of defence (Gruber et al., 2006). The primary control measure is insecticidal seed dressings.

Psylliodies chrysocephala – Cabbage stem flea beetle

This beetle is one of the most important pests of oilseed rape in Europe (Ekbom, 1995). Eggs, laid by adults in the soil at the base of seedlings, produce larvae that eat into leaf stocks and later into the stem and base of the biennial plants, where the larvae overwinter. Feeding damage results in weakened plants, resulting in reduced yield and winter kill. Control is dependent upon insecticide sprays.

Ceutorhynchus spp. – Stem weevils

Both *C. napi* and *C. quadridens*are are important pests in continental Europe. The weevils overwinter as adults and lay their eggs on leaf petioles of overwintered *Brassica* plants. The larvae eat into and feed in the stems resulting in weakened and broken plants. Insecticide sprays are used for control.

Aphid species

Three species of aphids can be of economic importance on Brassicaceous plants (Ekbom, 1995). *Lipaphis erysimi* and *Brevicoryne brassicae* prefer Brassicaceae hosts while *Myzus persicae* is polyphagous. On the Indian sub-continent *L. erysimi* is a very serious pest capable of reducing oilseed mustard yields by 50%. In temperate zones, *B. brassicae* is a common pest of vegetable Brassicas and occasionally of oilseed crops. Suction feeding causes a direct loss of vigour and yield. *M. persicae* also causes indirect damage as a vector of beet western yellow virus (BWYV) (Hill et al., 1991). Insecticide sprays can be used for control but care must be taken not to kill beneficial insects present during flowering.

Lepidoptera species

The lepidopteron pests occur sporadically and can have more than one generation per year. The eggs are laid on the leaves where the larvae feed. In Canada, a second generation of the diamondback moth (*Plutella xylostella*) may also attack pods of oilseed rape. The diamondback and *Pieris brassicae* (the large cabbage white butterfly) are also important pests of vegetable crops, where their leaf damage affects market value. Chemical control is applied where populations warrant.

Meligethes species

Pollen beetles are important pests of both spring and winter oilseed rape in Europe. Adult beetles move onto the crop from unrelated early flowering plants to feed on pollen from open flowers and to lay eggs in unopened buds. The larvae emerge and eat the stamens, causing buds to abort. The final instar larvae fall to the ground and pupate with the new generation emerging in July into August (Ekbom, 1995). Pyrethroids are the most commonly used chemical control.

Ceuthorhynchus assimilis – Seed pod weevil

This weevil occurs in both Europe and North America. It has one generation per year, emerging from over-wintering sites in the late spring to feed on the crop. The main damage is done to the seed pods. The adults make small holes in the pods to feed on the seeds within and to lay eggs. The larvae eat their way out of the pods, drop to the ground and pupate. The weevil has only recently invaded the oilseed rape growing area of western Canada. Control is by chemical sprays.

Dasineura brassicae – *Pod midge*

The pod midge is a European pest that uses the small holes in the pods made by the seed pod weevil for oviposition. The larvae eat the developing seed and cause the pods to open, losing their seed. The larvae over-winter in cocoons in the soil, pupate in the late spring and fly to the plants to oviposition, living only a few days (Ekbom, 1995). Early spraying for the pollen beetle can provide control of pod midge and other pod pests.

Beneficial insects

The interaction between bees, both farmed and wild, and *Brassica* plants are mutually beneficial. The bees aid fertilisation and receive nectar and pollen in return. Where grown, oilseed rape and mustard provide productive bee pasture while the fertilising activities of the bees are essential for the production of hybrid seed and tend to increase seed yields of commercial fields.

Animal interaction

Succulent *Brassica* plants attract many foraging animals including rabbits, rodents and deer to name a few. Winter oilseed rape is an important winter pasture for wild deer and other animals. Ruminant animals, both wild and domestic, under certain circumstances, can become ill from grazing kale or winter oilseed rape crops (Marquard and Walker, 1995). The toxic compound responsible is dimethyldisulphide that arises from the breakdown products of glucosinolates and S-methylcysteine sulphoxide (SMCO), also known as the kale anaemia factor (Maxwell, 1981). Birds often feed on fall-germinating seedlings and on the developing seed in the pod.

Soil microbial interaction

The genetic makeup of crop plants can influence the composition of the soil microbial community in which they grow. However, the interaction between plants and their residues with the soil microflora is not well understood (Dunfield and Germida, 2004). The soil microbial communities associated with the growing of conventional spring oilseed rape (both *B. napus* and *B. rapa*) and transgenic HR *B. napus* were investigated in western Canada plot trials. The soil microflora in the plots of the glyphosate resistant variety Quest differed significantly from that found in both conventional and transgenic glufosinate resistant varieties, particularly at the flowering stage. However, although the microbial diversity was altered, the effects varied by test site and plant growth stage. In addition, the change in the microbial community was temporary as no differences were found the following spring (Dunfield and Germida, 2004, 2003, 2001: Siciliano and Germida, 1999).

Allelopathy

There have been numerous reports of inhibitory affects by *Brassica* residues on the following planting of pasture, cereal and oilseed crops (Campbell, 1959; Bell and Muller, 1973; Rice, 1984; Mason-Sedun, Jesspo and Lovett, 1986; Horricks, 1969; Vera, McGregor and Downey, 1987). The allelopathic effects include germination inhibition, reductions in root growth, plant height, dry weight, tiller number and seed yield. Species involved in the inhibition included marrow stem kale (*B. oleracea*), oilseed and turnip rape (*B. napus*, *B. rapa*) and condiment and black mustard (*B. juncea* and *B. nigra*). The inhibiting compound(s) are leached by water from dead or decaying stems and leaves of *Brassica* vegetation. The compound(s) appear to reside in the upper soil layer for a short period and then dissipate. Mason-Sedun, Jesspo and Lovett (1986) compared the effect of water extracts from dry residues of four *Brassica* species on coleoptile growth of common wheat (*Triticum aestivum*). All residues significantly reduced grain yield, plant dry weight, plant height and tiller production, with the greatest level of inhibition resulting from *B. juncea* residues followed by *B. nigra*, *B. napus* and *B. rapa*.

Laboratory studies indicated that when stored, dry residues became less toxic over time. Waddington and Bowren (1978) found that rapeseed residue was no more toxic to barley, bromegrass or alfalfa than comparable amounts of wheat residue. Normally *Brassica* residue will have been rained on well before seeding, resulting in no inhibition. Indeed, there is good evidence that cereal crops are more productive following oilseed rape than another cereal (Almond, Dawkins and Askew, 1986). Vera, McGregor and Downey (1987) suggested that the primary cause of the observed inhibition in western Canada may be the release of a chemical compound from volunteer oilseed rape seedlings that are killed by cultivation at seeding time. The chemical was thought to be the indole glucosinolate, glucobrassicin, present in high concentrations in tissues of young seedlings (Röbbelen and Thies, 1980).

Pathogens

The *Brassica* crops and their wild allies are subject to a broad range of pathogens and adverse conditions or disorders associated with non-infectious causes. Although many of the *Brassica* species have many diseases in common, there are also significant differences in susceptibility among and within species. The *Compendium of Brassica Diseases* (Rimmer et al., 2007) provides an authorative and practical reference guide to disease problems in *Brassica* crops the world over. Colour plates and text describe the infectious diseases caused by fungi, oomycetes, bacteria, mollicutes, viruses and nematodes. In addition, non-infectious disorders such as those related to environmental effects, herbicide injury and nutritional deficiencies are also described. The American Phytopathological Society (APS) also provides a listing by common and scientific name of known *Brassica* diseases and conditions at its website as reproduced in Table 3.A1.2 in the annex (APS, 2001).

Of the many *Brassica* field crop diseases listed in Table 3.A1.2, three stand out as particularly troublesome as they are pandemic and have the potential to cause major crop injury: blackleg or stem canker (*Leptosphaeria maculans*); Sclerotinia stem rot (*Sclerotinia sclerotiorum*); and clubroot (*Plasmodiophora brassicae*). To date there are few control measures for these pathogens that are fully effective and economical.

Varieties with single race resistance have been developed, but the multi-race pathogenicity of these fungi has made it difficult to breed varieties with long-lasting resistance. However, it is anticipated that with the location of resistance genes on marker-saturated genome maps, breeders will be able to bring together multiple resistance genes from both within and outside the genus that will provide long-lasting disease resistance.

Breeding improved varieties

Introduction

The objective of all plant breeding programmes is to produce plants of greater value to the producer, the industry and the consumer. The objective is achieved by building on past advances, through the incorporation of desirable traits that impart increased yield, pest resistance, superior quality and/or utility to new varieties. To accomplish the task, many related disciplines are essential including genetics, biotechnology, agronomy, cytology, chemistry, pathology, entomology, physiology and statistics. Within the biotechnology component, gene transfer and the production of transgenic varieties has attracted public attention but the discipline is much broader and includes, among others, tissue culture, protoplast fusion, dihaploid production, gene identification and cloning.

The essential requirement for success is genetic variation for the trait or traits of interest. The breeder will normally search for the desired trait within adapted genotypes and then the crop's world germplasm collection. If it is not present within the species but present in a related species, interspecific and intergeneric crosses and/or protoplast fusion may be attempted. If those approaches fail, induced mutation may be explored. Generally gene transfer, because of regulatory hurdles, is the last resort.

Valuable, new gene-controlled traits are added with each improved variety. The breeder evaluates the need and the genetic variability available and stacks desirable traits, be they large or small advances, into the genetic base that previous breeders have built. Gene stacking is the very essence of plant breeding. Breeding techniques vary with the crop being bred and its mode of pollination and reproduction. Among the commercial *Brassica* crops, both self-compatible and self-incompatible species are present so that a wide array of techniques are employed, as described below, depending on the species and the trait or traits to be introduced.

The application of conventional genetic manipulation in plants can have major beneficial impact on the nutritional quality and quantity of the world's food supply. A very successful example, described below, is the conversion of *Brassica* oilseed crops from a problematic commodity to the high-quality productive crop we now define as canola.

Lipids not only make our food taste better but are required dietary ingredients. They are essential cell membrane components, regulating cell permeability and are responsible for vitamin transport as well as the starting point for hormone biosynthesis. Oils and fats are predominantly (~98%) triacylglycerols (TAGs) that consist of a three-carbon chain with fatty acids attached to each carbon. The fatty acid composition of an oil determines its value, use and nutritional worth.

Oils from *B. juncea*, *B. rapa* and later *B. napus* have been part of the Asian diet for centuries, but in Europe and the Americas they are relatively recent edible oil additions. Prior to and during the Second World War, rapeseed oil was primarily used as a lubricant for steam engines and as a lamp oil, but following the war, *B. napus* and *B. rapa* oils became an important constituent of margarine. Researchers became interested in the nutritional value of *Brassica* seed oils because they differed from most other edible oils in having a high percentage of long carbon chain monoenoic fatty acids, eicosenoic (C20:1) and erucic (C22:1) (Table 3.14).

Small animal feeding studies in the late 1950s and throughout the 1960s indicated that the nutritional value of rapeseed oil could be substantially improved if the long chain fatty acids could be reduced to <5% of the fatty acid total (Kramer et al., 1983). Breeding and selection within the world's germplasm was successful in developing plants of *B. napus* (Stefansson, Hougen and Downey, 1961), *B. rapa* (Downey, 1964) and later *B. juncea* (Kirk and Oram, 1981) that produced oils with less than 2% erucic acid. This oil was found nutritionally superior to the high erucic oil (Kramer et al., 1983) and proved to be an excellent liquid and salad oil, as well as a suitable ingredient for margarine and shortening manufacture. This new natural oil is called "canola oil" in most countries of the world and is defined as oils from *B. napus*, *B. rapa* or *B. juncea* containing less than 2% erucic acid of the fatty acid total. The genetic blocking of the biosynthesis of eicosenoic and erucic acids resulted in an increased percentage of oleic and linoleic acids (Table 3.14).

Table 3.14. **Fatty acid composition of rapeseed, canola, soybean, sunflower and linseed oils**

Fatty acid	Symbol[1]	Rapeseed	Canola	Soybean	Sunflower	Linseed
		Fatty acid composition (%)				
Palmitic	C16:0	4.0	4.7	11.5	7.5	7.0
Stearic	C18:0	1.5	1.8	3.5	4.5	4.0
Oleic	C18:1	17.0	61.5	23.0	16.0	20.0
Linoleic	C18:2	13.0	21.0	43.0	71.0	17.0
Linolenic	C18:3	9.0	11.0	8.0	1.0	52.0
Eicosenoic	C20:1	14.5	<1.0	0.0	0.0	0.0
Erucic	C22:1	41.0	<1.0	0.0	0.0	0.0

Note: 1. The first number denotes the number of carbon atoms in the fatty acid chain and the second numbe, the number of double bonds in the chain.

When nutritionists recommended that dietary intake of saturated fat be reduced, the nutritional value of canola oil gained widespread recognition since it contains the lowest level of saturated fatty acids of any edible oil (Grundy and Denke, 1990; Gurr, 1992; Hu et al., 1997; see Figure 3.41). Further, in 1985 Mattson and Grundy reported on the nutritional desirability of the so-called "Mediterranean diet", pointing out the health advantages of oils with a low level of saturates and high content of oleic acid. The fatty acid composition of canola oil met or exceeded the nutritional requirements of a superior edible oil, with the lowest saturate content (6-7%) of any edible oil and a high (58-60%) level of oleic (18:1n-9) that reduces the undesirable low-density lipoproteins (LDLs) without reducing the desirable high-density lipoproteins (HDLs).

Plant breeders have now developed varieties that produce canola oils with less than 3% α-linolenic acid which improves the oxidative stability of the oil and reduces the development of unpleasant flavours and cooking odours (Scarth, Rimmer and McVetty, 1995; Scarth et al., 1988; Eskin et al., 1989; Przybylski et al., 1993).

Figure 3.41. **Canola oil compared to other edible vegetable oils as to total saturated fat content and other fatty acids**

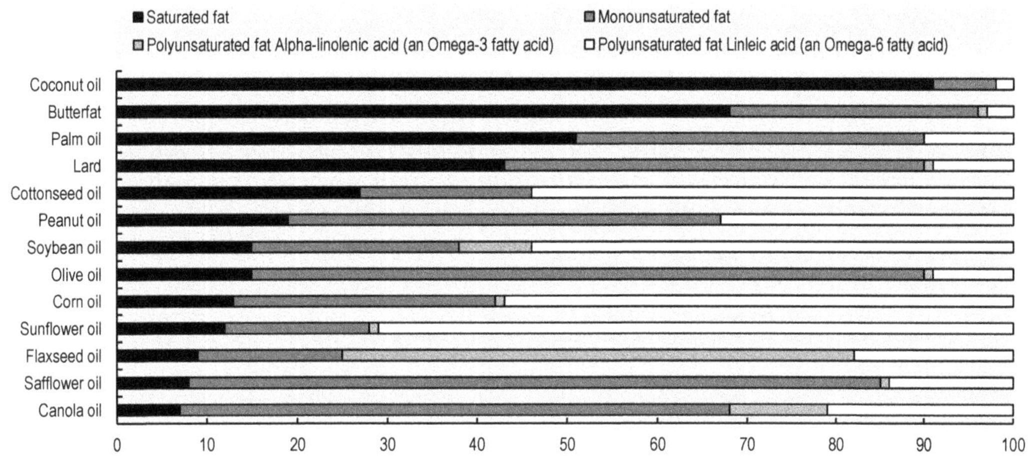

Source: Analyses conducted by POS Pilot Plant Corporation, Saskatoon, Canada, data courtesy of Canola Council of Canada.

More recently plant breeders have combined the low linolenic trait with a reduced level of linoleic acid to provide an oil with over 70% oleic acid (Table 3.15; Downey, 1996). The high oleic acid level further increases the oil's stability so that little or no hydrogenation of the oil is required, which would otherwise result in undesirable trans fatty acids. Canola varieties that produce this latter fatty acid composition now occupy about 10% of Canada's oilseed rape growing area.

Table 3.15. **Fatty acid composition of canola and specialty *B. napus* varieties grown in Canada**

Oil type	Fatty acid composition (%)							
	C16:0	C16:1	C18:0	C18:1	C18:2	C18:3	C20:1	C22:1
Canola	4.7	<1.0	1.8	61.5	21.0	11.0	<1.0	<1.0
High erucic	2.0	<1.0	2.0	13.0	12.0	9.0	7.0	54.0
Low linolenic	4.0	<1.0	2.0	64.0	27.0	2.0	1.0	<1.0
High oleic	4.0	<1.0	2.0	75.0	9.0	8.0	<1.0	<1.0

Oil extraction of *Brassica* oilseeds yields about 40% oil and some 60% high protein meal. The meal is used as a high-quality protein supplement in diets for animals, poultry and fish. Unfortunately, the plant translocates and concentrates the glucosinolates in the seed. As a result, rapeseed and mustard can contain over 120 mg/g of glucosinolates per whole seed. This high concentration of glucosinolates, and their breakdown products, greatly limited the amount of traditional rapeseed meal that could be fed to non-ruminant animals, such as swine and poultry. Glucosinolates and their breakdown products reduced the palatability of the meal but, more importantly, they interfered with the iodine uptake by the thyroid gland and are active goitrogens. Feeding rapeseed meal to non-ruminant animals frequently resulted in poor feed efficiency and weight gains as well as reproductive difficulties (Bell, 1993). Thus, the amount of seed that could be processed was determined by the limited size of the meal market.

A partial solution was the inactivation of the myrosinase enzyme as the first step in the oil extraction process but enzymes in the animal gut, although less efficient, were also able to hydrolyse the glucosinolates. The answer to this problem was to breed plants with little or no glucosinolates in their seed.

Analytical advances in the 1960s allowed breeders to identify plants with only 10-12 μmoles of aliphatic glucosinolate per gram oil free meal. These plants were crossed with low erucic acid varieties to produce "double low" or "canola quality" varieties of *B. napus* (Stefansson, 1983), *B. rapa* (Downey and Rakow, 1987) and *B. juncea* (Love et al., 1990). The reduction in glucosinolate levels allowed canola meal to be fed at maximum economic levels to non-ruminants and canola meal became the preferred protein supplement for dairy cattle. Canola is defined as seeds of the genus *Brassica* (*Brassica napus*, *Brassica rapa* or *Brassica juncea*) from which the oil shall contain less than 2% erucic acid in its fatty acid profile and the solid component shall contain less than 30.0 micromoles of any one or any mixture of 3-butenyl glucosinolate, 4-pentenyl glucosinolate, 2-hydroxy-3-butenyl, or 2-hydroxy-4-pentenyl glucosinolate, per gram of air-dried, oil free solid (Canola Council of Canada).

Breeding methods

The amphidiploids, *B. napus*, *B. juncea* and *B. carinata*, are largely self-pollinating with the self-pollinated progeny exhibiting very little, if any, loss in vigour. Thus, methods developed for highly inbred crops, such as the cereal grains, have been adapted for these partially outcrossing species. In the oilseed forms of these species, complete homozygosity is normally not the objective, although varietal distinctness, uniformity and stability are still a requirement. However, with cole crops and hybrids, high levels of homozygosity are required.

Regardless of the breeding technique employed, success is dependent upon the identification of suitable parents that, when crossed, will yield progenies that express the desirable traits of both parents.

Mass selection

This early plant-breeding technique relied on the identification and harvesting of seed from the most productive or desirable plants within a population for sowing in the following year. The system is one of population improvement based on plant phenotype and is best suited to self-fertilised crops and where gene action is additive. It lacks the efficiency of present-day techniques, but a variation is used today to preserve the identity of established varieties whereby off-types are removed from elite lines and breeder seed plots.

Pedigree method

In the past, most *B. napus* and *B. juncea* commercial varieties were developed using the pedigree method. Crosses are made between parents exhibiting the traits to be combined and the F_1s are selfed or intercrossed. The progeny are selfed or allowed to interpollinate and selection of the best F_4 rows is done within the best F_3 families. By the $F_{5\text{-}6}$, the vast majority of loci will be homozygous and the characteristics of the breeding line are fixed.

The pedigree method may be modified in various ways depending on the inheritance of the trait or traits being introduced or combined. The method is well suited to the mainly self-pollinating species *B. napus* and *B. juncea*, because the seed multiplication rate, unlike cereal grains, is high (*ca.* 1 000:1). In the self-incompatible *Brassica* vegetables and oilseed *B. rapa*, inbreeding leads to a rapid loss in vigour and reduced fertility. However, it is sometimes used to produce inbred lines destined for the production of hybrid vegetables. Some cauliflower varieties are exceptions, being natural self-pollinators that do not exhibit the usual vigour and fertility losses.

Single seed descent

As with the pedigree method, the first step in single seed descent (SSD) breeding is the careful choice of parents for hybridisation. However, unlike the pedigree method, selection is not practiced until a high degree of homozygosity is reached. The object is to advance generations as rapidly as possible and subsequently select among the randomly derived lines. The size of the segregating population is kept at a manageable level by planting only one randomly chosen single seed from each plant in the previous generation.

Since the degree of homozygosity is not as critical in *B. napus* and *B. juncea* as it is in cereals, this method has not been widely used in *Brassica* breeding programmes.

Backcross method

The backcross method is designed to introduce one or more specific trait(s) into an otherwise highly desirable parent or variety. The donor parent, containing the trait(s) to be incorporated, is crossed onto plants of an adapted, desirable, recurrent parent. Depending on the inheritance of the trait(s) and the ease or efficiency of selection, the F_1 or selected BCF_1 plants will be backcrossed to the recurrent parent. By the fourth to sixth backcross, the genetic makeup of the recurrent parent is expected to have been reconstituted with the new trait incorporated. However, linkage between the desirable trait and one or more undesirable characteristics may require selection within large populations to identify plants or lines with an uncoupled linkage.

Frequently in the self-pollinating species, only one or two backcrosses are made followed by pedigree selection.

Figure 3.42 illustrates the combined use of the backcross and pedigree methods.

In the self-incompatible species, backcrossing can also be effective for the incorporation of specific traits. However, crosses in oilseed *B. rapa* need to be made with sufficient numbers of recurrent parent plants to ensure that heterozygosity of the self-incompability alleles of the recurrent parent is maintained in the backcross generations. To overcome this potential problem, the "recurrent selection" breeding system is widely used.

Backcrossing is also effective in the self-incompatible vegetable species. Dickson and Wallace (1986) outline a complete backcross breeding programme for cabbage improvement.

Figure 3.42. **Breeding scheme combining the backcross and pedigree selection systems to develop a low erucic, low glucosinolate variety with high seed and oil yield**

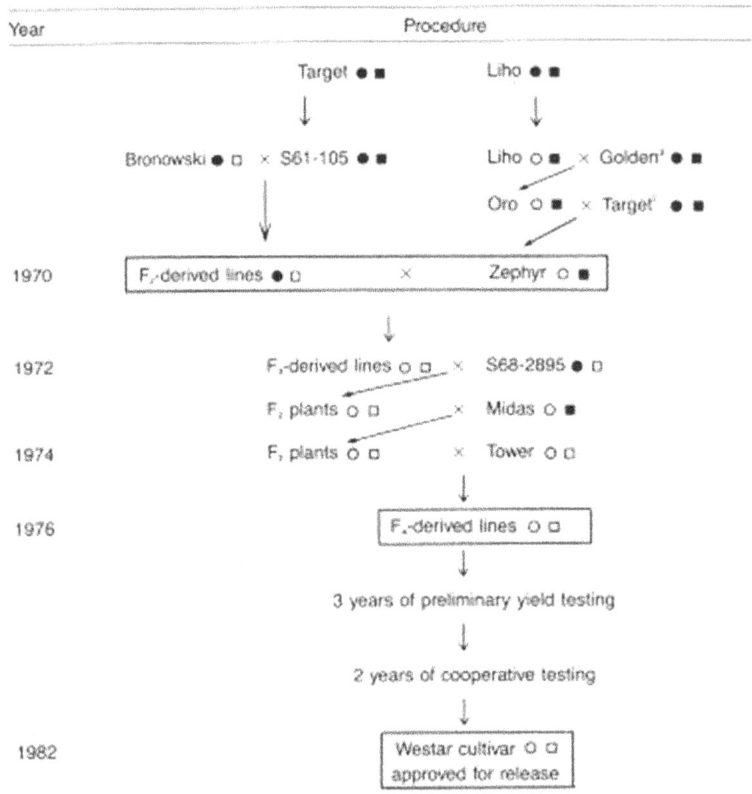

Notes: This breeding scheme uses agronomically superior parents that contribute either high (●) or low (○) erucic acid levels and high (■) or low (□) glucosinolate content.

Source: Downey and Rakow (1987).

Recurrent selection method

This method is standard procedure for improving populations of self-incompatible species. Any type of segregating population may be a candidate for improvement. Normally open-pollinated seed is harvested at random from individual plants within the population, and a progeny row or rows sown from each plant. However, some seed from each plant is held in reserve. The progeny rows are evaluated and the best performing identified. An equal amount of the reserve seed from the best single plants, based on the performance of their progeny, is bulked. The first cycle of recurrent selection is complete when the new seed composite is sown in an isolation plot, and the second cycle begins with the harvesting of random single plants within the new composite. A bulk sample from the remaining plants can be harvested and planted in replicated trials to measure the response to selection. Recurrent selection is continued as long as it is anticipated that there will be a reasonable response to selection.

With every additional trait under selection, the intensity of selection increases exponentially, thus it is difficult to improve a population for several traits simultaneously. This constraint is overcome by having specialised composites for different traits that are brought together after the original objectives for each have been met.

Synthetic varieties

Allard (1960) defines a synthetic variety as one "that is maintained from open pollinated seed following its synthesis by hybridisation in all combinations among a number of selected genotypes". This method, which is widely used in breeding forage crops, is also effective for the breeding of oilseed *B. rapa* (Falk et al., 1994). Equal amounts of seed from varieties or recurrent lines that arise from widely different gene pools are mixed and sown in Syn.-0 isolation plots. Seed harvested from the Syn.-0 plot constitutes the Syn.-1 generation. Syn.-1 seed from a two component synthetic will consist of 25% from each parental genotype and 50% hybrid seed. Thus, if the parental lines are good combiners, a significant amount of heterosis can be captured.

The method (Figure 3.43) has also been explored in *B. napus* (Becker, Löptien and Röbbelen, 1999) but breeding programmes in this species are now directed to F_1 hybrid varieties.

Normally, despite the high multiplication rate (1 000: 1), there is insufficient Syn.-1 seed for commercialisation so that Syn.-1 seed is sown to provide commercial Syn.-2 seed. This procedure has been used in Canada to produce the first commercial *B. rapa* synthetic varieties, Hysyn 100 and Hysyn 110. Because of the large number of genotypes within the parental lines, there is very little loss in heterosis between the Syn.-1 and Syn.-2 generations (Falk and Woods, 2003). If the market is very large a Syn.-3 generation could be added.

Figure 3.43. **Breeding scheme for development of commercial synthetic varieties of oilseed *Brassica* crops**

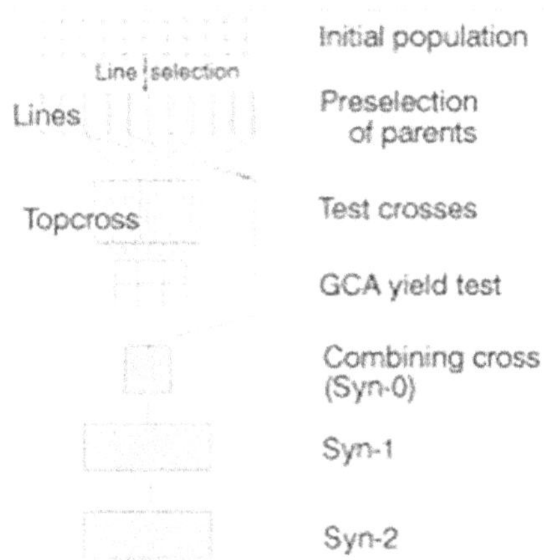

Source: Becker, Löptien and Röbbelen (1999).

Diallel and polycross methods

In vegetable crops, uniform maturity, head size and appearance are critical to the success of a variety and seed yield is of secondary importance. Further, the numbers of parents that make up a variety are few and the market price of seed is substantially greater than the commodity oilseed crops. Thus, breeding methods used for vegetables can be more intensive than the large population breeding methods used in oilseed improvement programmes. For example, if a deleterious trait is controlled by a recessive gene, it is difficult to completely eliminate it from a self-incompatible plant population.

However, within a small population of potential elite parents, diallel crossing, i.e. hand crossing each parent with all other potential parents, followed by progeny assessment, can eliminate the heterozygous parent(s). Although labour intensive, this technique is suitable for most vegetable Brassicas because individual plants (parents) can be vegetatively maintained over many generations. Vegetative propagation also makes possible the use of the polycross breeding method used to identify desirable parents with good general combining ability. In this method, parental clones are space planted in a field design that assures each parent is equally exposed to pollen from all the other parents in the nursery. Progeny evaluation then identifies the best parents for inter-pollination to produce seed of a new variety.

Hybrid varieties

The vigour, yield and uniformity advantages associated with hybrids in both oilseed and vegetable *Brassica* crops have been demonstrated by many breeders. The main constraint to their commercial exploitation has been an effective pollen control-fertility restoration system. Vegetable breeders have utilised the variations in SI alleles, which control the self-incompatible system, to produce single and double cross hybrids. Kuckuck (1979) illustrates how lines, selected for general combining ability and specific S alleles, are programmed to produce double-cross cabbage hybrids (Figure 3.44).

The self-incompatible parent can be maintained through bud pollination, micro propagation or by overcoming the SI barrier by exposing flowering plants to high CO_2 concentrations. Nuclear male sterility in oilseed rape has also been used commercially in China but the segregating male fertile progeny have to be removed by hand (Fu et al., 1997), thus making the system expensive in many regions.

Figure 3.44. **Self-incompatability scheme for breeding cabbage hybrid seed production**

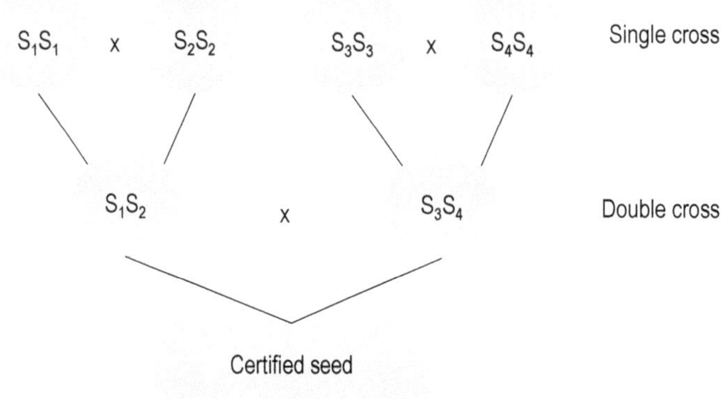

Source: Kuckuck (1979).

The most practical and efficient system is that of cytoplasmic male sterility (CMS). More than 17 different male sterile forms have been investigated in *Brassica* species (Stiewe et al., 1995; Prakash et al., 1995). Only a few have been developed to the commercial stage, but varietal development programmes worldwide are rapidly moving to the use of CMS-restorer systems for hybrid seed production. The CMS systems are based on genetic miscommunication between cytoplasmic mitochondria and nuclear genes, resulting in the disruption of normal anther and/or pollen development. There are three components to the system: the A line, carrying the cytoplasmic mitochondrial genome that results in male sterility, the B line that is fully fertile and maintains the

A line, and the R line with a nuclear gene that restores fertility. The R line should be highly heterotic to the A line to produce a high yielding, fully fertile F_1 commercial crop (Figure 3.45).

China developed the first *B. napus* commercial CMS system, known as the Polima system (Fu et al., 1997). However, the Polima system is rapidly being replaced in western breeding programmes with the *ogu*-INRA system, the Male Sterile Lembke (MSL) system and others under development (Stiewe et al., 1995; Prakash et al., 1995; Downey and Rimmer, 1993).

A transgenic pollen control-restorer system, developed by Plant Genetic Systems and commercialised by Bayer CropScience, is in widespread use in Canada and the United States. Details of how this system functions are outlined by Downey and Rimmer (1993).

Figure 3.45. **Production system for cytoplasmic male sterile hybrid seed of oilseed rape**

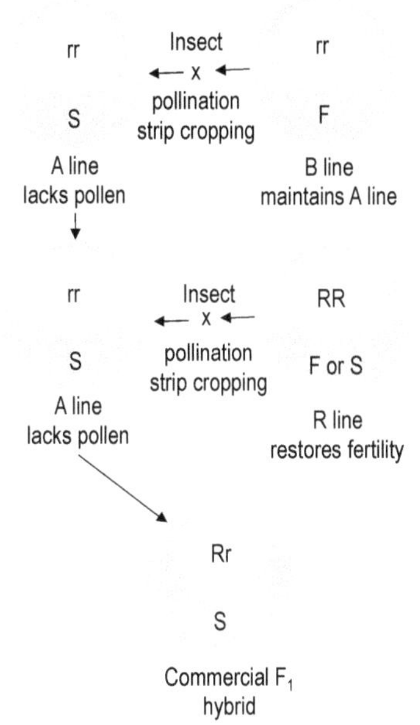

Note: The small circle represents the nucleus showing the fertility restorer genes r and R and the larger circle the cell with cytoplasm containing fertile (F) or sterile (S) mitochondrial genes.

Source: Modified from Buzza (1995).

Improvement through "interspecific hybrids" and "cybrids"

Interspecific and intergenomic crosses are important options for the introduction of desired traits that are not available, or cannot be found, within the primary gene pool of a crop species. Normally such crosses are difficult to make. As noted previously, there are many natural barriers, both pre- and post-fertilisation, that protect the integrity of a species. Further, even if such crossing is successful, chromosome pairing and alien gene introgression into the genome of the target species must occur.

However, in the Brassicaceae, a number of desirable nuclear genes from different genera and species have been transferred to targeted crop species. A list of traits that have been transferred to *B. napus*, *B. juncea* and/or *B. oleracea* from other Brassicaceae species is presented in Table 3.16 (Prakash et al., 2009).

The development of protoplast fusion technology has been highly successful in circumventing the natural sexual barriers that separate the Brassicaceae species and genera. The technology has the potential to access desirable genes present in distant relatives (Glimelius, 1999; Christey, 2004; Navrátilová, 2004; Liu, Xu and Deng, 2005). Prakash et al. (2009) have compiled a list of intertribal somatic hybrids in the Brassiceae and the desirable traits to be transferred (Table 3.17). Additional intergenomic hybrids have been produced but failed to establish in soil e.g. *Camelina sativa* + *B. carinata* (Narasimhulu et al., 1994), *C. sativa* + *B. oleracea* (Hansen, 1998) and *Barbarea vulgaris* + *B. napus* (Fahleson, Eriksson and Glimelius, 1994).

With some exceptions, the somatic hybrids so far obtained have exhibited a high degree of sterility and/or morphological abnormalities that have limited their use. However, the importance of somatic hybridisation is not so much the direct use of the resulting amphidiploids, containing both parental genomes, but rather to utilise the somatic hybrids as a bridge to transfer desirable traits to target species (Glimelius, 1999).

Cell fusion not only brings together the nuclear contents of both parents but also combines the cytoplasm and organellar content of fused cells. Frequently, to improve the outcome, the nucleus of one parent is eliminated by X-ray, centrifugation or chemical treatment before fusion but the fused cell contains the cytoplasm of both parents. The resulting plant is termed a "cybrid". This technique allows cytoplasmic substitution which frequently results in cytoplasmic male sterility (CMS). Cell fusion among the Brassicaceae, where the cytoplasm of both parents are combined, can also generate novel cytoplasmic variability, bringing about organellar reassortment and DNA rearrangement, which is not possible using sexual hybridisation.

Chloroplast segregation is independent of mitochondrial segregation and while mitochondrial recombination has been frequently observed in the Brassiceae (Glimelius, 1999), recombination is rarely found in the chloroplasts. It is also rare to have a mixture of the two chloroplasts occurring in the same hybrid. In general, the chloroplasts are usually contributed by crop species. This may occur because many of the fusions are with the allopolyploid crop species that contribute large numbers of chloroplasts per cell (Butterfass, 1989).

Table 3.16. **Introgression of nuclear genes conferring desirable traits to *Brassica* crops**

Trait	Donor species	Recipient species	Reference
Yellow seed coat	*B. rapa*	*B. napus*	Chen and Heneen (1991)
	B. juncea/B. carinata	*B. napus*	Rashid, Rakow and Downey (1994)
	B. carinata	*B. napus*	Qi et al. (1995)
	B. rapa/B. carinata	*B. napus*	Meng et al. (1998)
	B. rapa/B. juncea	*B. napus*	Rahman (2001); Potapov and Osipova (2003)
CMS fertility	*Raphanus sativus*	*B. napus* CMS (*Ogu*)	Heyn (1977); Rousselle et al. (1985)
Restoration	*Raphanus sativus*	*B. napus* CMS (*Kosena*)	Sakai et al. (1996)
	B. juncea	*B. napus* CMS (*Polima*)	Fan, Tai and Stefansson (1985)
	B. tournefortii	*B. napus* CMS (*Tour*)	Stiewe and Röbbelen (1994)
	Trachystoma ballii	*B. juncea* CMS (*Trachy*)	Kirti et al. (1997)
	Moricandia arvensis	*B. juncea* CMS (*Moricandia*)	Prakash et al. (1998)
	Erucastrum canariense	*B. juncea* CMS (*Canariense*)	Prakash et al. (2001)
		B. napus CMS (*Canariense*)	Banga et al. (2003)
	Enarthrocarpus lyratus	*B. rapa* CMS (*Lyratus*)	Deol et al. (2003)
		B. juncea/B. napus CMS (*Lyratus*)	Banga, Deol and Banga (2003)
Chlorosis removal	*Raphanus sativus*	*B. napus* CMS (*Ogu*)	Paulmann and Röbbelen (1988)
Beet cyst nematode	*Sinapis alba*	*B. napus*	Lelivelt et al. (1993)
Resistance	*Raphanus sativus*	*B. napus*	Lelivelt and Krens (1992); Voss, Snowdon and Lühs (2000); Peterka et al. (2004): Budahn et al. (2006)
Club root resistance	*B. napus*	*B. oleracea* var. *capitata*	Chiang, Chiang and Grant (1977); Chiang et al. (1980)
	B. rapa	*B. napus*	Gowers (1982)
Blackleg resistance	*B. juncea*	*B. napus*	Roy (1984); Dixelius (1999); Sacristan and Gerdemann (1986)
	B. nigra	*B. napus*	Struss et al. (1996); Chèvre et al. (1997b, 1996); Plieske, Struss and Röbbelen (1998); Dixelius (1999)
	Arabidopsis thaliana	*B. napus*	Bohman, Wang and Dixelius (2002); Ogbonnaya et al. (2003); Saal et al. (2004)
	Sinapis arvensis	*B. napus*	Snowdon et al. (2000); Winter et al. (2003)
	Coincya monensis	*B. napus*	Winter et al. (2003)
	B. rapa	*B. napus*	Chèvre et al. (2003)
Alternaria leaf spot	*Sinapis alba*	*B. napus*	Primard et al. (1988)
Resistance	*Diplotaxis erucoides*	*B. napus*	Klewer et al. (2003)
	Sinapis alba	*B. oleracea*	Sigareva, Ren and Earle (1999)
Black rot resistance	*B. juncea*	*B. napus*	Tonguc and Griffiths (2004)
Soft rot resistance	*B. oleracea* var. *italica*	*B. oleracea*; *B. rapa* subsp. *pekinensis*	Ren, Dickson and Earle (2001a, 2001b)
Altered oil quality	*Orychophragmus violaceus*	*B. napus*	Hu et al. (2002); Hua and Li (2006)
Earliness	*B. rapa*	*B. napus*	Shiga (1970); Namai, Sarashima and Hosoda (1980)
	B. oleracea	*B. napus*	Habman et al. (2010)
Low erucic acid	*B. juncea*	*B. carinata*	Getinet et al. (1994)
Low glucosinolate	*B. napus*	*B. carinata*	Getinet et al. (1994)

Source: Prakash et al. (2009).

Table 3.17. **Intertribal somatic hybrids in Brassiceae for the integration and incorporation of desirable traits into *Brassica* crops**

Somatic hybrid	Desirable trait for introgression	Reference
Arabidopsis thaliana (*n*=5) + *B. nigra*	Resistance to flea beetles, cold tolerance, short life cycle	Siemens and Sacristan (1995*)
Arabidopsis thaliana (*n*=5) + *B. oleracea*	Plastome transformation	Nitovskaya and Shakhovskyi (1998); Yamagishi and Nakagawa (2004); Nitovskaya et al. (2006a)
Arabidopsis thaliana (*n*=5) + *B. rapa*	Experimental demonstration	Gleba and Hoffmann (1980, 1979)
Arabidopsis thaliana (*n*=5) + *B. juncea*	Phosphinothricin resistance	Ovcharenko et al. (2004)
Arabidopsis thaliana (*n*=5) + *B. napus*	Herbicide resistance, Blackleg resistance	Bauer-Weston et al. (1993*); Forsberg, Landgren and K. Glimelius (1994*); Forsberg et al. (1998); Yamagishi et al. (2002*)
	Transposable element *Spm/dSpm*	Ovcharenko et al. (2005*)
Armoracia rusticana (*n*=16) + *B. oleracea*	Clubroot resistance	Navrátilová et al. (1997)
Barbarea vulgaris (*n*=8) + *B. oleracea*	Cold tolerance	Ryschka et al. (1999)
Barbarea vulgaris (*n*=8) + *B. rapa*	Cold tolerance	Oikarinen and Ryöppy (1992)
Barbarea vulgaris (*n*=8) + *B. napus*	Cold tolerance	Fahleson, Eriksson and Glimelius (1994)
Barbarea stricta (*n*=8) + *B. rapa*	Cold tolerance	Oikarinen and Ryöppy (1992)
Camelina sativa (*n*=20) + *B. oleracea*	Alternaria resistance	Hansen (1998); Sigareva and Earle (1999)
Camelina sativa (*n*=20) + *B. carinata*	Alternaria resistance	Narasimhulu et al. (1994)
Capsella bursa-pastoris (*n*=16) + *B. oleracea*	Resistance to flea beetles, alternaria blight	Nitovskaya et al. (1998); Sigareva and Earle (1999)
Crambe abyssinica (*n*=45) + *B. napus*	High erucic acid content, insect resistance	Wang, Sonntag and Rudloff (2003*); Wang et al. (2004*)
Lepidium meyenii (*n*=32) + *B. oleracea*	Glucosinolate content	Ryschka, Klocke and Schumann (2003)
Lesquerella fendleri (*n*=6) + *B. napus*	High lesquerolic acid content, drought tolerance	Skarzhinskaya, Landgren and Glimelius (1996**); Skarzhinskaya et al. (1998)
	Lesquerella chloroplasts	Schröder-Pontoppidan et al. (1999); Nitovskaya et al. (2006b)
Lunaria annua (*n*=14) + *B. napus*	High nervonic acid content	Craig and Millam (1995)
Matthiola incana (*n*=7) + *B. oleracea*	Oil quality	Ryschka et al. (1999)
Orychophragmus violaceus (*n*=12) + *B. napus*	High linoleic and palmitic acid content	Hu et al. (2002* ,1999)
	Phosphinothrin resistance	Sakhno et al. (2007)
	Chlorosis correction	Vasilenko et al. (2003)
Thlaspi perfoliatum (*n*=21) + *B. napus*	High nervonic acid content	Fahleson et al. (1994**)
Thlaspi caerulescens (*n*=7) + *B. napus*	Zinc and cadmium tolerance	Brewer et al. (1999)
Thlaspi caerulescens (*n*=7) + *B. juncea*	High metal accumulation	Dushenkov et al. (2002)

Note: * Denotes asymmetric hybrids; ** Both asymmetric and symmetric hybrids identified.

Source: Prakash et al. (2009).

Biotechnology in Brassica *breeding*

Introduction

Although the above breeding procedures have been very effective in combining important agronomic and nutritional traits in superior cultivars, the process of identifying the desired genotype in genetically stable, uniform and high-yielding varieties takes many years. Further, the small chromosome size plus their lack of distinctive features have been an additional limitation on the selection of superior genotypes. However, beginning in the mid- to late 1980s, developments in tissue culture, embryo rescue, cell fusion, molecular markers and genetic mapping have not only reduced the time from cross to market but have given breeders powerful tools to quickly identify and assemble desirable traits in a single genotype. In addition, these biotech tools have greatly expanded the size and variation of the available gene pool, well beyond species boundaries.

Doubled haploid breeding

The doubled haploid (DH) breeding technique is now widely used in *B. napus* and *B. juncea* breeding programmes (Ferrie and Keller, 2004). This breeding tool not only eliminates the several generations needed to attain genetic stability and uniformity in breeding lines, but also significantly reduces the size of populations needed to find a desired genotype. For example, in *B. napus*, two genes code for the level of the fatty acid erucic in the seed oil, and an additional six genes code for the content of glucosinolates in the seed. Thus, when making a high by low cross, to produce progeny that have both low erucic acid and low glucosinolate (double low or canola quality), large segregating populations must be examined since the desired genotype must have all eight genes in the recessive state.

Table 3.18 illustrates the DH technique's increased selection efficiency, particularly when the selected plants are completely homozygous individuals that can be used directly as pure breeding varieties or as hybrid parents.

Table 3.18. **Minimum population size required to select the least frequent homozygote at 95% probability**

Number. of genes	Minimum F_2 population	
	Diploid	Haploid
1	11	5
2	47	11
4	766	47
5	3 067	95
6	12 269	191
7	49 077	382
8	**196 259**	**766**
10	3 123 923	3 067

Source: Rajhathy (1976).

The technique involves inducing large numbers of immature pollen grains (microspores) from *Brassica* species to develop into plants with the gametic or half the somatic chromosome number. Such plants are termed haploids and are sterile. By applying colchicine to the developing haploid plant, cell division is temporarily arrested, bringing about chromosome duplication. The result is a doubled haploid (DH) or dihaploid plant that is fully fertile and totally homozygous. Thus, complete homozygosity is reached in a single generation, and all seeds arising from self-fertilisation of that plant will be genetically identical. It is this single step to homozygosity that reduces the number of generations and time required to develop a new variety or hybrid parent. However, in a breeding programme, large populations of DH lines must be generated and evaluated since no prior selection has taken place. DH lines are usually derived from F_1 donors, although the use of F_2 and F_3 donor plants allows for more recombination and some preselection.

Molecular markers and their application

Marker-assisted selection and chromosome mapping came into general use in the 1980s with the development of restriction fragment length polymorphisms (RFLP) techniques that resulted in the first linkage maps for *B. oleracea* (Slocum et al., 1990), *B. rapa* (Song et al., 1991) and *B. napus* (Landry et al., 1991). This technique was important in identifying genomes and their chromosomes, locating genes and qualitative trait loci (QTLs), which are DNA regions containing a gene or genes that regulate traits of agronomic or quality interest.

The discovery of the polymerase chain reaction (PCR) by Mullis and Faloona (1987) resulted in new types of genetic markers such as amplified fragment length polymorphisms (AFLPs) that are more sensitive than RFLPs and simultaneously detect various polymorphisms in different genomic regions.

Additional marker systems have since been added to the toolbox including: random amplified polymorphic DNAs (RAPDs); sequence tagged sites (STS); simple sequence repeats (SSRs) or microsatellites and single nucleotide polymorphisms (SNPs). Breeders use these molecular markers to produce densely marked chromosome maps that can then be used to: 1) characterise germplasm and its genetic variability; 2) estimate the genetic distance between gene pools, inbreds and populations; 3) detect and locate QTLs and monogenic traits of interest; 4) select genotypes based on the presence or absence of specific markers; 5) identify useful candidate genes for sequencing (for more detailed information on genome mapping and molecular breeding in *B. napus*, see Snowdon, Lühs and Friedt, 2007; Snowdon et al., 2007). The marker systems differ in their ease of use, cost and other characteristics. It is expected that the SNPs system will become the marker system of preference, dispite its initial high cost, due to its ease of use, low cost per analysis and high level of reproducibility (Korzun, 2003).

Comparative genomic gene identification

The distantly related and intensively studied species *Arabidopsis thaliana* provides information that is highly relevant for gene isolation and characterisation in *Brassica* crops. However, the genomes of *Brassica* species are much more complex (Snowdon, Lühs and Friedt, 2007).

A comprehensive comparative RFLP linkage map of *A. thaliana* and *B. napus* genomes indicated the 5 *Arabidopsis* chromosomes could be allocated to a minimum of 22 conserved, duplicated and rearranged blocks throughout the *B. napus* genome (Parkin et al., 2005).

Such information highlights the complexity of genome rearrangements between the two species, but also the great potential the model genome offers for comparative genetic analysis of the *Brassica* crops (Snowdon, Lühs and Friedt, 2007).

TILLING technique

The technique of TILLING (targeted induced local lesions in genomes) can be used to identify a series of mutations (alleles) in a target gene by heteroduplex analysis (McCallum et al., 2000). This method combines a standard technique of mutagenesis with a chemical mutagen such as ethyl methanesulfonate (EMS),[6] with a sensitive DNA screening technique that identifies single-base mutations (also called point mutations) in a target gene.

This technique is available from the Canadian TILLING Initiative (CAN-TILL) at the University of British Columbia on a fee-for-service basis. The CAN-TILL facility is currently developing a large-scale mutant population for *Brassica napus* as part of a Genome Canada project and has completed projects on *B. oleracea* and *Arabidopsis thaliana*. *B. rapa* TILLING services are available from RevGenUK in the United Kingdom (John Innes Centre).

Gene transfer

The transfer of a gene(s) from an unrelated species is undertaken only when the desired trait cannot be found or induced by traditional methods. Because of the huge costs and time required to comply with multiple regulations in multiple countries, only those traits that have a potentially large and valuable market are considered for commercial exploitation.

Notes

1. The authority for the scientific names used in this chapter is given in Tables 3.2 and 3.3. The nomenclatural authority for genus and species names not listed in the tables will be included in the text where they first appear.
2. Autosyndesis is defined as the pairing of completely or partially homologous chromosomes during prophase of the first meiotic division.
3. This section is drawn from Wang, Guan and Zhang (2007).
4. This section is drawn from Downey, Klaasen and Stringham (1980); Dickson and Wallace (1986).
5 This citation has been added for update in January 2016.
6. Ethyl methanesulfonate (EMS): *http://en.wikipedia.org/wiki/Ethyl_methanesulfonate*.

References

AAFC (1996a), "Decision document DD96-07: Determination of environmental safety of Monsanto Canada Inc.'s Roundup® herbicide-tolerant *Brassica napus* canola line GT200", Agriculture and Agri-Food Canada, Canadian Food Inspection Agency, Nepean, Ontario, *www.cera-gmc.org/files/cera/GmCropDatabase/docs/decdocs/01-290-006.pdf.*

AAFC (1996b), "Decision document DD96-011: Determination of environmental safety of AgrEvo Canada Inc.'s glufosinate-ammonium-tolerant canola line HCN28", Agriculture and Agri-Food Canada.

AAFC (1995a), "Decision document DD95-01: Determination of environmental safety of AgrEvo Canada Inc.'s glufosinate-ammonium-tolerant canola", Agriculture and Agri-Food Canada, Canadian Food Inspection Agency, *www.inspection.gc.ca/plants/plants-with-novel-traits/approved-under-review/decision-documents/dd1995-01/eng/1331753432250/1331754555841.*

AAFC (1995b), "Decision document DD95-02: Determination of environmental safety of Monsanto Canada Inc.'s Roundup® herbicide-tolerant *Brassica napus* canola line GT73", Agriculture and Agri-Food Canada, Canadian Food Inspection Agency, *www.inspection.gc.ca/plants/plants-with-novel-traits/approved-under-review/decision-documents/dd1995-02/eng/1303706149157/1303751841490.*

AAFC (1995c), "Decision document DD95-03: Determination of environmental safety of pioneer Hi-Bred International Inc.'s Imidazolinone-tolerant canola", Agriculture and Agri-Food Canada, Canadian Food Inspection Agency, *www.inspection.gc.ca/plants/plants-with-novel-traits/approved-under-review/decision-documents/dd1995-03/eng/1303707055806/1303751917598.*

AAFC (1995d), "Decision document DD95-04: Determination of environmental safety of Plant Genetic Systems Inc (PGS) novel hybridization system for canola (*Brassica napus* L.)", Agriculture and Agri-Food Canada, Canadian Food Inspection Agency, *www.inspection.gc.ca/plants/plants-with-novel-traits/approved-under-review/decision-documents/dd1995-04/eng/1303707629454/1303752000144.*

AAFC (1994), "Regulatory Directive DIR 94-09: The biology of *Brassica napus* L. (canola/rapeseed)", Agriculture and Agri-Food Canada.

Abel, S. and H.C. Becker (2007), "The effect of autopolyploidy on biomass production in homozygous lines of *Brassica rapa* and *Brassica oleracea*", *Plant Breeding*, Vol. 126, No. 6, pp. 642-643, December, *http://dx.doi.org/10.1111/j.1439-0523.2007.01405.x.*

Adler, L.S. et al. (1993), "Potential for persistence of genes escaped from canola: Germination cues in crop, wild, and crop-wild hybrid *Brassica rapa*", *Functional Ecology*, Vol. 7, No. 6, pp. 736-745, December, *http://dx.doi.org/10.2307/2390196.*

Agrisearch (2001), *A Physical Survey of Representative Australian Roadside Vegetation to Evaluate the Incidence and Distribution of Canola and Key Brassicaceae Weeds*, pp. 54.

Alam, Z. (1945), "Nomenclature of oleiferous Brassicas cultivated in Punjab", *Indian Journal of Agricultural Science*, Vol. 15, No. 4, pp. 173-181.

Alam, M. et al. (1992), "Cross compatibility studies within the genus *Brassica* 1. Amphidiploid combinations", *Scientific Khyber*, Vol. 5, No. 2, pp. 89-92.

Al Mouemar, A. and H. Darmency (2004), "Lack of stable inheritance of introgressed transgene from oilseed rape in wild radish", *Environmental Biosafety Research*, Vol. 3, No. 4, October-December, pp. 209-214.

Al-Shehbaz, I.A. (1984), "The tribes of Cruciferae (Brassicaceae) in the southeastern United States", *Journal of Arnold Arboretum*, Vol. 65, pp. 343-373, July.

Al-Shehbaz, L.A., M.A. Beilstein and E.A. Kellogg (2006), "Systematics and phylogeny of the Brassicaceae (Cruciferae): An overview", *Plant Systematics and Evolution*, Vol. 259, No. 2, pp. 89-120, July, *http://dx.doi.org/10.1007/s00606-006-0415-z*.

Allainguillaume, J. et al. (2006), "Fitness of hybrids between rapeseed (*Brassica napus*) and wild *Brassica rapa* in natural habitats", *Molecular Ecology*, Vol. 15, No. 4, pp. 1 175-1 184, April.

Allard, R.W. (1960), *Principles of Plant Breeding*, J. Wiley, New York.

Allender, C.J. and G.J. King (2010), "Origins of the amphiploid species *Brassica napus* L. investigated by chloroplast and nuclear molecular markers", *BMC Plant Biology*, Vol. 10, No. 54, *http://dx.doi.org/10.1186/1471-2229-10-54*.

Almond, J.A., T.C.K. Dawkins and M.F. Askew (1986), "Aspects of crop husbandry", in: Scarisbrick, R.D.H. and R.W. Daniels (eds.), *Oilseed Rape*, Collins, London, pp. 127-175.

Ammitzbøll, H. and R.B. Jørgensen (2006), "Hybridization between oilseed rape (*Brassica napus*) and different populations and species of *Raphanus*", *Environmental Biosafety Research*, Vol. 5, No. 1, pp. 3-13, January-March.

Andersson, M.S. and M.C. de Vicente (2010), "Canola, oilseed rape", in: Andersson, M.S. and M.C. de Vicente (eds.), *Gene Flow between Crops and their Wild Relatives*, John Hopkins University Press, pp. 73-123.

Aono, M. et al. (2006), "Detection of feral transgenic oilseed rape with multiple-herbicide resistance", *Environmental Biosafety Research*, Vol. 5, No. 2, pp. 77-87, April-June.

AOSCA (n.d.), "Crop certification", Association of Official Seed Certifying Agencies, *www.aosca.org/page/Crop_Certification.aspx* (accessed 27 October 2011).

Appelqvist, L.-Å., and R. Ohlson (1972), *Rapeseed: Cultivation, Composition, Processing and Utilization*, Elsevier, Amsterdam and New York.

APS (2001), "Diseases of rapeseed = Canola (*B. napus* L. and *Brassica rapa* L. (= *B. campestris* L.))", The American Phytopathological Society (APS), *www.apsnet.org/publications/commonnames/Pages/Rapeseed.aspx* (accessed 27 October 2011).

Arabidopsis Genome Initiative (2000), "Analysis of the genome sequence of the flowering plant *Arabidopsis thaliana*", *Nature*, Vol. 408, No. 6 814, pp. 796-815, December.

Armstrong, S.J. et al. (1998), "Physical mapping of DNA repetitive sequences to mitotic and meiotic chromosomes of *Brassica oleracea* var. *alboglabra* by fluorescence *in situ* hybridization", *Heredity*, Vol. 81, pp. 666-673, *http://dx.doi.org/10.1046/j.1365-2540.1998.00432.x*.

Arumuganathan, K., and D.E. Earle (1991), "Nuclear DNA content of some important species", Plant Molecular Biology Reporter, Vol. 9(3), pp. 208-2018, *https://www.researchgate.net/profile/Elizabeth_Earle2/publication/225167844_Nuclear_DNA_content_of_some_important_plant_species/links/00b7d52ffc9cf93d3f000000.pdf*.

Australian Bureau of Statistics (2002), *Agricultural Census 2001*, Canola Statistics available at the ABS website: *www.abs.gov.au/websitedbs/D3310114.nsf/home/Home?opendocument* (accessed October 2011).

Axelsson, T. et al. (2000), "Amphidipoid *Brassica juncea* contains conserved progenitor genomes", *Genome*, Vol. 43, pp. 679-688.

Ayotte, R., P.M. Harney and V. Souza Machado (1987), "The transfer of triazine resistance from *Brassica napus* L. to *B. oleracea* L. I. Production of F_1 hybrids through embryo rescue", *Euphytica*, Vol. 36, No. 2, pp. 615-624, January.

Bailey, L.H. (1930), "The cultivated Brassicas: Second paper", *Gentes Herbarum*, No. 2, pp. 211-267.

Baker, J and C. Preston (2008), "Canola (*Brassica napus* L.) seedbank declines rapidly in farm managed fields in South Australia", *Australian Journal of Agricultural Research*, Vol. 59, No. 8, pp. 780-784, *http://dx.doi.org/10.1071/AR07436*.

Banga, S.S., J.S. Deol and S.K. Banga (2003), "Alloplasmic male sterile *Brassica juncea* with *Enarthrocarpus lyratus* cytoplasm and the introgression of gene(s) for fertility restoration from cytoplasmic donor species", *Theoretical and Applied Genetics*, Vol. 106, pp. 1 390-1 395.

Banga, S.S. et al. (2003), "Alloplasmic line of *Brassica napus* L. with *Erucastrum canariense* cytoplasm is male sterile", *Proceedings of the 11th International Rapeseed Congress*, Copenhagen, 6-10 July, pp. 324-325.

Baranger, A. et al. (1995), "Effect of oilseed rape genotype on the spontaneous hybridization rate with a weedy species: An assessment of transgene dispersal", *Theoretical and Applied Genetics*, Vol. 91, Nos. 6-7, pp. 956-963, November, *http://dx.doi.org/10.1007/BF00223906*.

Barber, S. (1999), "Transgenic plants: Field testing and commercialisation including a consideration of novel herbicide resistant oilseed rape (*Brassica napus* L.)", in: Lutmann, P.J.W. (ed.), *Gene Flow and Agriculture: Relevance for Transgenic Crops*, British Crop Protection Council Symposium Proceedings, Vol. 72, pp. 49-56.

Bateman, A.J. (1974a), "Contamination in seed crops: I. Insect pollination", *Journal of Genetics*, Vol. 48, No. 2, pp. 257-275, August.

Bateman, A.J. (1974b), "Contamination in seed crops: III. Relation with isolation distance", *Heredity*, Vol. 1, pp. 303-336.

Batra, V., K.R. Shivanna and S. Prakash (1989), "Hybrids of wild species *Erucastrum gallicum* and crop brassicas", *Proceedings of the 6th International Congress of SABRAO*, Tsukuba, pp. 443-446.

Bauer-Weston, W.B. et al. (1993), "Production and characterization of asymmetric somatic hybrids between *Arabidopsis thaliana* and *Brassica napus*", *Theoretical and Applied Genetics*, Vol. 86, Nos. 2-3, pp. 150-158, April, *http://dx.doi.org/10.1007/BF00222073*.

Becker, H.C., C. Damgaard and B. Karlsson (1992), "Environmental variation for outcrossing rate in rapeseed (*Brassica napus*)", *Theoretical and Applied Genetics*, Vol. 84, No. 3, pp. 303-306, July, *http://dx.doi.org/10.1007/BF00229487*.

Becker, H.C., H. Löptien and G. Röbbelen (1999), "Breeding: An overview", in: Gómez-Campo, C. (ed.), *Biology of Brassica Coenospecies*, Elsevier Science, pp. 413-460.

Becker, H.C. et al. (1991), "Genotypic and environmental variation for outcrossing rate in rapeseed (*Brassica napus*)", in McGregor, D.I. (ed.), *Proceedings of the 8th International Rapeseed Congress*, Saskatoon, Canada, pp. 1 455-1 459.

Beckie, H.J. (2011), "Herbicide-resistant weed management: Focus on glyphosate", *Pest Management Science*, Vol. 67, No. 9, pp. 1 037-1 048, *http://onlinelibrary.wiley.com/doi/10.1002/ps.2195/epdf*.

Beckie, H.J. and M.D.K. Owen (2007), "Herbicide-resistant crops as weeds in North America", *CAB Reviews: Perspectives in Agriculture, Veterinary Science, Nutrition and Natural Resources*, Vol. 2, No. 044, *www.weeds.iastate.edu/weednews/2007/PAV2044.pdf*.

Beckie, H.J. and S.I. Warwick (2010), "Persistence of an oilseed rape transgene in the environment", *Crop Protection*, Vol. 29, No. 5, pp. 509-512, May, *http://dx.doi.org/10.1016/j.cropro.2009.11.013*.

Beckie, H.J., L.M. Hall and S.I. Warwick (2001), "Impact of herbicide-resistant crops as weeds in Canada", *Proceedings of the Brighton Crop Protection Conference: Weeds*, British Crop Protection Council, pp. 135-142.

Beckie, H.J. et al. (2006), "A decade of herbicide-resistant crops in Canada", *Canadian Journal of Plant Science*, Vol. 86, No. 4, pp. 1 243-1 264, *http://dx.doi.org/10.4141/P05-193*.

Beckie, H.J. et al. (2004), "Multiple herbicide-resistant canola can be controlled by alternative herbicides", *Weed Science*, Vol. 52, No. 1, pp. 152-157, *www.jstor.org/stable/4046805*.

Beckie, H.J. et al. (2003), "Gene flow in commercial fields of herbicide-resistant canola (*Brassica napus*)", *Ecological Applications*, Vol. 13, No. 5, pp. 1 276-1 294, October.

Beismann, H. and A. Roller (2003), "Assessing the number of transgenic oilseed rape seeds in the soil seedbank of former release sites", *Aspects of Applied Biology*, Vol. 69, pp. 209-215.

Bell, J.M. (1993), "Factors affecting the nutritional value of canola meal: A review", *Canadian Journal of Animal Science*, Vol. 73, No. 4, pp. 679-697, *http://dx.doi.org/10.4141/cjas93-075*.

Bell, D.T. and C.H. Muller (1973), "Dominance of California annual grasslands by *Brassica nigra*", *The American Midland Naturalist*, Vol. 90, No. 2, pp. 277-299, October, *http://dx.doi.org/10.2307/2424453*.

Benabdelmouna, A. et al. (2003), "Genome discrimination in progeny of interspecific hybrids between *Brassica napus* and *Raphanus raphanistrum*", *Genome*, Vol. 46, No. 3, pp. 469-472.

Bengtsson, L. et al. (1972), "Botany of rapeseed", in: Appelqvist, L.-Å. and R. Olson (eds.), *Rapeseed: Cultivation, Composition, Processing and Utilization*, Elsevier, Amsterdam and New York, pp. 36-48.

Bennett, M.D. and I.J. Leitch (2012), *Angiosperm DNA C-values database* (release 8.0, December 2012), *http://data.kew.org/cvalues/*.

Beschorner, M., B. Plümper and W. Odenbach (1995), "Analysis of self-incompatability interactions in 30 resynthesized *Brassica napus* lines. I. Fluorescence microscopic studies", *Theoretical and Applied Genetics*, Vol. 90, pp. 665-670, *http://dx.doi.org/10.1007/BF00222131*.

Bielikova, L. and S. Rakousky (2001), "Survey on oilseed rape cultivation and weed relatives in the Czech Republic", *European Science Foundation Meeting of a Working Group on Interspecific Gene Flow from Oilseed Rape to Weedy Species*, June, Rennes, France.

Bijral, J.S. and T.R. Sharma (1998), "Production and cytology of intergeneric hybridization between *Brassica napus* and *Diplotaxis catholica*", *Cruciferae Newsletter Eucarpia*, Vol. 20, No. 15.

Bijral, J.S. and T.R. Sharma (1996a), "Cytogenetics of intergeneric hybrids between *Brassica napus* L. and *Eruca sativa* Lam", *Cruciferae Newsletter Eucarpia*, Vol. 18, pp. 12-13.

Bijral, J.S. and T.R. Sharma (1996b), "Intergeneric hybridization between *Brassica napus* and *Diplotaxis muralis*", *Cruciferae Newsletter Eucarpia*, Vol. 18, pp. 10-11.

Bijral, J.S., T.R. Sharma and K.S. Kanwal (1993), "Morphocytogenetics of *Brassica napus* L. × *Sinapis alba* L. sexual hybrids", *Indian Journal of Genetics of Plant Breeding*, Vol. 53, No. 4, pp. 442-444.

Bijral, J.S. et al. (1995), "Interspecific hybrids of *Brassica maurorum* with *Brassica* crops, and their cytology", *Cruciferae Newsletter Eucarpia*, Vol. 17, pp. 18-19.

Bilsborrow, P.E. et al. (1998), "Contamination of edible double-low oilseed rape cultivars via pollen transfer from high erucic cultivars", *J; Sci; of Food and Agriculture*, Vol. 76, No. 1, pp. 17-22, *http://dx.doi.org/10.1002/(SICI)1097-0010(199801)76:1<17::AID-JSFA903>3.0.CO;2-9*.

Bing, D.J., R.K. Downey and G.F.W. Rakow (1991), "Potential of gene transfer among oilseed *Brassica* and their weedy relatives", in: McGregor, D.I. (ed.), *Proceedings of the 8th International Rapeseed Congress*, Saskatoon, Canada, pp. 1 022-1 027.

Bing, D.J. et al. (1996), "Assessment of transgene escape from *Brassica rapa* (*B. campestris*) into *B. nigra* or *Sinapis arvensis*", *Plant Breeding*, Vol. 115, No. 1, pp. 1-4, April, *http://dx.doi.org/10.1111/j.1439-0523.1996.tb00861.x*.

Bohman, S., M. Wang and C. Dixelius (2002), "*Arabidopsis thaliana*-derived resistance against *Leptosphaeria maculans* in a *Brassica napus* genomic background", *Theoretical and Applied Genetics*, Vol. 105, No. 4, pp. 498-504.

Bonnemaison, L. (1965), "Insect pests of crucifers and their control", *Annual Reviews of Entomology*, Vol. 10, pp. 233-256, January, *http://dx.doi.org/10.1146/annurev.en.10.010165.001313*.

Booth, E.J. et al. (1996), "Assessment of the ecological consequences of introducing transgenic rapeseed", *4th ESA Congress Book of Abstracts (Persistence of Oil-modified Oilseed Rape, Sinapis arvensis and Brassica nigra)*, pp. 144-145.

Boulter, G.S. (1983), "The history and marketing of rapeseed oil in Canada", in: Kramer, J.K. et al. (eds.), *High and Low Erucic Acid Rapeseed Oils*, Academic Press, Toronto, pp. 62-83.

Brewer, E.P. et al. (1999), "Somatic hybridization between the zinc accumulator *Thlaspi caerulescens* and *Brassica napus*", *Theoretical and Applied Genetics*, Vol. 99, No. 5, pp. 761-771.

Briggs, F.N. and P.F. Knowles (1967), *Introduction to Plant Breeding*, Reinhold Publishing.

Brown, A.P. et al. (1996), "Gene transfer between canola (*Brassica napus*) and related weed species", *Proceedings of the 8th Symposium on Environmental Releases of Biotechnology Products: Risk Assessment Methods and Research Progress*, Ottawa, Ontario, Canada.

Brown, J. and A.P. Brown (1996), "Gene transfer between canola (*Brassica napus* L. and *B. campestris* L.) and related weed species", *Annals of Applied Biology*, Vol. 129, No. 3, pp. 513-522, December, *http://dx.doi.org/10.1111/j.1744-7348.1996.tb05773.x*.

Brown, P.D. and M.J. Morra (1997), "Control of soil-borne plant pests using glucosinolate-containing plants", *Advances in Agronomy*, Vol. 61(C), pp. 167-213, *http://dx.doi.org/10.1016/S0065-2113(08)60664-1*.

Brown, J. et al. (1997), "Intergeneric hybridization between *Sinapis alba* and *Brassica napus*", *Euphytica*, Vol. 93, No. 2, pp. 163-168, January.

Budahn, H. et al. (2006), "Intergeneric transfer of nematode resistance from *Raphanus* to *Brassica* using a series of rape-radish chromosome addition lines", *Acta Horticulturae 706: IV International Symposium on Brassicas and XIV Crucifer Genetics Workshop*, *http://dx.doi.org/10.17660/ActaHortic.2006.706.16*.

Burkill, I.H. (1930), "The Chinese mustards in the Malay peninsula", *Gad's Bulletin*, No. 5, pp. 99-117.

Butterfass, T.H. (1989), "Nuclear control of plastid division", in: Boffey, S.A. and D. Lloyd (eds.), *Division and Segregation of Organelles*, Society for Experimental Biology, Seminar Series 25, pp. 21-38.

Buzza, G.C. (1995), "Plant breeding", in: Kimber, D.S. and D.I. McGregor (eds.), *Brassica Oilseeds, Production and Utilization*, CAB International, Wallingford, Oxon, United Kingdom, pp. 153-175.

Campbell, A.G. (1959), "A germination inhibitor and root growth retarder in chou Moellier (*Brassica oleracea* var.)", *Nature*, Vol. 183, pp. 1 263-1 264, http://dx.doi.org/10.1038/1831263a0.

Canadian Food Inspection Agency (2009), "5.9 Legal Reference Samples", *Procedures for the Registration of Crop Varieties in Canada*, Government of Canada, *www.inspection.gc.ca/english/plaveg/variet/proced/regproe.shtml#aii59* (accessed 27 October 2011).

Canadian Food Inspection Agency (2007), "The biology of Brassica juncea (canola/mustard), *Biology Document BIO2007-01: A Companion Document to the Directive 94-08 (Dir94-08), Assessment Criteria for Determining Environmental Safety of Plants with Novel Traits*, *www.inspection.gc.ca/plants/plants-with-novel-traits/applicants/directive-94-08/biology-documents/brassica-juncea/eng/1330727837568/1330727899677*.

Canadian Seed Growers Association (CSGA) (2009), "Canadian regulations and procedures for breeder seed crop production", *http://seedgrowers.ca/wp-content/uploads/Breeder-Seed-Crop-Regulations_ENGLISH_20100322.pdf* (accessed 13 January 2016).

Canola Council of Canada, "What is canola? – Official definition of canola", *http://www.canolacouncil.org/oil-and-meal/what-is-canola/* (accessed 14 January 2016).

CAN-TILL (The Canadian TILLING Initiative), "CAN-TILL", University of British Columbia, Vancouver, BC, Canada, *www3.botany.ubc.ca/can-till/* (accessed14 January 2016).

CETIOM (2000), "Introduction of genetically modified rapeseed tolerant to various herbicides in the French agriculture system: Evaluation of the agro-environmental impact and development of management scenarios", *Summary of GMO 2000 report prepared in the framework of the Moratorium*, Ver. 4.2, pp. 29, Centre technique interprofessionnel des oléagineux métropolitains.

Chadoeuf, R., H. Darmency and J. Maillet (1998), "Survival of buried seeds of interspecific hybrids between oilseed rape, hoary mustard and wild radish", *Field Crops Research*, Vol. 58, No. 3, pp. 197-204, *http://dx.doi.org/10.1016/S0378-4290(98)00095-1*.

Champolivier, J. et al. (1999), "Management of transgenic crops within the cropping system", in: Lutmann, P.J.W. (ed.), *Gene Flow and Agriculture: Relevance for Transgenic Crops*, British Crop Protection Council Symposium Proceedings, Vol. 72, pp. 233-240.

Chang, K.C. (1970), "The beginnings of agriculture in the Far East", *Antiquity*, Vol. 44, No. 175, pp. 175-185, September, *http://dx.doi.org/10.1017/S0003598X0004148X*.

Chapman, G.P. and Y.W. Wang (2002), *The Plant Life of China: Diversity and Distribution*, Springer-Verlag, Berlin.

Chen, B.Y. and W.K. Heneen (1991), "The basic number of *Brassica* genomes: *x*=3?", *Cruciferae Newsletter*, Vol. 14-15, pp. 20-21.

Chen, B.Y., W.K. Heneen and V. Simonsen (1989), "Comparative and genetic studies of isozymes in resynthesized and cultivated *Brassica napus* L., *B. campestris* L. and *B. alboglabra* Bailey", *Theoretical and Applied Genetics*, Vol. 77, No. 5, pp. 673-679, May, *http://dx.doi.org/10.1007/BF00261242*.

Chen, L. et al. (2005), "Production of interspecific somatic hybrids between tuber mustard (*Brassica juncea*) and red cabbage (*Brassica oleracea*)", *Plant Cell, Tissue and Organ Culture*, Vol. 80, No. 3, pp. 305-311.

Chèvre, A.M. et al. (2004), "A review on interspecific gene flow from oilseed rape to wild relatives", Chapter 18, in: den Nijs, H.C.M. et al. (eds.), *Introgression from Genetically Modified Plants into Wild Relatives*, CABI Pub, pp. 235-251.

Chèvre, A.M. et al. (2003), "Introduction of black leg resistance from *Brassica rapa* into *B. napus*", *Proceedings of the 11th International Rapeseed Congress*, Copenhagen, 6-10 July, Vol. 1, pp. 32-35.

Chèvre, A.M. et al. (2001), "Impacts agro-environnementaus de la mise en culture de variétés de colza transgéniques tolérantes à des herbicides", *Comptes-rendu de l'Académie d'Agriculture de France*, Vol. 87, No. 5, pp. 11-20.

Chèvre, A.M. et al. (2000), "Assessment of interspecific hybridization between transgenic oilseed rape and wild radish under normal agronomic conditions", *Theoretical and Applied Genetics*, Vol. 100, No. 8, pp. 1 233-1 239, June.

Chèvre, A.M. et al. (1999), "Gene flow from oilseed rape to weeds", in: Lutmann, P.J.W. (ed.), *Gene Flow and Agriculture: Relevance for Transgenic Crops*, British Crop Protection Council Symposium Proceeding, Vol. 72, pp. 125-130.

Chèvre, A.M. et al. (1998), "Characterisation of backcross generations obtained under field conditions from oilseed rape-wild radish F1 interspecific hybrids: An assessment of transgene dispersal", *Theoretical and Applied Genetics*, Vol. 97, pp. 90-98.

Chèvre, A.M. et al. (1997a), "Gene flow from transgenic crops", *Nature*, Vol. 389, No. 924, October, *http://dx.doi.org/10.1038/40054*.

Chèvre, A.M. et al. (1997b), "Selection of stable *Brassica napus-B. juncea* recombinant lines resistant to black leg (*Leptosphhaeria maculans*). 1. Identification of molecular markers, chromosomal and genomic origin of the introgression", *Theoretical and Applied Genetics*, Vol. 95, pp. 1 104-1 111.

Chèvre, A.M. et al. (1996), "Interspecific gene flow as a component of risk assessment for transgenic *Brassicas*", *ISHS Acta Horticulturae 407: ISHS Brassica Symposium: IX Crucifer Genetics Workshop*, pp. 169-179, *http://dx.doi.org/10.17660/ActaHortic.1996.407.21*.

Chèvre, A.M. et al. (1994), "Comparison of somatic and sexual *B. napus-Sinapis alba* hybrids and their progeny by cytogenetic and molecular characterization", *Genome*, Vol. 37, No. 3, pp. 367-374, June.

Chiang, M.S. (1974), "Cabbage pollen germination and longevity", *Euphytica*, Vol. 23, No. 3, pp. 579-584.

Chiang, M.S., B.Y. Chiang and W.F. Grant (1977), "Transfer of resistance to race 2 of *Plasmodiophora brassicae* from *Brassica napus* to cabbage (*B. oleracea* var. *capitata*). I. Interspecific hybridization between *B. napus* and *B. oleracea* var. *capitata*", *Euphytica*, Vol. 26, No. 2, pp. 319-336, June.

Chiang, M.S. et al. (1980), "Transfer of resistance to race 2 of *Plasmodiophora brassiceae* from *Brassica napus* to cabbage (*B. oleracea* var. *capitata*): IV. A resistant 18-chromosome B1 plant and its B2 progeny", *Euphytica*, Vol. 29, pp. 47-55.

Choudhary, B.R. and P. Joshi (1999), "Interspecific hybridization in *Brassica*", *Proceedings of the 10th International Rapeseed Congress*, Australia, Contribution No. 516, *www.regional.org.au/au/gcirc/4/516.htm*.

Christey, M.C. (2004), "*Brassica* protoplast culture and somatic hybridization", in: Pua, E.C. and C.J. Douglas (eds.), *Biotechnology in Agriculture and Forestry*, Vol. 54, Springer, New York, pp. 119-148.

Chyi, Y.S., M.E. Hoenecke and J.L. Sernyk (1992), "A genetic linkage map of restriction fragment length polymorphism loci for *Brassica rapa* (syn. *campestris*)", *Genome*, Vol. 35, No. 5, pp. 746-757, *http://dx.doi.org/10.1139/g92-115*.

Colbach, N. et al. (2005), "Spatial aspects of gene flow between rapeseed varieties and volunteers", *Agronomy for Sustainable Development*, Vol. 25, No. 3, pp. 355-368, *https://hal.archives-ouvertes.fr/hal-00886294*.

Coulthart, M.B. and KE. Denford (1982), "Isozyme studies in *Brassica*. I. Electrophoretic techniques for leaf enzymes and comparison of *B. napus*, *B. campristris* and *B. oleracea* using phosphoglucomutase", *Canadian Journal of Plant Science*, Vol. 62, pp. 621-630.

Craig, A. and S. Millam (1995), “Modification of oilseed rape to produce oils for industrial use by means of applied tissue culture methodology”, *Euphytica*, Vol. 85, No. 1, pp. 323-327.

Crawford, J.W., G.R. Squire and D. Burn (1999), “Modelling spread of herbicide resistance in oilseed rape”, in: Gray, A.J. et al. (eds.), *Environmental Impact of Genetically Modified Crops*, DETR Research Report No. 10, pp. 97-106, DETR, London.

Crawley, M.J. and S.L. Brown (1995), “Seed limitation and the dynamics of feral oilseed rape on the M25 motorway”, *Proceedings B of the Royal Society of London*, Vol. 259, No. 1 354, pp. 49-54, *http://dx.doi.org/10.1098/rspb.1995.0008*.

Crawley, M.J. et al. (2001), “Transgenic crops in natural habitats”, *Nature*, Vol. 409, No. 6 821, pp. 682-683, February, *http://dx.doi.org/10.1038/35055621*.

Crawley, M.J. et al. (1993), “Ecology of transgenic oilseed rape in natural habitat”, *Nature*, Vol. 363, pp. 620-623, June, *http://dx.doi.org/10.1038/363620a0*.

Cresswell, J.E. (1994), “A method for quantifying the gene flow that results from a single bumblebee visit using transgenic oilseed rape, *Brassica napus* L. cv. Westar”, *Transgenic Research*, Vol. 3, No. 2, pp. 134-137, March.

CropScience (2009), “Bayer CropScience first to sequesce the entire genome of rapeseed/canola”, *www.cropscience.bayer.ca/en/News/2009/News98.aspx?overviewId=4B8E36F2-C34D-42B1-8573-8894864C8229* (accessed 14 January 2016).

Daniels, R. et al. (2005), “The potential for dispersal of herbicide tolerance genes from genetically-modified, herbicide-tolerant oilseed rape crops to wild relatives”, Final Report to DEFRA, July.

Darmency, H. and A. Fleury (2000), “Mating system in *Hirschfeldia incana* and hybridization to oilseed rape”, *Weed Research*, Vol. 40, No. 2, pp. 231-238, April, *http://dx.doi.org/10.1046/j.1365-3180.2000.00186.x*.

Darmency, H., A. Fleury and E. Lefol (1995), “Effect of transgenic release on weed biodiversity: Oilseed rape and wild radish”, *Crop Protection*, Vol. 5a2, pp. 433-438.

Darmency, H., E. Lefol and A. Fleury (1998), “Spontaneous hybridizations between oilseed rape and wild radish”, *Molecular Ecology*, Vol. 7, No. 11, pp. 1 467-1 473, November, *http://dx.doi.org/10.1046/j.1365-294x.1998.00464.x*.

Dass, H. and N. Nybom (1967), “The relationships between *Brassica nigra*, *B. campestris*, *B. oleracea*, and their amphidiploid hybrids studied by means of numerical chemotaxonomy”, *Canadian J. of Genetic Cytology*, Vol. 9, No. 4, pp. 880-890, *http://dx.doi.org/10.1139/g67-093*.

Daun, J.K. (1983), “The introduction of low erucic acid rapeseed varieties into Canadian production”, Chapter 7, in: Kramer, J.K.G. et al. (eds.), *High and Low Erucic Acid Rapeseed*, Academic Press, Toronto, pp. 161-180, *http://dx.doi.org/10.1016/B978-0-12-425080-2.50012-8*.

Davey, V.M. (1959), “Cultivated *Brassiceae*: Information available to the breeder”, *Report of the Scottish Plant Breeding Station*, pp. 23-62.

De Candolle, A. (1885), *Origin of Cultivated Plants*, The International Scientific Series, D. Appleton and Comp., London, *https://archive.org/details/origincultivate03candgoog*.

De Nettancourt, D. (1971), *Incompatibility in Angiosperms*, Springer-Verlag, Berlin.

Deol, J.S. et al. (2003), “*Enarthrocarpus lyratus*-based cytoplasmic male sterility and fertility restorer system in *Brassica rapa*”, *Plant Breeding*, Vol. 122, No. 5, pp. 438-440, October, *http://dx.doi.org/10.1046/j.1439-0523.2003.00836.x*.

Devaux, C. et al. (2007), “Modelling and estimating pollen movement in oilseed rape (*Brassica napus*) at the landscape scale using genetic markers”, *Molecular Ecology*, Vol. 16, No. 3, pp. 487-499, February.

Dhillon, S.S. et al. (1985), “Root tumours in interspecific crosses of *Brassica*”, *Cruciferae Newsletter*, No. 10.

Dickinson, H.G. (1995), "Dry stigmas, water and self-incompatibility in *Brassica*", *Sexual Plant Reproduction*, Vol. 8, No. 1, pp. 1-10, January.

Dickson, M.H. and D.H. Wallace (1986), "Cabbage breeding", Chapter 11, in: Bassett, M.J. (ed.), *Breeding Vegetable Crops*, AVI Publishing Co., pp. 395-432, available at: *http://faculty.agron.iastate.edu/fehr/BVC/11BVC.PDF*.

Diederichsen, E. and M.D. Sacristan (1988), "Interspecific hybridizations in the genus *Brassica* followed by in-ovule embryo culture", *Cruciferae Newsletter Eucarpia*, No. 13, pp. 20-21.

Dietz-Pfeilstetter, A. and P. Zwerger (2009), "In-field frequencies and characteristics of oilseed rape with double herbicide resistance", *Environmental Biosafety Research*, Vol. 8, pp. 100-111, *http://dx.doi.org/10.1051/ebr/2009006*.

Dietz-Pfeilstetter, A. and P. Zwerger (2004), "Dispersal of herbicide resistance genes during the large scale cultivation of different transgenic, herbicide resistant oilseed rape varieties", *Journal of Plant Diseases and Protection*, Vol. XIX, pp. 831-838.

Dietz-Pfeilstetter, A., K. Metge and J. Schönfeld (2006), "Assessment of transgene spread from oilseed rape by population dynamic and molecular analyses of feral oilseed rape", Chapter 2, *Journal of Plant Diseases and Protection*, Special Issue XX, pp. 39-47 (in German).

Dietz-Pfeilstetter, E. et al. (1998), "Untersuchungen zum Auskreuzungsverhalten von Basta-tolerant Winterraps auf Nicht Transgenem Raps (*Brassica napus*)", *Mitt. a. d. Biol. Bundesanst.*, H.257, pp. 121 (in German).

Dignam, M. (2001), *Bush, Parks and Road and Rail Weed Management Survey*, Mark Dignam and Associates, Australia.

Dixelius, C. (1999), "Inheritance of the resistance to *Leptosphaeria maculans* of *Brassica nigra* and *B.juncea* in near-isogenic lines of *B. napus*", *Plant Breeding*, Vol. 118, No. 2, pp. 151-156, April, *http://dx.doi.org/10.1046/j.1439-0523.1999.118002151.x*.

Dixon, G.R. (2007), *Vegetable Brassicas and Related Crucifers*, Crop Production Science in Horticulture Series, No. 14, CABI, Oxfordshire, United Kingdom.

Downey, R.K. (1999a), "Risk assessment of outcrossing of transgenic *Brassica*, with focus on *B. rapa* and *B. napus*", *Proceedings of the 10th International Rapeseed Congress*, Australia, Contribution No. 61, *www.regional.org.au/au/gcirc/4/61.htm*.

Downey, R.K. (1999b), "Gene flow and rape – the Canadian experience", in: Lutmann, P.J.W. (ed.), *Gene Flow and Agriculture: Relevance for Transgenic Crops*, British Crop Protection Council Symposium Proceedings, Vol. 72, pp. 109-116.

Downey, R.K. (1996), "Diversification of canola/rapeseed fatty acid supply for the year 2000", *OCL*, Vol. 22, No. 3, pp. 9-13, *http://dx.doi.org/10.1051/ocl/2015011*.

Downey, R.K. (1983), "The origin and description of the Brassica oilseed crops", in: Kramer, J.K.G., F.D. Sauer and W.J. Pigden (eds.), *High and Low Erucic Acid Rapeseed Oils: Production, Usage, Chemistry and Toxicological Evaluation*, Academic Press, Toronto, pp. 1-20.

Downey, R.K. (1964), "A selection of *Brassica campestris* L. containing no erucic acid and in its seed oil", *Canadian Journal of Plant Science*, Vol. 44, No. 3, pp. 295, *http://dx.doi.org/10.4141/cjps64-057*.

Downey, R.K. and J.M. Armstrong (1962), "Inter-compatibility and performance of diploid and derived tetraploid rape (*Brassica campestris* L.)", *Canadian Journal of Plant Science*, Vol. 42, No. 4, pp. 672-680, *http://dx.doi.org/10.4141/cjps62-114*.

Downey, R.K. and H. Beckie (2002), "Isolation effectiveness in canola pedigree seed production", Agriculture and Agri-Food Canada, Saskatoon Research Centre, Saskatoon, Canada, *http://oapf.saskorganic.com/pdf/canola-study.pdf*.

Downey, R.K. and J. Buth (2003), "Transgenic rapeseed – Grower adoption and consumer acceptance", *Proceedings of the 11th Rapeseed Congress*, Copenhagen, pp. 1 190-1 194.

Downey, R.K. and G.F.W. Rakow (1987), "Rapeseed and mustard", in: Fehr, W. (ed.), *Principles of Cultivar Development Vol. 2 Crop Species*, Macmillan, New York, pp. 437-486.

Downey, R.K. and S.R. Rimmer (1993), "Agronomic improvement in oilseed Brassicas", *Advances in Agronomy*, Vol. 50, pp. 1-66.

Downey, R.K. and G. Röbbelen (1989), "Brassica species", in: Röbbelen, G. et al. (eds.), *Oil Crops of the World*, McGraw-Hill, New York, pp. 339-362.

Downey, R.K., A.J. Klaasen and G.R. Stringham (1980), "Rapeseed and mustard", in: Fehr, W.R. and H.H. Hadley (eds.), *Hybridization of Crop Plants*, American Society of Agronomy, Crop Science Society of America, pp. 495-509.

Dunfield, K.E. and J.J. Germida (2004), "Impact of genetically modified crops on soil- and plant-associated microbial communities", *Journal of Environmental Quality*, Vol. 33, No. 3, pp. 806-815, May-June.

Dunfield, K.E. and J.J. Germida (2003) "Seasonal changes in the rhizophere microbial communities associated with field-grown genetically modified canola (*Brassica napus*)", *Applied and Environmental Microbiology*, Vol. 69, No. 12, pp. 7 310-7 318, December, *http://dx.doi.org/10.1128/AEM.69.12.7310-7318.2003.*

Dunfield, K.E. and J.J. Germida (2001), "Diversity of bacterial communities in the rhizophere and root-interior of field-grown genetically modified *Brassica napus*", *FEMS Microbiology Ecology*, Vol. 38, No. 1, pp. 1-9, December, *http://dx.doi.org/10.1111/j.1574-6941.2001.tb00876.x.*

Dushenkov, S. et al. (2002), "Bioengineering of a phytoremediation plant by means of somatic hybridization", *International Journal of Phytoremediation*, Vol. 4, No. 2, pp. 117-126.

Eastham, K. and J. Sweet (2002), "Genetically modified organisms (GMOs): The significance of gene flow through pollen transfer", *Environmental Issue Report*, No. 28, European Environment Agency, Copenhagen, *www.pedz.uni-mannheim.de/daten/edz-bn/eua/02/28.pdf.*

Eber, F. et al. (1994), "Spontaneous hybridization between a male sterile oilseed rape and two weeds", *Theoretical and Applied Genetics*, Vol. 88, Nos. 3-4, pp. 362-368, June, *http://dx.doi.org/10.1007/BF00223646.*

Ekbom, B. (1995), "Insect pests", in: Kimber, D.S. and D.I. McGregor (eds.), *Brassica Oilseeds, Production and Utilization*, CAB International, Oxon, United Kingdom, pp. 141-152.

Elliott, R.H., C. Franke and G.F.W. Rakow (2008), "Effects of seed size and seed weight on seedling establishment, vigor and tolerance of Argentine canola (*Brassica napus*) to flea beetles, *Phyllotreta* spp", *Canadian Journal of Plant Science*, Vol. 88, No. 1, pp. 207-217, *http://dx.doi.org/10.4141/CJPS07059.*

Ellis, R.H. et al. (1994), "The long-term storage of seeds of seventeen crucifers at very low moisture contents", *Plant Genetics Resources Newsletter*, No. 98.

Erickson, L.R., N.A. Straus and W.B. Beversdorf (1983), "Restriction patterns reveal origins of chloroplast genomes in *Brassica* amphidiploids", *Theoretical and Applied Genetics*, Vol. 65, pp. 201-206.

Eskin, N.A.M. et al. (1989), "Stability of low linolenic acid canola oil to frying temperature", *Journal of the American Oil Chemists' Society*, Vol. 66, pp. 1 081-1 084.

European Commission (2000), "Opinion regarding submission for placing on the market of glufosinate tolerant oilseed rape transformation event liberator PHOE 6/AC notified by the Hoechst schering Agrevo Company [Now Aventis CropScience] (Notification C/DE/98/6), Directorate general "Health & Consumer Protection", available at: *http://ec.europa.eu/food/fs/sc/scp/out88_gmo_en.html.*

European Commission (1999), "Opinion of the Scientific Committee on Plants, adopted on 18 May 1999, on the invocation by France of Article 16 ('Safeguard Clause') of Council Directive 90/220/EEC with respect to a genetically modified oilseed rape notification C/UK/94/M1/1 (Plant Genetic Systems N.V.)-(SCP/GM/150-final)", *http://ec.europa.eu/food/fs/sc/scp/out37_en.html*.

European Commission (1998a), "Opinion of the Scientific Committee on Plants regarding the glufosinate tolerant hybrid rape derived from genetically modified parental lines (MS8xRF3) notified by plant genetic systems (Notification C/B/96/01)", submitted by the Scientific Committee on Plants 19 May 1998.

European Commission (1998b), "Opinion of the Scientific Committee on Plants regarding submission for placing on the market of glufosinate tolerant oilseed rape transformation event liberator pHOE 6/AC notified by the AgrEvo Company [Now Aventis CropScience] (Notification C/DE/98/6)", *http://ec.europa.eu/food/fs/sc/scp/out88_gmo_en.html*.

Fahleson, J., I. Eriksson and K. Glimelius (1994), "Intertribal somatic hybrids between *Brassica napus* and *Barbarea vulgaris* – Production of in vitro plantlets", *Plant Cell Reproduction*, Vol. 13, No. 7, pp. 411-416, *http://dx.doi.org/10.1007/BF00234149*.

Fahleson, J. et al. (1994), "Intertribal somatic hybrids between *Brassica napus* and *Thlaspi perfoliatum* with high content of the *T. perfoliatum*-specific nervonic acid", *Theoretical and Applied Genetics*, Vol. 87, No. 7, pp. 795-804, February, *http://dx.doi.org/10.1007/BF00221131*.

Falk, K.C. and D.L. Woods (2003), "Seed yield of successive synthetic generations in summer turnip rape", *Canadian Journal of Plant Science*, Vol. 83, No. 2, pp. 271-274, *http://dx.doi.org/10.4141/P01-182*.

Falk, K.C. et al. (1994), "Performance of inter-cultivar summer turnip rape hybrids in Saskatchewan", *Canadian Journal of Plant Science*, Vol. 74, No. 3, pp. 441-445, *http://dx.doi.org/10.4141/cjps94-082*.

Fan, Z., W. Tai and B.R. Stefansson (1985), "Male sterility in *Brassica napus* L. associated with an extra chromosome", *Canadian Journal of Genetics and Cytology*, Vol. 27, No. 4, pp. 467-471, *http://dx.doi.org/10.1139/g85-068*.

FAOSTAT (2013), FAO Statistics online database, "Production / Crops, – rapeseed, Year 2013", Food and Agriculture Organization of the United Nations, *http://faostat3.fao.org/home/E* (accessed 5 January 2016).

Farnham, M.W. (2007), "Vegetable crucifers – Status report", USDA Crucifer Crop Germplasm Committee, *www.ars-grin.gov/npgs/cgc_reports/crucifer1201.htm*.

Fenwick, J.W. et al. (1983), "Glucosinolates and their breakdown products in food and food plants", *Critical Reviews in Food Science and Nutrition*, Vol. 18, No. 2, pp. 123-201, *http://dx.doi.org/10.1080/10408398209527361*.

Fernandez-Escobar. J. et al. (1988), "Genetics of erucic acid content in interspecific hybrids of Ethiopian mustard (*Brassica carinata* Braun) and rapeseed (*B. napus* L.)", *Plant Breeding*, Vol. 100, pp. 310-315.

Ferrie, A.M. and W.A. Keller (2004), "*Brassica* improvement through microsporculture", in: Pua, E.C. and C.J. Douglas (eds.), *Biotechnology in Agriculture and Forestry*, Vol. 54: Brassica, pp. 148-168, Springer.

Finkelstein, R.R. et al. (1985), "Role of ABA in maturation of rapeseed embryos", *Plant Physiology*, Vol. 78, No. 3, pp. 630-636, July.

Flannery, M.L. et al. (2006), "Plastid genome characterization in *Brassica* and Brassicaceae using a new set of nine SSRs", *Theoretical and Applied Genetics*, Vol. 113, pp. 1 221-1 231.

Ford, C.S. et al. (2006), "Spontaneous gene flow from rapeseed (*Brassica napus*) to wild *Brassica oleracea*", *Proceedings of the Royal Society B*, Vol. 273, No. 1 605, pp. 3 111-3 115.

Forsberg, J., M. Landgren and K. Glimelius (1994), "Fertile somatic hybrids between *Brassica napus* and *Arabidopsis thaliana*", *Plant Science*, Vol. 95, No. 2, pp. 213-223, *http://dx.doi.org/10.1016/0168-9452(94)90095-7.*

Forsberg, J. et al. (1998), "UV dose-dependent DNA elimination in asymmetric hybrids between *Brassica napus* and *Arabidopsis thaliana*", *Plant Science*, Vol. 131, pp. 65-76.

Förster, K. and W. Diepenbrock (2002), "Use of genetically modified plants – Consequences for crop production", *Schwellenwerte für Produkte aus Gentechnisch Veränderten Pflanzen*, Mitt. DFG, Vol. 7, pp. 50-58.

Frandsen, K.J. (1947), "The experimental formation of *Brassica napus* L. var. *oleifera* DC. and *Brassica carinata* Braun", *Dansk Botanisk Arkiv Udgivet af Dansk Botanisk Forening*, Vol. 12, pp. 1-16.

Frandsen, K.J. (1943), "The experimental formation of *Brassica juncea* Czern. et Coss.", *Dansk Botanisk Arkiv Udgivet af Dansk Botanisk Forening*, Vol. 11, pp. 1-17.

Frello, S. et al. (1995), "Inheritance of rapeseed (*Brassica napus*) – Specific RAPD markers and a transgene in the cross *B. juncea* × (*B. juncea* × *B. napus*)", *Theoretical and Applied Genetics*, Vol. 91, No. 2, pp. 236-241, July, *http://dx.doi.org/10.1007/BF00220883.*

Friesen, L.F., A.G. Nelson and R.C. Van Acker (2003), "Evidence of contamination of pedigreed canola (*Brassica napus*) seed lots in western Canada with genetically engineered herbicide resistance traits", *Agronomy Journal*, Vol. 95, No. 5, pp. 1 342-1 347, *http://dx.doi.org/10.2134/agronj2003.1342.*

Fu, T.D. et al. (1997), "Rapeseed varieties and improvement in China", *GCIRC Bulletin*, Vol. 14, pp. 90-95, available at: *www.gcirc.org/fileadmin/documents/Bulletins/B14/GCIRCBUL14_23.pdf.*

Fujimoto, R. et al. (2006), "Suppression of gene expression of a recessive *SP11/SCR* allele by an intranscribed *SP11/SCR* allele in *Brassica* self-incompatibility", *Plant Molecular Biology*, Vol. 61, pp. 577-587.

Fukui, K.S. et al. (1998), "Quantitative karyotyping of three diploid *Brassica* species by imaging methods and localization of 45S rDNA loci on the identified chromosomes", *Theoretical and Applied Genetics*, Vol. 96, Nos. 3-4, pp. 325-330, *http://dx.doi.org/10.1007/s001220050744.*

George, R.A.T. (2009), *Vegetable Seed Production 3rd Edition*, CABI Publishing, Oxon, United Kingdom.

Getinet, A. et al. (1994), "Development of zero erucic acid Ethiopian mustard through an interspecific cross with zero erucic oriental mustard", *Canadian Journal of Plant Science*, Vol. 74, No. 4, pp. 793-795, *http://dx.doi.org/10.4141/cjps94-141.*

Gleba, Y.Y. and F. Hoffmann (1980), "*Arabidobrassica*: A novel plant obtained by protoplast fusion", *Planta*, Vol. 149, No. 2, pp. 112-117, July, *http://dx.doi.org/10.1007/BF00380870.*

Gleba, Y.Y. and F. Hoffmann (1979), "*Arabidobrassica*: Plant genome engineering by protoplast fusion", *Naturwissenschaften*, Vol. 66, pp. 547-554.

Glimelius, K. (1999), "Somatic hybridization", in: Gómez-Campo, C. (ed.), *Biology of Brassica Coenospecies*, Elsevier Science, Amsterdam, Netherlands.

Gómez-Campo, C. (1999), "Taxonomy", in: Gómez-Campo, C. (ed.), *Biology of Brassica Coenospecies*, Elsevier, Amsterdam, Netherlands, pp. 3-32.

Gómez-Campo, C. (1983), "Studies on Cruciferae: X. Concerning some west Mediterranean species of *Erucastrum*", *Anales Jardín Botánico de Madrid*, Vol. 40, No. 1, pp. 63-72, *www.rjb.csic.es/jardinbotanico/ficheros/documentos/pdf/anales/1983/Anales_40(1)_063_072.pdf.*

Gómez-Campo, C. (1980), "Morphology and morpho-taxonomy of the tribe *Brassiceae*", in: Tsunoda, S., K. Hinata and C. Gómez-Campo (eds.), *Brassica Crops and Wild Allies*, Japan Scientific Societies Press, Tokyo, pp. 3-31.

Gómez-Campo, C. and M.E. Tortosa (1974), "The taxonomic and evolutionary significance of some juvenile characters in Brassiceae", *Botanical Journal of the Linnean Society*, Vol. 69, pp. 105-124.

GoshDastidar, N. and N.S. Varma (1999), "A study on intercrossing between transgenic *B. juncea* and other related species", *Proceedings of the 10th International Rapeseed Congress*, Australia, Contribution No. 244, *www.regional.org.au/au/gcirc/4/244.htm*.

Department of Primary Industries, Water and Environment (2003), "Tasmanian rural and marine industry profiles, 2004", Government of Tasmania, Australia.

Gowers, S. (1982), "The transfer of characters from *Brassica campestris* L. to *Brassica napus* L.: Production of clubroot-resistant oil-seed rape (*Brassica napus* subsp. *oleifera*)", *Euphytica*, Vol. 31, No. 3, pp. 971-976, December.

Graziano Ceddia, M., M. Bartlett and C. Perrings (2007), "Landscape gene flow, coexistence and threshold effect: The case of genetically modified herbicide tolerant oilseed rape (*Brassica napus*)", *Ecological Modelling*, Vol. 205, Nos. 1-2, pp. 169-180, July, *http://dx.doi.org/10.1016/j.ecolmodel.2007.02.025*.

Gruber, M.Y. et al. (2006), "'HAIRY CANOLA' – Arabidopsis *GL3* induces a dense covering of trichomes on *Brassica napus* seedlings", *Plant Molecular Biology*, Vol. 60, No. 5, pp. 679-698, March.

Gruber, S. and W. Claupein (2007), "Fecundity of volunteer oilseed rape and estimation of potential gene dispersal by a practice-related model", *Agriculture, Ecosystems & Environment*, Vol. 119, Nos. 3-4, pp. 401-408, March, *http://dx.doi.org/10.1016/j.agee.2006.08.007*.

Gruber, S., K. Emrich and W. Claupein (2009), "Classification of canola (*Brassica napus*) winter cultivars by secondary dormancy", *Canadian Journal of Plant Science*, Vol. 89, No. 4, pp. 613-619, *http://dx.doi.org/10.4141/CJPS08190*.

Gruber, S., C. Pekrun and W. Claupein (2004), "Seed persistence of oilseed rape (*Brassica napus*): Variation in transgenic and conventionally bred cultivars", *Journal of Agricultural Science*, Vol. 142, No. 1, pp. 29-40, February, *http://dx.doi.org/10.1017/S0021859604003892*.

Gruber, S., C. Pekrun and W. Claupein (2003), "Life cycle and gene dispersal of oilseed rape volunteers (*Brassica napus* L.)", *Proceedings of the BCPC International Congress: Crop Science and Technology*, Glasgow, Scotland, pp. 1 093-1 098.

Grundy, S.M. and M. Denke (1990), "Dietary influences on serum lipids and lipoproteins", *Journal of Lipid Research*, Vol. 31, No. 7, pp. 1 149-1 172, July.

Guéritaine, G., S. Bazot and H. Darmency (2003), "Emergence and growth of hybrids between *Brassica napus* and *Raphanus raphanistrum*", *New Phytology*, Vol. 158, No. 3, pp. 561-567, June, *http://dx.doi.org/10.1046/j.1469-8137.2003.00755.x*.

Guéritaine, G. et al. (2002), "Fitness of backcross six of hybrids between transgenic oilseed rape (*Brassica napus*) and wild radish (*Raphanus raphanistrum*)", *Molecular Ecology*, Vol. 11, No. 8, pp. 1 419-1 426, August.

Gulden, R.H., S.J. Shirtliffe and A.G. Thomas (2003a), "Harvest losses of canola (*Brassica napus*) cause large seedbank inputs", *Weed Science*, Vol. 51, No. 1, pp. 83-86, January-February, *www.jstor.org/stable/4046629*.

Gulden, R.H. S.J. Shirtliffe and A.G. Thomas (2003b), "Secondary seed dormancy prolongs persistence of volunteer canola in Western Canada", *Weed Science*, Vol. 51, No. 6, pp. 904-913, November, December, *www.jstor.org/stable/4046744*.

Gulden, R.H., A.G. Thomas and S.J. Shirtliffe (2004a), "Relative contribution of genotype, seed size and environment to secondary seed dormancy Potential in Canadian spring oilseed rape (*Brassica napus*)", *Weed Research*, Vol. 44, No. 2, pp. 97-106, April, *http://dx.doi.org/10.1111/j.1365-3180.2003.00377.x*.

Gulden, R.H., A.G. Thomas and S.J. Shirtliffe (2004b), "Secondary dormancy, temperature and burial depth regulate seedbank dynamics in canola", *Weed Science*, Vol. 52, pp. 382-388, May, *http://dx.doi.org/10.1614/WS-03-123R1*.

Gupta, S.K. (1997), "Production of interspecific and intergeneric hybrids in *Brassica* and *Raphanus*", *Cruciferae Newsletter Eucarpia*, No. 19, pp. 21-22.

Gurr, M.I. (1992), "Dietary lipids and coronary heart disease: Old evidence, new perspective", *Progress in Lipid Research*, Vol. 31, No. 3, pp. 195-243.

Habman, M.H. et al. (2010), "Exploitation of a late flowering species *Brassica oleracea* L. for the improvement of earliness in *B. napus* L.: An untraditional approach".*Euphytica*, Vol. 177, pp. 365-374, *http://dx.doi.org/10.1007/s10681-010-0253-5*.

Hails, R.S. et al. (1997), "Burial and seed survival in *Brassica napus* subsp. *oleifera* and *Sinapis arvensis* including a comparison of transgenic and non-transgenic lines of the crop", *Proceedings of the Royal Society B*, Vol. 264, No. 1 378, pp. 1-7, January, *http://dx.doi.org/10.1098/rspb.1997.0001*.

Håkansson, A. (1956), "Seed development of *Brassica oleracea* and *B. rapa* after certain reciprocal pollinations", *Hereditas*, Vol. 42, Nos. 3-4, pp. 373-395, December, *http://dx.doi.org/10.1111/j.1601-5223.1956.tb03024.x*.

Halfhill, M.D. et al. (2002), "*Bt*-transgenic oilseed rape hybridization with its weedy relative, *Brassica rapa*", *Environmental Biosafety Research*, Vol. 1, No. 1, pp. 19-28, October.

Hall, L.M. et al. (2005), "Volunteer oilseed rape – Will herbicide resistance traits assist ferality?", Chapter 5, in: Gressel, J. (ed.), *Crop Ferality and Volunteerism*, CRC London Press, pp. 59-79, *http://dx.doi.org/10.1201/9781420037999.ch5*.

Hall, L. et al. (2000), "Pollen flow between herbicide-resistant *Brassica napus* volunteers", *Weed Science*, Vol. 48, No. 6, pp. 688-694, November-December, *www.jstor.org/stable/4046338*.

Hansen, L.B. et al. (2001), "Introgression between oilseed rape (*Brassica napus* L.) and its weedy relative *B. rapa* L. in a natural population", *Genetic Resources and Crop Evolution*, Vol. 48, No. 6, pp. 621-627, December.

Hansen, L.N. (1998), "Intertribal somatic hybridization between rapid cycling *Brassica oleracea* L. and *Camelina sativa* (L.) Crantz", *Euphytica*, Vol. 104, No. 3, pp. 173-179, December.

Harberd, D.J. (1976), "Cytotaxonomic studies of *Brassica* and related genera", in: Vaughan, J.G. et al. (eds.), *The Biology and Chemistry of the Cruciferae*, Academic Press, London, pp. 47-68.

Harberd, D.J. (1972), "A contribution to the cyto-taxonomy of Brassica (Cruciferae) and its allies", *Botanical Journal of the Linnean Society*, Vol. 65, No. 1, pp. 1-23, January, *http://dx.doi.org/10.1111/j.1095-8339.1972.tb00920.x*.

Harberd, D.J. and E.D. McArthur (1980), "Meiotic analysis of some species and genus hybrids in the Brassiceae", in: Tsunoda, S., K. Hinata and C. Gómez-Campo (eds.), *Brassica Crops and Wild Allies*, Japan Scientific Societies Press, Tokyo, pp. 65-67.

Harberd, D.J. and E.D. McArthur (1972), "Cytotaxonomy of *Rhynchosinapis* and *Hutera* (Cruciferae-Brassiceae)", *Heredity*, Vol. 28, pp. 254-257.

Harker, K.N. et al. (2006), "Persistence of glyphosate-resistant canola in Western Canadian cropping systems", *Agronomy Journal*, Vol. 98, No. 1, pp. 107-119, *http://dx.doi.org/10.2134/agronj2005.0168*.

Harper, F.R. and B. Berkenkamp (1975), "Revised growth-stage key for *Brassica camprestris* and *B. napus*", *Canadian Journal of Plant Science*, Vol. 55, No. 2, pp. 657-658, *http://dx.doi.org/10.4141/cjps75-103*.

Hatakeyama K. et al. (1998), "Dominance relationships between *S*-alleles in self-incompatible *Brassica campestris* L.", *Heredity*, Vol. 80, pp. 241-247, *http://dx.doi.org/10.1046/j.1365-2540.1998.00295.x*.

Hauser, T.P., R. Jorgensen and H. Ostergard (1997), "Preferential exclusion of hybrids in mixed pollinations between oilseed rape (*Brassica napus*) and weedy *B. campestris* (Brassiceae)", *American Journal of Botany*, Vol. 84, No. 6.

Hauser, T.P. et al. (2001), "Environmental effects on male and female fitness of *Brassica* hybrids", *European Science Foundation Meeting of a Working Group on Interspecific Gene Flow from Oilseed Rape to Weedy Species*, June 2001, Rennes, France.

Hayter, K.E. and J.E. Cresswell (2006), "The influence of pollinator abundance on the dynamics and efficiency of pollination in agricultural *Brassica napus*: Implications for landscape-scale gene dispersal", *Journal of Applied Ecology*, Vol. 43, No. 6, pp. 1 196-1 202, December, *http://dx.doi.org/10.1111/j.1365-2664.2006.01219.x*.

Heather, D.W. and J.B. Sieczka (1991), "Effect of seed size and cultivar on emergence and stand establishment of broccoli in crusted soil", *Journal of the American Society for Horticultural Science*, Vol. 116, No. 6, pp. 946-949, November.

Hedge, I.C. (1976), "A systematic and geographical survey of the Old World Cruciferae", in: Vaughan, J.G., A.J. MacLeod and B.M.G. Jones (eds.), *The Biology and Chemistry of the Cruciferae*, Academic Press, New York, pp. 1-45.

Helm, J. (1963), "Morphologisch-taxonomische Gleiderung der Kultursippen von *Brassica oleracea*", *Kulturpflanze*, Vol. 11, pp. 92-210.

Hemingway, J.S. (1995), "Mustards: *Brassica spp.* and *Sinapis alba* (Cruciferae)", in: Smartt, J. and N.W. Simmons (eds.), *Evolution of Crop Plants*, Longman, London, pp. 82-86.

Herklots, G.A.C. (1972), *Vegetables in South-East Asia: Asiatic Cabbages*, George Allen & Unwin, London, pp. 190-224.

Hertz, B. (1999), *Expert Testimony in Monsanto Canada and Monsanto Company vs. Percy Schmeiser and Schmeiser Enterprises*, Federal Court-Trial Division.

Heyn, F.W. (1977), "Transfer of restorer genes from *Raphanus* to cytoplasmic male sterile *Brassica napus*", *Cruciferae Newsletter*, No. 1, pp. 15-16.

Heywood, V.H. (1964), "Brassica", in: Tutin, T.G. et al. (eds.), *Flora Europaea*, Vol. 1, Cambridge University Press, pp. 335-339.

Hilhorst, H.W.M. and P.E. Toorop (1997), "Review on dormancy, germinability and germination in crop and weed species", *Advances in Agronomy*, Vol. 61, pp. 111-165.

Hill, S.A. et al. (1991), "The incidence and importance of beet western yellow virus in oilseed rape", *IOBC/WPRS Bulletin*, Vol. 14, pp. 36-45.

Hinata, K. and T. Nishio (1980), "Self-incompatability in Crucifera", in: Tsunoda, S., K. Hinata and C. Gómez-Campo (eds.), *Brassica Crops and Wild Allies*, pp. 223-234.

Hinata, K., A. Isogai and K. Isuzugawa (1994), "Manipulation of sporophytic self-incompatibilities on plant breeding", Chapter 6, in: Williams, E.G. et al. (eds.), *Genetic Control of Self-incompatibility and Reproductive Development in Flowering Plants*, Volume 2 of the series Advances in Cellular and Molecular Biology of Plants, Kluwer, Dordrecht, Netherlands, pp. 102-115.

Holm, L. et al. (1997), *World Weeds, Natural Histories and Distribution*, John Wiley & Sons, New York.

Horricks, J.S. (1969), "Influence of rape residue on cereal production", *Canadian Journal of Plant Science*, Vol. 49, No. 5, pp. 632-634, *http://dx.doi.org/10.4141/cjps69-110*.

Hosaka, K. et al. (1990), "Development and chromosomal localization of genome specific DNA markers of *Brassica* and the evolution of amphidiploids and *n*=9 diploid species", *Genome*, Vol. 33, No. 1, pp. 131-142, *http://dx.doi.org/10.1139/g90-021*.

Hu, F.B. et al. (1997), "Dietary fat intake and the risk of coronary heart disease in women", *New England Journal of Medecine*, Vol. 337, pp. 1 491-1 499, *http://dx.doi.org/10.1056/NEJM199711203372102*.

Hu, J. and C.F. Quiros (1991), "Molecular and cytological evidence of deletions in alien chromosomes for two monosomic addition lines of *Brassica campestris-oleracea*", *Theoretical and Applied Genetics*, Vol. 81, No. 2, pp. 221-226, *http://dx.doi.org/10.1007/BF00215726*.

Hu, Q. et al. (2002), "Intergeneric hybrid between *Brassica napus* and *Orychophragmus violaceus* containing traits of agronomic importance for oilseed rape breeding", *Theoretical and Applied Genetics*, Vol. 105, pp. 834-840.

Hu, Q. et al. (1999), "Intergeneric hybridization by protoplast fusion aiming at modification of fatty acid composition in *Brassica napus* L.", *Proceedings of the 10th International Rapeseed Congress*, Canberra, Contribution No. 332, pp. 26-29, *www.regional.org.au/au/gcirc/4/332.htm*.

Hua, Y.W. and Z.Y. Li (2006), "Genomic in situ hybridization analysis of *Brassica napus* × *Orychophragmus violaceus* hybrids and production of *B. napus* aneuploids", *Plant Breeding*, Vol. 125, pp. 144-149.

Huang, B. et al. (2002), "Production and cytogenetics of intergeneric hybrids between Ogura CMS *Brassica napus* and *Raphanus raphanistrum*", *Cruciferae Newsletter Eucarpia*, No. 24, pp. 25-27.

Hüsken, A. and A. Dietz-Pfeilstetter (2007), "Pollen-mediated intraspecific gene flow from herbicide resistant oilseed rape (*Brassica napus* L.)", *Transgenic Research*, Vol. 16, No. 5, pp. 557-569, October.

Inaba, R. and T. Nishio (2002), "Phylogenetic analysis of *Brassiceae* based on the nucleotide sequences of the S-locus related gene, SLR1", *Theoretical and Applied Genetics*, Vol. 105, No. 8, pp. 1 159-1 165, December.

Ingram, J. (2000), "The separation distances required to ensure cross-pollination is below specified limits in non-seed crops of sugar beet, maize and oilseed rape", *Plant Varieties & Seeds*, Vol. 13, No. 3, pp. 1 981-1 999.

Inomata, N. (2007), "*Brassica* vegetable crops", in: Singh, R. (ed.), *Genetic Resources, Chromosome Engineering, and Crop Improvement*, CRC Press, New York, pp. 115-146.

Inomata, N. (1988), "Intergeneric hybridization between *Brassica napus* and *Sinapis arvensis* and their crossability", *Cruciferae Newsletter*, No. 13, pp. 22-23.

Jackson, S.A. et al. (2000), "Comparative fluorescence *in situ* hybridization mapping of a 431-kb *Arabidopsis thaliana* bacterial artificial chromosome contig reveals the role of chromosomal duplications in the expansion of the *Brassica rapa* genome", *Genetics*, Vol. 156, No. 2, pp. 833-838, October.

Jenkins, T.E., A.J. Conner and C.M. Frampton (2001), "Investigating gene introgression from rape to wild turnip", *New Zealand Plant Protection*, Vol. 54, pp. 101-104.

John Innes Centre (n.d.), "RevGenUK", *http://revgenuk.jic.ac.uk* (accessed 27 October 2011).

Johnston, J.S. et al. (2005), "Evolution of genome size in Brassicaceae", *Annals of Botany*, Vol. 95, No. 1, pp. 229-235, January.

Jørgensen, R.B. (1999), "Gene flow from oilseed rape (*Brassica napus*) to related species", in: Lutmann, P.J.W. (ed.), *Gene Flow and Agriculture: Relevance for Transgenic Crops*, British Crop Protection Council Symposium Proceedings, Vol. 72, pp. 117-124.

Jørgensen, R.B. and B. Andersen (1994), "Spontaneous hybridization between oilseed rape (*Brassica napus*) and weedy *B. campestris* (Brassicaceae): A risk of growing genetically modified oilseed rape", *Journal of Botany*, Vol. 81, pp. 1 620-1 626.

Jørgensen, R.B. et al. (1998), "Introgression of crop genes from oilseed rape (*Brassica napus*) to related wild species – An avenue for the escape of engineered genes", *Acta Horticulturae*, Vol. 459, pp. 211-217.

Jørgensen, R.B. et al. (1996), "Spontaneous hybridization between oilseed rape (*Brassica napus*) and weedy relatives", *Acta Horticulturae*, Vol. 407, pp. 193-200.

Jørgensen, T., T. Pavlo Hauser and R. Bagger Jørgensen (2007), "Adventitious presence of other varieties in oilseed rape (*Brassica napus*) from seed bank and certified seeds", *Seed Science Research*, Vol. 17, pp. 115-125, *http://dx.doi.org/10.1017/S0960258507708103*.

Kamler, F. (2000), "Transgene rape and honey bees", *Proceedings of the 8th International Pollination Symposium*, Mosonmagyarovar, Hungary, July.

Karssen, C.M. (1980/81), "Environmental conditions and endogenous mechanisms involved in secondary dormancy of seeds", *Israel Journal of Botany*, Vol. 29, Nos. 1-4, pp. 45-64, *http://dx.doi.org/10.1080/0021213X.1980.10676875*.

Kerlan, M.C., A.M. Chèvre and F. Eber (1993), "Interspecific hybrids between a transgenic rapeseed (*Brassica napus*) and related species – Cytogenetical characterization and detection of the transgene", *Genome*, Vol. 36, No. 6, pp. 1 099-1 106, December.

Kerlan, M.C. et al. (1992), "Risk assessment of outcrossing of transgenic rapeseed to related species: I. Interspecific hybrid production under optimal conditions with emphasis on pollination and fertilization", *Euphytica*, Vol. 62, pp. 145-153.

Khanna, V.K. and J.B. Chowdhury (1974), "Studies on pollen tube growth and seed set in some interspecific crosses in genus Brassica", *Haryana Agricultural University Journal of Research*, Vol. 4, No. 3, pp. 231-235.

Kirk, J.T.O. and R.N. Oram (1981) "Isolation of erucic acid-free lines of *Brassica juncea*: Indian mustard now a potential oilseed crop in Australia", *Journal of the Australian Institute of Agricultural Science*, Vol. 47, pp. 51-52.

Kirti, P.B. et al. (1997), "Introgression of a gene restoring gertility to CMS *(Trachystoma) Brassica juncea* and the genetics of restoration", *Plant Breeding*, Vol. 116, No. 3, pp. 259-262, July, *http://dx.doi.org/10.1111/j.1439-0523.1997.tb00992.x*.

Kirti, P.B. et al. (1995), "Transfer of *Ogu* cytoplasmic male sterility to *Brassica juncea* and improvement of male sterile through somatic cell fusion", *Theoretical and Applied Genetics*, Vol. 91, pp. 517-521.

Kjaer, A. (1960), "Naturally derived isothocyanates (mustard oils) and their parent glucosides", in: Zechmeister, L. (ed.), *Progress in the Chemistry of Organic Natural Products*, Springer-Verlag, Vienna, pp. 122-176.

Klein, E.K. et al. (2006), "Pollen dispersal of oilseed rape: Estimation of the dispersal function and effects of field dimension", *Journal of Applied Ecology*, Vol. 43, No. 1, pp. 141-151, February, *http://dx.doi.org/10.1111/j.1365-2664.2005.01108.x*.

Klewer, A. et al. (2003), "Incorporation of blackspot resistance from different origins into oilseed rape", *Proceedings of the 11th International Rapeseed Congress*, Copenhagen, 6-10 July, pp. 65-67.

Knispel, A.L. et al. (2008), "Gene flow and multiple herbicide resistance in escaped canola populations", *Weed Science*, Vol. 56, No. 1, pp. 72-80, *http://dx.doi.org/10.1614/WS-07-097.1*.

Korzun, V. (2003), "Molecular markers and their applications in cereals breeding", in: *Proceedings of "Marker Assisted Selection: A Fast Track to Increase Genetic Gain in Plant and Animal Breeding?", Session I: MAS in Plants*, 17-18 October 2003, Turin, Italy, pp. 18-22, *www.fao.org/biotech/docs/Korzun.pdf*.

Koul, K.K., R. Nagpal and S.N. Raina (2000), "Seed coat microsculpturing in *Brassica* and allied genera (subtribes Brassicinae, Raphaninae, Moricandiinae)", *Annals of Botany*, Vol. 86, No. 2, pp. 385-397, *http://dx.doi.org/10.1006/anbo.2000.1197*.

Kramer, J.K.G. et al. (1983), *High and Low Erucic Acid Rapeseed Oils*, Academic Press, Toronto.

Kuckuck, H. (1979), *Gartenbauliche Pflanzenzüchtung, 2nd ed*, Parey, Berlin, Hamburg, Germany, pp. 64-72.

Kumazawa, S. and S. Able (1955), "Studies on varieties of mustard", *Journal of the Japanese Society for Horticultural Science*, Vol. 24, No. 2, pp. 69-84, *http://doi.org/10.2503/jjshs.24.69*.

Kusaba, M. et al. (2001), "Self-incompatibility in the genus *Arabidopsis*: Characterization of the *S* locus in the outcrossing *A. lyrata* and its autogamous relative *A. thaliana*", *Plant Cell*, Vol. 13, No. 3, pp. 627-643, March.

Lagercrantz, U. and D. Lydiate (1996), "Comparative genome mapping in *Brassica*", *Genetics*, Vol. 144, No. 4, pp. 1 903-1 910.

Lal, G. (1993), "Seed production technology for tropical and subtropical cauliflower", in: *Breeding of Solanaceous and Cole Crops*, AVRDC, Chinese Taipei, pp. 268-279.

Lamb, R.J. (1989), "Entomology of oilseed *Brassica* crops", *Annual Reviews of Entomology*, Vol. 34, pp. 211-229, January, *http://dx.doi.org/10.1146/annurev.en.34.010189.001235*.

Landbo, L. and R.B. Jørgensen (1997), "Seed germination in weedy *Brassica campestris* and its hybrids with *Brassica napus*: Implications for risk assessment of transgenic oilseed rape", *Euphytica*, Vol. 97, pp. 209-218.

Landbo, L., B. Andersen and R.B. Jørgensen (1996), "Natural hybridisation between oilseed rape and a wild relative: Hybrids among seeds from weedy *B. campestris*", *Hereditas*, Vol. 125, No. 1, pp. 89-91, November, *http://dx.doi.org/10.1111/j.1601-5223.1996.00089.x*.

Landry, B.S. et al. (1991), "Agenetic map for *Brassica napus* based on restriction fragment length polymorphisms detected with expressed DNA sequences", *Genome*, Vol. 34, No. 4, pp. 543-552, *http://dx.doi.org/10.1139/g91-084*.

Lavigne, C. et al. (1998), "A pollen dispersal experiment with transgenic oilseed rape. Estimation of the average pollen dispersal of an individual plant within a field", *Theoretical and Applied Genetics*, Vol. 96, No. 6, pp. 886-896, *http://dx.doi.org/10.1007/s001220050816*.

Leadlay, E.A. and V.H. Heywood (1990), "The biology and systematics of the genus *Coincyc* Porta and Rigo ex Rouy (Cruciferae)", *Botanic Journal of the Linnean Society*, Vol. 102, pp. 313-398.

Leckie, D., A. Smithson and I.R. Crute (1993), "Gene movement from oilseed rape to weedy populations – A component of risk assessment for transgenic cultivars", *Aspects of Applied Biology*, Vol. 35, pp. 61-66.

Leeson, J.Y. et al. (2005), *Prairie Weed Surveys of Cereal, Oilseed and Pulse Crops from the 1970s to the 2000s*, Weed Survey Series Publication, 05-1, Agriculture and Agri-Food Canada, Saskatoon Research Centre, Canada.

Leflon, M. et al. (2006), "Pairing and recommendation at meiosis of *Brassica rapa* (AA) × *Brassica napus* (AACC) hybrids", *Theoretical and Applied Genetics*, Vol. 113, No. 8, pp. 1 467-1 480, November.

Lefol, E., V. Danielou and H. Darmency (1996), "Predicting hybridization between transgenic oilseed rape and wild mustard", *Field Crops Research*, Vol. 45, No. 1, pp. 153-161, *http://dx.doi.org/10.1016/0378-4290(95)00067-4*.

Lefol, E., A. Fleury and H. Darmency (1996), "Gene dispersal from transgenic crops. II. Hybridization between oilseed rape and wild hoary mustard", *Sexual Plant Reproduction*, Vol. 9, pp. 189-196.

Lefol, E., G. Seguin-Swartz and R.K. Downey (1997), "Sexual hybridization in crosses of cultivated Brassica species with the crucifers *Erucastrum gallicum* and *Raphanus raphanistrum*: Potential for gene introduction", *Euphytica*, Vol. 95, pp. 127-139.

Lefol, E. et al. (1995), "Gene dispersal from transgenic crops. I. Growth of interspecific hybrids between oilseed rape and the wild hoary mustard", *Journal of Applied Ecology*, Vol. 32, No. 4, pp. 803-808, November, *http://dx.doi.org/10.2307/2404819*.

Lefol, E. et al. (1991), "Escape of engineered genes from rapeseed to wild Brassiceae", *Proceedings of the Brighton Crop Protection Conference Weeds*, Vol. 3, pp. 1 049-1 056.

Légère, A. et al. (2001), "Presence and persistence of volunteer canola in Canadian cropping systems", *Proceedings of the Brighton Crop Protection Conference: Weeds*, Brighton Crop Protection Council, pp. 143-148.

Lelivelt, C.L.C. and F.K. Krens (1992), "Transfer of resistance to the beet cyst nematode (*Heterodera schachtii* Schm) into the *Brassica* gene pool through intergeneric somatic hybridization with *Raphanus sativus* L.", *Theoretical and Applied Genetics*, Vol. 83, Nos. 6-7, pp. 887-894, April, *http://dx.doi.org/10.1007/BF00226712*.

Lelivelt, C.L.C. et al. (1993), "Transfer of resistance to the beet cyst nematode (*Heterodera schachtii* Schm.) from *Sinapis alba* L. (white mustard) to the *Brassica napus* L. gene pool by means of sexual and somatic hybridization", *Theoretical and Applied Genetics*, Vol. 85, Nos. 6-7, pp. 688-696, February, *http://dx.doi.org/10.1007/BF00225006*.

Lim, K.B. et al. (2005), "Characterization of rDNAs and tandem repeats in the heterochromatin of *Brassica rapa*", *Molecular Cell*, Vol. 19, No. 3, pp. 436-444, June.

Linder, R. (1998), "Potential persistence of transgenes: Seed performance of transgenic canola and wild relatives × canola hybrids", *Ecological Applications*, Vol. 8, No. 4, pp. 1 180-1 195.

Linder, R. and J. Schmitt (1995), "Potential persistence of escaped transgenes: Performance of transgenic oil-modified *Brassica* seeds and seedlings", *Ecological Applications*, Vol. 5, pp. 1 056-1 068, *http://dx.doi.org/10.2307/2269354*.

Liu, H. (1985), *Genetics and Breeding of Rapeseed*, Shanghai Science and Technology Press, Shanghai, China (in Chinese).

Liu, J.H., M. Landgren and K. Glimelius (1996), "Transfer of the *Brassica tournefortii* cytoplasm to *B. napus* for the production of cytoplasmic male sterile *B. napus*", *Physiologia Plantarum*, Vol. 96, No. 1, pp. 123-129, January, *http://dx.doi.org/10.1111/j.1399-3054.1996.tb00192.x*.

Liu, J., X. Xu and X. Deng (2005), "Intergeneric somatic hybridization and its application to crop genetic improvement", *Plant Cell, Tissue and Organ Culture*, Vol. 82, No. 1, pp. 19-44.

Lizgunova, T.V. (1959), "The history of botanical studies of the cabbage, *Brassica oleracea* L.", *Bulletin of Applied Botany, Genetics and Plant Breeding*, Vol. 32, pp. 37-70.

Lokanadha, R.D. and N. Saria (1994), "Hybridization of *Brassica tournefortii* and cultivated Brassicas", *Cruciferae Newsletter Eucarpia*, No. 16, pp. 32-33.

López-Granados, F. and P.J.W. Lutman (1998), "Effect of environmental conditions on the dormancy and germination of volunteer oilseed rape seed (*Brassica napus*)", *Weed Science*, Vol. 46, No. 4, pp. 419-423, August, *www.jstor.org/stable/4046058*.

Love, H.K. et al. (1990), "Development of low glucosinolate mustard", *Canadian Journal of Plant Science*, Vol. 70, No. 2, pp. 425-429, *http://dx.doi.org/10.4141/cjps90-049*.

Lutman, P.J.W., S.E. Freeman and C. Pekrun (2003), "The long-term persistence of seeds of oilseed rape (*Brassica napus*) in arable fields", *Journal of Agricultural Science*, Vol. 141, No. 2, pp. 231-240, September, *http://dx.doi.org/10.1017/S0021859603003575*.

Lutman, P.J.W. et al. (2005), "Persistence of seeds from crops of conventional and herbicide tolerant oilseed rape (*Brassica napus*)", *Proceedings of the Royal Society B.*, Vol. 272, No. 1 575, pp. 1 909-1 915, September, *http://dx.doi.org/10.1098/rspb.2005.3166*.

Lysak, M.A. and C. Lexer (2006), "Towards the era of comparative evolutionary genomics in *Brassicaceae*", *Plant Systematics and Evolution*, Vol. 259, No. 2, pp. 175-198, *http://dx.doi.org/10.1007/s00606-006-0418-9*.

Lysak, M.A. et al. (2009), "The dynamic ups and downs of genome size evolution in *Brassicaceae*", *Molecular Biology and Evolution*, Vol. 26, No. 1, pp. 85-98, *http://dx.doi.org/10.1093/molbev/msn223*.

Lysak, M.A. et al. (2007), "Ancestral chromosomal blocks are triplicated in Brassiceae species with varying chromosome number and genome size", *Plant Physiology*, Vol. 145, No. 2, pp. 402-410, October, *http://dx.doi.org/10.1104/pp.107.104380*.

Lysak, M.A. et al. (2006), "Mechanisms of chrosome number reduction in *Arabidopsis thaliana* and related Brassicaceae species", *Proceedings of the National Academy of Sciences USA*, Vol. 103, No. 13, pp. 5 224-5 229, *http://dx.doi.org/10.1073/pnas.0510791103*.

Lysak, M.A. et al. (2005), "Chromosome triplication found across the tribe *Brassiceae*", *Genome Research*, Vol. 15, No. 4, pp. 516-525, April, *http://dx.doi.org/10.1101/gr.3531105*.

MacDonald, R.L. and G.J. Kuntz (2000), "Monitoring program to assess the occurrence and fate of SeedLink canola volunteers following the 1999 growing season in Western Canada", *Aventis CropScience Report*, No. ACC00-03.

Manasse, R. and P. Kareiva (1991), "Quantifying the spread of recombinant genes and organisms", in: Ginzburgh, L. (ed.), *Assessing Ecological Risks of Biotechnology*, Butterworth-Heinemann, Boston, pp. 215-231.

Marquard, R. and K.C. Walker (1995), "Environmental impact of rapeseed production", in: Kimber, D.S. and D.I. McGregor (eds.), *Brassica Oilseed: Production and Utilization*, CAB International, Oxon, United Kingdom, pp. 195-213.

Martens, G. (2001), "From Cinderella to Cruela: Volunteer canola", *2nd Annual Manitoba Agronomists Conference 2001*, University of Manitoba, Canada, pp. 151-154, *www.umanitoba.ca/faculties/afs/MAC_proceedings/2001/pdf/martens.pdf*.

Masden, S.B. (1962), "Germination of buried and dry stored seed III. 1934-1960", *Proceedings of the International Seed Testing Association*, Vol. 27, pp. 920-928.

Mason-Sedun, W., R.S. Jesspo and J.V. Lovett (1986), "Differential phytotoxicity among species and cultivars of the genus *Brassica* to wheat. I. Laboratory and field screening of species", *Plant and Soil*, Vol. 93, No. 1, pp. 3-16, February.

Mathias, R. (1991), "Improved embryo rescue technique for intergeneric hybridization between *Sinapis* species and *B. napus*", *Cruciferae Newsletter*, No. 14/15, pp. 90-91.

Mattson, F.H. and S.M. Grundy (1985), "Comparison of effects of dietary saturated, monounsaturated, and polyunsaturated fatty acids on plasma lipids and lipoproteins in man", *Journal of Lipid Research*, Vol. 26, No. 2, pp. 194-201, February.

Maxwell, M. (1981), "Production of a heinz body anaemia in the domestic fowl after ingestion of dimethyl disulphide: A haematological and ultrastructural study", *Reviews of Veterinary Science*, Vol. 30, pp. 233-238.

McCallum, C.M. et al. (2000), "Targeting induced local lesions in genomes (TILLING) for plant functional genomics", *Plant Physiology*, Vol. 123, No. 2, pp. 439-442, *http://www.plantphysiol.org/content/123/2/439.full*.

McCartney, H.A. and M.E. Lacey (1991), "Wind dispersal of pollen from crops of oilseed rape (*Brassica napus* L.)", *Journal of Aerosol Science*, Vol. 22, No. 4, pp. 467-477, *http://dx.doi.org/10.1016/0021-8502(91)90005-3*.

McGrath, J.M. et al. (1990), "Identification of *Brassica oleracea* monosomic alien chromosome addition lines with molecular markers reveals extensive gene duplication", *Molecular and General Genetics*, Vol. 223, No. 2, pp. 198-204, September.

McNaughton, I.H. (1995a), "Turnip and relatives *Brassica campestris* (Cruciferae)", in: Smartt, J. and N.W. Simmonds (eds.), *Evolution of Crop Plants*, Longman, London, pp. 62-68.

McNaughton, I.H. (1995b), "Swedes and rapes *Brassica napus* (Cruciferae)", in: Smartt, J. and N.W. Simmonds (eds.), *Evolution of Crop Plants*, Longman, London, pp. 68-75.

Meng, J. et al. (1998), "The production of yellow-seeded *Brassica napus* (AACC) through crossing interspecific hybrids *B. campestris* (AA) and *B. carinata* (BBCC) with *B. napus*", *Euphytica*, Vol. 103, No. 3, pp. 329-333, October.

Mesquida, J. and M. Renard (1982), "Study of the pollen dispersal by wind and of the importance of wind pollination in rapeseed (*Brassica napus* var. *oleifera* Metzger)", *Apidologie*, Vol. 4, pp. 353-366, English summary.

Messéan, A. et al. (2007), "Occurrence of genetically modified oilseed rape in the harvest of subsequent conventional oilseed rape over time", *European Journal of Agronomy*, Vol. 27, No. 1, pp. 115-122, July, *http://dx.doi.org/10.1016/j.eja.2007.02.009*.

Metz, P.L.J., J.-P. Nap and W.J. Stiekema (1995), "Hybridization of radish (*Raphanus sativus* L.) and oilseed rape (*Brassica napus* L.) through a flower-culture method", *Euphytica*, Vol. 83, No. 2, pp. 159-168, June.

Metz, P.L.J. et al. (1997), "The impact on biosafety of the phosphinothricin-tolerance transgene in inter-specific *B. rapa* × *B. napus* hybrids and their successive backcrosses", *Theoretical and Applied Genetics*, Vol. 95, No. 3, pp. 442-450, August.

Mikkelsen, T.R., J. Jensen and R.B. Jørgensen (1996), "Inheritance of oilseed rape (*Brassica napus*) RAPD markers in a backcross progeny with *Brassica campestris*", *Theoretical and Applied Genetics*, Vol. 92, Nos. 3-4, pp. 492-497, *http://dx.doi.org/10.1007/BF00223698*.

Mizushima, U. (1980), "Genome analysis in *Brassica* and allied genera", in: Tsunoda, S., K. Hinata and C. Gómez-Campo (eds.), *Brassica Crops and Wild Allies, Biology and Breeding*, Japan Scientific Societies Press, Tokyo, pp. 89-108.

Mizushima, U. (1972), "Evolution of species in Brassiceae and their breeding", *Kagaku to Seibutsu*, Vol. 10, pp. 78-85.

Mizushima, U. and S. Tsunoda (1967), "A plant exploration in Brassica and allied genera", *Tohoku Journal of Agricultural Research*, Vol. 17, pp. 249-276.

Mohammad, A. (1935), "Pollination studies on toria (*Brassica napus* L. var *dichotoma* Prain) and sarson (*Brassica camprestris* Prain)", *Indian Journal of Agricultural Science*, Vol. 5, pp. 125-154.

Momoh, E.J.J. et al. (2002), "Variation in the development of secondary seed dormancy in oilseed rape genotypes under conditions of stress", *Weed Research*, Vol. 42, pp. 446-455.

Morinaga, T. (1934a), "Interspecific hybridation in *Brassica*: VI. The cytology of F1 hybrids of *B. juncea* and *B. nigra*", *Cytologia*, Vol. 6, pp. 62-67.

Morinaga, T. (1934b), "On the chromosome number of *Brassica juncea* and *B. napus*, on the hybrid between the two, and on offspring line of the hybrid", *The Japanese Journal of Genetics*, Vol. 9, No. 3, pp. 161-163, *http://dx.doi.org/10.1266/jjg.9.161*.

Morinaga, T. (1933), "Interspecific hybridization in *Brassica*. V. The cytology of F1 hybrids of *B. carinata* and *B. alboglabra*", *Japan Journal of Botany*, Vol. 6, pp. 467-475.

Morinaga, T. (1931), "Interspecific hybridization in *Brassica*. IV. The cytology of F1 hybrids of *B. carinata* and some other species with 10 chromosomes", *Cytologia*, Vol. 3, No. 1, pp. 77-83, *http://doi.org/10.1508/cytologia.3.77*.

Morinaga, T. (1929), "Interspecific hybridization in *Brassica*. II. The cytology of F1 hybrids of *B. cernua* and various other species with 10 chromosomes", *The Japanese Journal of Botany*, Vol. 4, pp. 277-289.

Morinaga, T. (1928), "Preliminary note on interspecific hybridization in *Brassica*", *Proceedings of the Imperial Academy*, Vol. 4, No. 10, pp. 620-622, *http://doi.org/10.2183/pjab1912.4.620*.

Morris, W.K. et al. (1994), "Do barren zones and pollen traps reduce gene escape from transgenic crops?", *Ecological Applications*, Vol. 4, No. 1, pp. 157-165, *http://dx.doi.org/10.2307/1942125*.

Moyes, C.L. et al. (2002), "Barriers to gene flow from oilseed rape (*Brassica napus*) into populations of *Sinapis arvensis*", *Molecular Ecology*, Vol. 11, No. 1, pp. 103-112, January.

Mullis, K. and F. Faloona (1987), "Specific synthesis of DNA *in vitro* via a polymerase-catalyzed chain reaction", *Methods in Enzymology.*, Vol. 155, pp. 335-350.

Mun, J.-H. et al. (2009), "Genome-wide comparative analysis of the *Brassica rapa* gene space reveals genome shrinkage and differential loss of duplicate genes after whole genome triplication", *Genome Biol.* Vol. 10 No. 10, pp. 1-18, *http://dx.doi.org/10.1186/gb-2009-10-10-r111*.

Nagpal, R. et al. (1996), "Transfer of *Brassica tournefortii* (TT) genes to allotetraploid oilseed *Brassica* species (*B. juncea* AABB, *B. napus* AACC, *B. carinata* BBCC): Homoeologous pairing is more pronounced in the three-genome hybrids (TACC, TBAA, TCAA, TCBB) as compared to allodiploids (TA, TB, TC)", *Theoretical and Applied Genetics*, Vol. 92, pp. 566-571.

Namai, H., M. Sarashima and T. Hosoda (1980), "Interspecific and intergeneric hybridization breeding in Japan", in: Tsunoda, S., K. Hinata and C. Campo- Gómez (eds.), *Brassica Crops and Wild Allies: Biology and Breeding*, Japan Scientific Societies Press, Tokyo, pp. 191-204.

Narasimhulu, S.B. et al. (1994), "Intergeneric protoplast fusion between *Brassica carinata* and *Camelina sativa*", *Plant Cell Reports*, Vol. 13, No. 11, pp. 657-660, August, *http://dx.doi.org/10.1007/BF00232940*.

Nasrallah, J.B., T. Nishio and M.E. Nasrallah (1991), "The self-incompatibility genes of *Brassica*: Expression and use in genetic ablation of floral tissues", *Annual Review Plant Physiol. and Plant Mol. Biology*, Vol. 42, pp. 393-422, *http://dx.doi.org/10.1146/annurev.pp.42.060191.002141*.

Navabi, Z. K. et al. (2010) "*Brassica* B-genome resistance to stem rot (*Sclerotinia sclerotiorum*) in a doubled haploid population of *Brassica napus* x *Brassica carinata*", *Canadian Journal of Plant Pathology*, Vol. 32, No. 2, pp. 237-246.

Navrátilová, B. (2004), "Protoplast cultures and protoplast fusion focused on Brassicaceae – A review", *Horticultural Science*, Vol. 31, pp. 140-157.

Navrátilová, B. et al. (1997), "Construction of intergeneric somatic hybrids between *Brassica oleracea* and *Armoracia rusticana*", *Biologia Plantarium*, Vol. 39, No. 4, pp. 531-541.

Neutrofal, F. (1927), "Zytologische Studien uber die Kulturrassen von *Brassica oleracea*", *Österreichische Botanische Zeitschrift*, Vol. 76, No. 2, pp. 101-115, *www.jstor.org/stable/23652239*.

Nieuwhof, M. (1969), *Cole Crops*, World Crop Books, London.

Nishi, S. (1980), "Differentiation of *Brassica* crops in Asia and the breeding of 'Hakuran', a newly synthesized leafy vegetable", in: Tsunoda, S., K. Hinata and C. Gómez-Campo (eds.), *Brassica Crops and Wild Allies*, Japan Scientific Societies Press, Tokyo, pp. 133-150.

Nishi, S. (1968), "'HAKURAN'. An artificially synthesized heading *Brassica napus", Japan Agricultural Research Quarterly*, Vol. 3, No. 4, pp. 18-22, *www.jircas.affrc.go.jp/english/publication/jarq/03-4/03-4index.html*.

Nishiyama, I., M. Sarashima and Y. Matsuzawa (1991), "Critical discussion on abortive interspecific crosses in *Brassica*", *Plant Breeding*, Vol. 107, No. 4, pp. 288-302, December, *http://dx.doi.org/10.1111/j.1439-0523.1991.tb00552.x*.

Nitovskaya, I.O. and A.M. Shakhovskyi (1998), "Obtaining of asymmetrical somatic hybrids between *Brassica oleracea* L. and *Arabidopsis thaliana* L.", *Tsitol. Genet.*, Vol. 32, pp. 72-81 (in Russian: English abstract).

Nitovskaya, I.O. et al. (2006a), "Production of *Brassica oleracea* (+*Arabidopsis thaliana*) and *Brassica napus* cell lines resistant to spectinomycin/streptomycin as a result of plastome genetic transformation", *Tsitol Genet.*, Vol. 40, pp. 3-10 (in Ukrainian: English abstract).

Nitovskaya, I.O. et al. (2006b), "Construction of cybrid transplastomic *Brassica napus* plants containing *Lesquerella fendlei* chloroplasts", *Tsitol Genet.*, Vol. 40, pp. 11-21 (in Ukrainian: English abstract).

Nitovskaya, I.O. et al. (1998), "Production of "Brassicapsella" somatic hybrids based on the double inactivation of donor protoplasts", *Dopov. Nats. Akad. Nauk. Ukr.*, Vol. 0(9), pp. 181-186 (in Ukrainian: English abstract).

Norris, C.E. and J. Sweet (2002), "Monitoring large scale releases of genetically modified crops (EPG 1/5/84)", *Final Report of Monitoring Studies of Field Scale Releases of GM Oilseed Rape Crops in England from 1994-2000 (NIAB)*, NIAB, Cambridge, *www.gmo-safety.eu/pdf/aktuell/NIAB-monitoring.pdf*.

Norris, C.E. et al. (1999), "Monitoring weediness and persistence of genetically modified oilseed rape (*Brassica napus*) in the UK", in: Lutmann, P.J.W. (ed.), *Gene Flow and Agriculture: Relevance for Transgenic Crops*, British Crop Protection Council Symposium Proceedings, No. 72, pp. 255-260.

Nou, I.S. et al. (1993), "Comparison of *S*-alleles and *S*-glycoproteins between two wild populations of *Brassica campestris* in Turkey and Japan", *Sexual Plant Reproduction*, Vol. 6, No. 2, pp. 79-86, May.

Ockendon, D. (2000), "The *S*-allele collection of *Brassica oleracea*", *ISHS Acta Horticulturae 539: III. International Symposium on Brassicas and XII Crucifer Genetics Workshop*, pp. 25-30, *http://dx.doi.org/10.17660/ActaHortic.2000.539.1*.

OECD (2011), "Revised consensus document on compositional considerations for new varieties of low erucic acid rapeseed (canola): Key food and feed nutrients, anti-nutrients and toxicants", *Series on the Safety of Novel Foods and Feeds*, No. 24, OECD, Paris, *www.oecd.org/officialdocuments/publicdisplaydocumentpdf/?cote=ENV/JM/MONO(2011)55&docLanguage=En*.

OECD (1997), "Consensus document on the biology of *Brassica napus* L. (oilseed rape)", *Series on Harmonisation of Regulatory Oversight of Biotechnology*, No. 7, OECD, Paris.

OECD (n.d.), "OECD seed schemes website", OECD, Paris, *www.oecd.org/tad/seed* (accessed 17 November 2011).

Ogbonnaya, F.C. et al. (2003), "Progress in the utilization of *Brassica nigra* in breeding for resistance to black leg (*Leptosphaeria maculans*)", *Proceedings of the 11th International Rapeseed Congress*, Copenhagen, Vol. 1, pp. 39-41.

Ohio State University (2009), "Vegetable seed production – 'dry' seeds", Department of Horticulture and Crop Science.

Ohkawa, Y. (1986), "Comparison of *B. napus* to *B. campestris* and *B. oleracea* based on the cytoplasmic characters, cytoplasmic male sterility and chloroplast DNA", *Japan Agricultural Research Quarterly*, Vol. 19, pp. 253-258.

Oikarinen, S. and P.H. Ryöppy (1992), "Somatic hybridization of *Brassica campestris* and *Barbarea* species", *Proceedings of the XIII Eucarpia Congress: Reproductive Biology and Plant Breeding*, Angers, France, 6-11 July, pp. 261-262.

Okamoto, S. et al. (2007), "Self-compatibility in *Brassica napus* is caused by independent mutations in the *S*-locus genes", *The Plant Journal*, Vol. 50, No. 3, pp. 391-400.

Olsson, G. (1960), "Species crosses within the genus *Brassica*. II. Artificial *Brassica juncea* Coss.", *Hereditas*, Vol. 46, pp. 171-222.

Opeña, R.T., C.G. Kuo and J.Y. Yoon (1988), "Breeding and seed production of Chinese cabbage in the tropics and suntropics", Vol. 17, Asian Vegetable Research Development Center, Chinese Taipei.

Orson, J. (2002), "Gene stacking in herbicide tolerant oilseed rape: Lessons from the North American experience", *English Nature Research Reports*, No. 443, *http://publications.naturalengland.org.uk/file/81043*.

Østergaard, L. et al. (2006), "Pod shatter-resistant *Brassica* fruit produced by ectopic expression of the *FRUITFULL* gene", *Plant Biotechnology Journal*, Vol. 4, No. 1, pp. 45-51, January.

Ovcharenko, O.O. et al. (2005), "Creation and analysis of *Brassica napus* + *Arabidopsis thaliana* somatic hybrids possessing maize Spm/dspm heterologous transposable system", *Tsitol Genet.*, Vol. 39, pp. 50-56.

Ovcharenko, O.O. et al. (2004), "Obtaining of intertribal *Brassica juncea* + *Arabidopsis thaliana* somatic hybrids and study of transgenic trait behaviour", *Tsitol Genet.*, Vol. 38, pp. 3-8.

Palmer, J.D. (1988), "Intraspecfic variation and multicircularity in *Brassica* mitochondrial DNAs", *Genetics*, Vol. 118, pp. 341-351.

Palmer, J.D. et al. (1983), "Chloroplast DNA evolution and the origin of amphiploid *Brassica* species", *Theoretical and Applied Genetics*, Vol. 65, No. 3, pp. 181-189, *http://dx.doi.org/10.1007/BF00308062*.

Palmer, T.P. (1962), "Population structure, breeding system, interspecific hybridization and alloploidy", *Heredity*, Vol. 17, pp. 278-283.

Parkin, I.A., A.G. Sharpe and D.J. Lydiate (2003), "Patterns of genome duplication within *Brassica napus* genome", *Genome*, Vol. 46, pp. 291-303.

Parkin, I.A. et al. (2005), "Segmental structure of the *Brassica napus* genome based on comparative analysis with *Arabidopsis thaliana*", *Genetics*, Vol. 171, No. 2, pp. 765-781, October, *http://dx.doi.org/10.1534/genetics.105.042093*.

Parkin, I.A. et al. (1995), "Identification of the A and C genomes of amphidloid *Brassica napus* (oilseed rape)", *Genome*, Vol. 38, No. 6, pp. 1 122-1 131.

Pascher, K.P. et al. (2006), "Feral oilseed rape – Investigations on its potential for hybridization", *Final Report to the Commission of the Federal Ministry of Health and Women*, Austria.

Paul, E.M., C. Thompson and J.M. Dunwell (1995), "Gene dispersal from genetically modified oilseed rape in the field", *Euphytica*, Vol. 81, pp. 283-289.

Paulmann, W. and G. Röbbelen (1988), "Effective transfer of cytoplasmic male fertility from radish (*Raphanus sativus* L.) to rape (*Brassica napus* L.)", *Plant Breeding*, Vol. 100, pp. 299-309.

PBO (Plant Biotechnology Office, Canada) (1998), "Decision document 98-25: Determination of environmental safety of Rhône Poulenc's oxynil herbicide-tolerant *Brassica napus* canola line Westar Oxy-235", Canadian Food Inspection Agency, Plant Biosafety Office, available at: *www.cera-gmc.org/files/cera/GmCropDatabase/docs/decdocs/01-290-019.pdf*.

Pekrun, C. and P.J.W. Lutman (1998), "The influence of post-harvest cultivation on the persistence of volunteer oilseed rape", *Aspects of Applied Biology*, Vol. 51, pp. 113-118.

Pekrun, C., J.D.J. Hewitt and P.J.W. Lutman (1998), "Cultural control of volunteer oilseed rape (*Brassica napus*)", *Journal of Agricultural Science*, Vol. 130, No. 2, pp. 155-163, March.

Pekrun, C., P.J.W. Lutman and K. Baeumer (1998), "Research on volunteer rape: A review", *Pflanzenbauwissenschaften*, Vol. 2, pp. 84-90.

Pekrun, C., P.J.W. Lutman and K. Baeumer (1997), "Induction of secondary dormancy in rape seeds (*Brassica napus* L.) by prolonged imbibitions of water stress or oxygen deficiency in darkness", *European Journal of Agronomy*, Vol. 6, Nos. 3-4, pp. 245-255, *http://dx.doi.org/10.1016/S1161-0301(96)02051-5*.

Pekrun C., T.C. Potter and P.J.W. Lutman (1997), "Genotypic variation in the development of secondary dormancy in oilseed rape and its impact on the persistence of volunteer rape", *Proceedings of the Brighton Crop Protection Conference: Weeds*, Vol. 1, pp. 243-248.

Pessel, F.D. et al. (2001), "Persistence of oilseed rape (*Brassica napus* L.) outside of cultivated fields", *Theoretical and Applied Genetics*, Vol. 102, No. 6, pp. 841-846, May.

Peterka, H. et al. (2004), "Transfer of resistance against the beet cyst nematode from radish (*Raphanus sativus*) to rape (*Brassica napus*) by monosomic chromosome addition", *Theoretical and Applied Genetics*, Vol. 109, No. 1, pp. 30-41, June.

Pinder, R. et al. (1999), "Evaluating the risk of transgene spread from *Brassica napus* to related species", in: Lutman, P.J.W. (ed.), *Gene Flow and Agriculture: Relavance for Transgene Crops*, Proceedings of the British Crop Protection Council Symposium, Vol. 72, pp. 275-280.

Pivard, S. et al. (2008), "Where do the feral oilseed rape populations come from? A large scale study of their possible origin in a farmland area", *Journal of Applied Ecology*, Vol. 45, No. 2, pp. 476-485, April, *http://dx.doi.org/10.1111/j.1365-2664.2007.01358.x*.

Plieske, J., D. Struss and G. Röbbelen (1998), "Inheritance of resistance derived from the B-genome of *Brassica* against *Phoma lingam* and the development of molecular markers", *Theoretical and Applied Genetics*, Vol. 97, No. 5, pp. 929-936, October.

Plümper, B. (1995), "Somatische und Sexuelle Hybridisierung für den Transfer von Krankheitsresistenzen auf *Brassica napus* L.", PhD Thesis, Free University of Berlin, Berlin.

Potapov, D.A. and G.M. Osipova (2003), "Development of yellow seeded *Brassica napus* in Siberia", *Proceedings of the 11th International Rapeseed Congres*s, Copenhagen, pp. 250-252.

Prain, D. (1898), "The mustards cultivated in Bengal", *Agricultural Ledger*, Vol. 5, pp. 1-80.

Prakash, O. (1961), *Food and Drinks in Ancient India*, Munshi Ram Manohar, Delhi, pp. 265-266.

Prakash, S. and V. L. Chopra (1991), "Cytogenetics of crop Brassicas and their allies", in: Tsuchiya, T. and P.K. Gupta (eds.), *Chromosome Engineering in Plants: Genetics, Breeding, Evolution*, Elsevier, Amsterdam, pp. 161-180.

Prakash, S. and K. Hinata (1980), "Taxonomy, cytogenetics and origin of crop Brassicas, a review", *Opera Botanica*, Vol. 55, pp. 3-57.

Prakash, S. et al. (2009), "*Brassica* and its close allies: Cytogenetics and evolution", *Plant Breeding Reviews*, Vol. 31, pp. 21-187, *http://dx.doi.org/10.1002/9780470593783.ch2*.

Prakash, S. et al. (2001), "Expression of male sterility in alloplasmic *Brassica juncea* with *Erucastrum canariense* cytoplasm and development of fertility restoration system", *Plant Breeding*, Vol. 120, pp. 178-182.

Prakash, S. et al. (1998), "A *Moricandia arvensis* based cytoplasmic male sterility and fertility restoration system in *Brassica juncea*", *Theoretical and Applied Genetics*, Vol. 97, No. 3, pp. 488-492, August.

Prakash, S. et al. (1995), "Cytoplasmic male sterility (CMS) systems other than *ogu* and polima in *Brassicae:* Current status", *Proceedings of the 9th International Rapeseed Congress*, Cambridge, United Kingdom, Vol. 1, pp. 44-48.

Price, J.S. et al. (1996), "Seed losses in commercial harvesting of oilseed rape", *J. of Agricultural Engineering Research*, Vol. 65, No. 3, pp. 183-191, *http://dx.doi.org/10.1006/jaer.1996.0091*.

Primard, C. et al. (1988), "Interspecific somatic hybridization between *Brassica napus* and *Brassica hirta* (*Sinapis alba* L.)", *Theoretical and Applied Genetics*, Vol. 74, No. 4, pp. 546-552, April.

Przybylski, R. et al. (1993), "Stability of low linolenic acid canola oil to accelerated storage at 60 degrees C", *Lebensmittel- Wissenschaft & Technologie*, Vol. 26, No. 3, pp. 205-209.

Purugganan, M.D., A.L. Boyles and J.I. Suddith (2000), "Variation and selection at the cauliflower floral homeotic gene accompanying the evolution of domesticated *Brassica oleracea*", *Genetics*, Vol. 155, No. 2, pp. 855-862, June.

Qi, C.K. et al. (1995), "A successful transfer of yellow-seeded trait from *Brassica carinata* to *B. napus*", *Proceedings of the 9th International Rapeseed Congress*, Cambridge, United Kingdom, 4-7 July, Vol. 4, pp. 1 137-1 140.

Qi, X.-H., M.-F. Zhang and J.-H. Yang (2007), "Molecular phylogeny of Chinese vegetable mustard (*Brassica juncea*) based on the internal transcribed spacers (ITS) of nuclear ribosomal DNA", *Genetic Resources and Crop Evolution*, Vol. 54, No. 8, pp. 1 709-1 716, December, *http://dx.doi.org/10.1007/s10722-006-9179-0*.

Quazi, M.H. (1988), "Interspecific hybrids between *Brassica napus* L. and *B. oleracea* L. developed by embryo culture", *Theoretical and Applied Genetics*, Vol. 75, No. 2, pp. 309-318, January.

Quiros, C.F., O. Ochoa and D.S. Douches (1988), "Exploring the role of x=7 species in *Brassica* evolution: Hybridization with *B. nigra* and *B. oleracea*", *Journal of Heredity*, Vol. 79, No. 5, pp. 351-358.

Rahman, M.H. (2001), "Production of yellow seeded *Brassica napus* through interspecific crosses", *Plant Breeding*, Vol. 120, No. 6, pp. 463-472, December, *http://dx.doi.org/10.1046/j.1439-0523.2001.00640.x*.

Rajhathy, T. (1976), "Haploid flax revisited", *Zeitschrift fuer Pflanzenzuechtung*, Vol. 67, pp. 1-10.

Rakow, G. (2004), "Species origin and economic importance of *Brassica*", *Biotechnology in Agriculture and Forestry*, Vol. 54, pp. 6-7, available at: *www.springer.com/cda/content/document/cda_downloaddocument/9783540202646-c1.pdf?SGWID=0-0-45-115127-p23861264*.

Rakow, G. and D. Rode (2009), "AC vulcan oriental mustard", *Canadian Journal of Plant Science*, Vol. 89, No. 2, pp. 325-329, *http://dx.doi.org/10.4141/CJPS08147*.

Rakow, G. and D.L. Woods (1987), "Outcrossing in rape and mustard under Saskatchewan prairie conditions", *Canadian Journal of Plant Science*, Vol. 67, No. 1, pp. 147-151, *http://dx.doi.org/10.4141/cjps87-017*.

Rakow, G. et al. (2009), "Centennial brown brown condiment mustard", *Canadian Journal of Plant Science*, Vol. 89, No. 2, pp. 337-340, *http://dx.doi.org/10.4141/CJPS08146*.

Ramanujam, S. and D. Srinivasachar (1943), "Cytogenetic investigations in the genus *Brassica* and the artificial synthesis of *B. juncea*", *Indian Journal of Genetic Plant Breeding*, Vol. 3, pp. 73-88.

Ramsay, G., C. Thompson and G. Squire (2003), "Quantifying landscape-scale gene flow in oilseed rape", DEFRA Project RG0216, Department for Environment, Food & Rural Affairs, London, *http://webarchive.nationalarchives.gov.uk/20081023141438/http://www.defra.gov.uk/environment/gm/research/pdf/epg_rg0216.pdf*.

Ramsay, G. et al. (1999), "Honeybees as vectors of GM oilseed rape pollen", in: Lutmann, P.J.W. (ed.), *Gene Flow and Agriculture: Relevance for Transgenic Crops*, British Crop Protection Council Symposium Proceedings, Vol. 72, pp. 209-214.

Rana, D. et al. (2004), "Conservation of the microstructure of genome segments on *Brassica napus* and its diploid relatives", *The Plant Journal*, Vol. 40, No. 5, pp. 725-733, December.

Ranito-Lehtimäki, A. (1995), "Aerobiology of pollen and pollen antigens", Chapter 15, in: Cox, C.S. and C.M. Wathes (eds.), *Bioaerosols Handbook*, CRC Lewis Press, Boca Raton, Florida, pp. 387-406.

Rasche, E. and M. Gadsby (1997), "Glufosinate-ammonium tolerant crops – International commercial developments and experience", *Proceedings of the Brighton Crop Protection Conference Weeds*, pp. 941-946.

Rashid, A., G. Rakow and R.K. Downey (1994), "Development of yellow seeded *Brassica napus* through interspecific crosses", *Plant Breeding*, Vol. 112, No. 2, pp. 127-134, March, *http://dx.doi.org/10.1111/j.1439-0523.1994.tb00660.x*.

Raybould, A.F. and A.J. Gray (1993), "Genetically modified crops and hybridization with wild relatives: A UK perspective", *Journal of Applied Ecology*, Vol. 30, No. 2, pp. 199-219, *http://dx.doi.org/10.2307/2404623*.

Reboud, X. (2003), "Effect of gap on gene flow between otherwise adjacent transgenic *Brassica napus* crops", *Theoretical and Applied Genetics*, Vol. 106, pp. 1 048-1 058.

Ren, J.P., M.H. Dickson and E.D. Earle (2001a), "Improved resistance to bacterial softrot by protoplast fusion between *Brassica rapa* and *B. oleracea*", *Theoretical and Applied Genetics*, Vol. 100, No. 5, pp. 810-819, March.

Ren, J.P. M.H. Dickson and E.D. Earle (2001b), "CC-14-1 and CC-18-2 progenies of Chinese cabbage derived from somatic hybridization for resistance to bacterial soft rot", *Horticultural Science*, Vol. 36, pp. 990-991.

Renfrew, J.M. (1973), *Palaoethnobotany: The Prehistoric Food Plants of the Near East and Europe*, Columbia University Press, New York.

Reuter, H. et al. (2008), "Hazard mitigation or mitigation hazard? Would genetically modified dwarfed oilseed rape (*Brassica napus*) increase feral survival?", *Environmental Science and Pollution Research*, Vol. 15, No. 7, pp. 529-535, *http://dx.doi.org/10.1007/s11356-008-0049-5*.

Rice, E.L. (1984), *Allelopathy. 2nd Edition*, Academic Press, New York.

Rich, T.C.G. (1991), *Crucifers of Great Britain and Ireland*, B.S.B.I. Handbook, No. 6, Botanical Society of the British Isles, London.

Rieger, M.A. et al. (2002), "Pollen-mediated movement of herbicide resistance between commercial canola fields", *Science*, Vol. 296, No. 5 577, pp. 2 386-2 388, *http://dx.doi.org/10.1126/science.1071682*.

Rieger, M.A. et al. (2001), "Hybridization between *Brassica napus* L. and *Raphanus raphanistrum* L. under agronomic field conditions", *Theoretical and Applied Genetics*, Vol. 103, pp. 555-560.

Rieger, M.A. et al. (1999), "Risks of gene flow from transgenic herbicide-resistant canola (*Brassica napus*) to weedy relatives in Southern Australian cropping system", *Australian Journal of Agricultural Research*, Vol. 50, pp. 115-128.

Rimmer, S.R. et al. (2007), *Compendium of Brassica Diseases*, APS Press, St. Paul, Minnesota.

Ringdahl, E.A., P.B.E. McVetty and J.L. Sernyk (1987), "Intergeneric hybridization of *Diplotaxis* spp. with *Brassica napus*: A source of new CMS systems?", *Canadian Journal of Plant Science*, Vol. 67, No. 1, pp. 239-243, *http://dx.doi.org/10.4141/cjps87-030*.

Ripley, V.L. and P.G. Arnison (1990), "Hybridization of *Sinapis alba* L. and *Brassica napus* L. via embryo rescue", *Plant Breeding*, Vol. 104, No. 1, pp. 26-33, March, *http://dx.doi.org/10.1111/j.1439-0523.1990.tb00398.x*.

Röbbelen, G. (1960), "Beiträge zur Analyse des Brassica-Genomes", *Chromosoma*, Vol. 11, pp. 205-228.

Röbbelen, G. and W. Thies (1980), "Variation in rapeseed glucosinolates and breeding for improved meal quality", in: Tsunoda, S., K. Hinata and C. Campo- Gómez (eds.), *Brassica Crops and Wild Allies: Biology and Breeding*, Japan Scientific Societies Press , Tokyo, pp. 285-300.

Rousselle, P. et al. (1985), "Restauration de la fertilité pour l'androsterilite genocytoplasmique chez le colza (*Brassica napus* L.) : Utilisation des Raphano-Brassica", *Agronomie: Sciences des productions végétales et de l'environnement*, Vol. 5, No. 5, pp. 431-437.

Roy, N.N. (1984), "Interspecific transfer of *Brassica juncea*-type high blackleg resistance to *Brassica napus*", *Euphytica*, Vol. 33, No. 2, pp. 295-303, June.

Roy, N.N. (1980), "Species crossability and early generation plant fertility in interspecific crosses of *Brassica*", *SABRAO Journal*, Vol. 12, pp. 43-54.

Roy, N.N. (1977), "Interspecific cross compatibility and F1 sterile oilseed Brassicas", *Proceedings of the 3rd International SABRAO Congress*, Vol. 2, pp. 31-33.

Ryschka, U., E. Klocke and G. Schumann (2003), "High frequency recovery of intergeneric fusion products of *Brassica oleracea* (+) *Lepidium meyenii* and their molecular characterization by RAPD and AFLP", *ISHS Acta Horticulturae 625: XXVI International Horticultural Congress: Biotechnology in Horticultural Crop Improvement: Achievements, Opportunities and Limitations*, pp. 145-149, *http://dx.doi.org/10.17660/ActaHortic.2003.625.15*.

Ryschka, U. et al. (1999), "Somatic cell hybridization for transfer of disease resistance in *Brassica*", in: Altman, A. et al. (eds.), *Plant Biotechnology and In Vitro Biology in the 21st Century*, Vol. 36 of the series Current Plant Science and Biotechnology in Agriculture, Kluwer Academic Publishers, Amsterdam, pp. 205-208.

Saal, B. et al. (2004), "Identification of a *Brassica juncea*-derived recessive gene conferring resistance to *Leptosphaeria maculans* in oilseed rape", *Plant Breeding*, Vol. 123, No. 6, pp. 505-511, December, *http://dx.doi.org/10.1111/j.1439-0523.2004.01052.x*.

Sacristan, M.D. and M. Gerdemann (1986), "Different behavior of *Brassica juncea* and *B. carinata* as sources of *Phoma lingam* resistance in experiments of interspecific transfer to *B. napus*", *Plant Breeding*, Vol. 97, No. 4, pp. 304-314, December, *http://dx.doi.org/10.1111/j.1439-0523.1986.tb01071.x*.

Sadik, S. (1962), "Morphology of the curd of cauliflower", *American Journal of Botany*, Vol. 49, pp. 290-297.

Saji, H. et al. (2005), "Monitoring the escape of transgenic oilseed rape around Japanese ports and roadsides", *Environmental Biosafety Research*, Vol. 4, No. 4, pp. 217-222, December.

Sakai, T. et al. (1996), "Introduction of a gene from fertility restored radish (*Raphanus sativus*) into *Brassica napus* by fusion of X-irradiated protoplasts from a radish restorer line and iodacetoamide-treated protoplasts from a cytoplasmic male-sterile cybrid of *B. napus*", *Theoretical and Applied Genetics*, Vol. 93, No. 3, pp. 73-379, August, *http://dx.doi.org/10.1007/BF00223179*.

Sakhno, L.O. et al. (2007), "Phosphinothricin-resistant *Brassica napus* + *Orychophragmus violaceus* somatic hybrids", *Cytology and Genetics*, Vol. 41, No. 1, pp. 1-5.

Salisbury, P. (2002), *Genetically Modified Canola in Australia: Agronomic and Environmental Considerations*, Downey, R.K. (ed.), Australian Oilseeds Federation.

Sarwar, M. and J.A. Kirkegaard (1998), "Biofumigation potential of Brassicas II. Effect of environment and ontogeny on glucosinolate production and implications for screening", *Plant and Soil*, Vol. 201, No. 1, pp. 91-101.

Sauer, J.D. (1993), *Historical Geography of Crop Plants*, CRC Press, Boca Raton, Florida.

Sauermann, W. (1987), "Entwicklung des Glucosinolatgehalts von 00-Raps in den Nängigkeit vom Fremddurchwuchs", *Raps*, Vol. 5, pp. 12-13.

Scarth, R., S.R. Rimmer and P.B.E. McVetty (1995), "Apollo low linolenic summer rape", *Canadian J. of Plant Science*, Vol. 75, No. 1, pp. 203-204, *http://dx.doi.org/10.4141/cjps95-037*.

Scarth, R. et al. (1988), "Stellar low linolenic high linoleic acid summer rape", *Canadian Journal of Plant Science*, Vol. 68, No. 2, pp. 509-511, *http://dx.doi.org/10.4141/cjps88-061*.

Schafer, E.H. (1977), "T'ang", in: Chang, K.C. (ed.), *Food in Chinese Culture*, Yale University Press, New Haven, Connecticut.

Scheffler, J.A. and P.J. Dale (1994), "Opportunities for gene transfer from transgenic oilseed rape (*Brassica napus*) to related species", *Transgenic Research*, Vol. 3, No. 5, pp. 263-278.

Scheffler J.A., R. Parkinson and P.J. Dale (1995), "Evaluating the effectiveness of isolation distances for field plots of oilseed rape (*Brassica napus*) using a herbicide-resistance transgene as a selectable marker", *Plant Breeding*, Vol. 114, No. 4, pp. 317-321, August, *http://dx.doi.org/10.1111/j.1439-0523.1995.tb01241.x*.

Scheffler, J.A., R. Parkinson and P.J. Dale (1993), "Frequency and distance of pollen dispersal from transgenic oilseed rape (*Brassica napus*)", *Transgenic Research*, Vol. 2, No. 6, pp. 356-364, November.

Schiemann, E. (1932), *Entstehung der Kulurpflanzen*, Hadb. Vererbwis. Lfg., Vol. 15.

Schlink, S. (1998), "10 year survival of rape seed (*Brassica napus* L.) in soil", *Zeitschrift für Pflanzenkrankheiten und Pflanzenschutz, Sonderheft*, Vol. XVI, pp. 169-172.

Schlink, S. (1995), "Überdauerungsvermögen und Dormanz von Rapssamen (*Brassica napus* L.) im Boden", *9th European Weed Research Society Symposium*, Budapest, Doorwerth, Netherlands, European Weed Research Society, pp. 65-73.

Schröder-Pontoppidan, M. et al. (1999), "Very long chain and hydroxylated fatty acids in offspring of somatic hybrids between *Brassica napus* and *Lesquerella fendleri*", *Theoretical and Applied Genetics*, Vol. 99, No. 1, pp. 108-114, July.

Schulz, O.E. (1936), "Cruciferae", in: Engler, A. and H. Harms (eds.), *Die Natürlichen Pflanzenfamilien*, 2nd edition, 17B, Leipzig, pp. 227-658.

Schulz, O.E. (1919), "IV. 105 Cruciferae-Brassiceae. Part 1. Subtribes I. Brassicinae and II. Raphaninae", in: Engler, A. (ed.), *Das Pflanzenreich,* Heft 68-70, Wilhelm Engelmann, Leipzig, pp. 1-290.

Scott, S.E. and M.J. Wilkinson (1998), "Transgene risk is low", *Nature*, Vol. 393, No. 320, May, *http://dx.doi.org/10.1038/30642*.

Sharma, T.R. and B.M. Singh (1992), "Transfer of resistance to *Alternaria brassicae* in *Brassica juncea* through interspecific hybridization among *Brassica*", *Journal of Genetic Breeding*, Vol. 46, No. 4, pp. 373-378.

Sharpe, A.G. et al. (1995), "Frequent nonreciprocal translocations in the amphidiploids genome of oilseed rape (*Brassica napus*)", *Genome*, Vol. 38, pp. 1 112-1 121.

Shiga, T. (1970), "Rapeseed breeding by interspecific crossing between *Brassica napus* and *Brassica campestris* in Japan", *JARQ*, Vol. 5, pp. 5-10.

Shpota, V.I. and V.E. Podkolzina (1986), "Development of erucic acid-free forms of Indian mustard", *Selektsiia I Semenovodstvo*, Vol. 4, pp. 19-21.

Siciliano, S.D. and J.J. Germida (1999), "Taxonomic diversity of bacteria associated with the roots of field-grown transgenic *Brassica napus* cv. Quest, compared to the non-transgenic *B. napus* cv. Excel and *B. rapa* Parkland", *FEMS Microbiology*, Vol. 29, No. 3, pp. 263-272, July, *http://dx.doi.org/10.1016/S0168-6496(99)00019-7*.

Siemens, J. (2002), "Interspecific hybridisation between wild relatives and *Brassica napus* to introduce new resistance traits into the oilseed rape gene pool", *Czech Journal of Genetics of Plant Breeding*, Vol. 38, pp. 155-157.

Siemens, J. and M.D. Sacristán (1995), "Production and characterization of somatic hybrids between *Arabidopsis thaliana* and *Brassica nigra*", *Plant Science*, Vol. 111, No. 1, pp. 95-106, October, *http://dx.doi.org/10.1016/0168-9452(95)04221-F*.

Sigareva, M.A. and E.D. Earle (1999), "Regeneration of plants from protoplasts of *Capsella bursa-pastoris* and somatic hybridization with rapid cycling *Brassica oleracea*", *Plant Cell Reports*, Vol. 18, No. 5, pp. 412-417, January.

Sigareva, M., J. Ren and E.D. Earle (1999), "Introgression of resistance to *Alternaria brassicicola* from *Sinapis alba* to *Brassica oleracea* via somatic hybridization and back-crosses", *Cruciferae Newsletter*, No. 21, pp. 135-136.

Simard, M.J. et al. (2002), "The frequency and persistence of canola (*Brassica napus*) volunteers in Québec cropping systems", *Weed Technology*, Vol. 16, pp. 433-439.

Simpson, E.C. et al. (1999), "Gene flow in genetically modified herbicide tolerant oilseed rape (*Brassica napus*) in the UK", in: Lutmann, P.J.W. (ed.), *Gene Flow and Agriculture: Relevance for Transgenic Crops*, British Crop Protection Council Symposium Proceedings, Vol. 72, pp. 75-81.

Sinskaia, E.N. (1928), "The oleiferous plants and root crops of the family Cruciferae Bull", *Bulletin of Applied Botany, Genetics and Plant Breeding*, Vol. 19, pp. 1-648.

Skarzhinskaya, M., M. Landgren and K. Glimelius (1996), "Production of intertribal somatic hybrids between *Brassica napus* L. and *Lesquerella fendleri* (Gray) Wats", *Theoretical and Applied Genetics*, Vol. 93, No. 8, pp. 1 242-1 250, December, available at: *http://dx.doi.org/10.1007/BF00223456*.

Skarzhinskaya, M. et al. (1998), "Genome organization of *Brassica napus* L. and *Lesquerella fendleri* and analysis of their somatic hybrids using genomic *in situ* hybridization", *Genome*, Vol. 41, pp. 691-701.

Slocum, M.K. (1989), "Analyzing the genomic structure of *Brassica* species using RFLP analysis", in: Helentjaris, T. and B. Burr (eds.), *Development and Application of Molecular Markers to Problems in Plant Genetics*, Cold Sping Harbor Lab Press, New York, pp. 73-80.

Slocum, M.K. et al. (1990), "Linkage arrangement of restriction fragment length polymorphisms in *Brassica oleracea*", *Theoretical and Applied Genetics*, Vol. 80, pp. 57-64.

Smith, L.B. and G.J. King (2000), "The distribution of *BoCAL-a* alleles in *Brassica oleracea* is consistent with a genetic model for curd development and domestication of the cauliflower", *Molecular Breeding*, Vol. 6, No. 6, pp. 603-613, December.

Snogerup, S. (1980), "The wild forms of the *Brassica oleracea* group (2n = 18) and their possible relations to the cultivated ones", in: Hinata, S.T. and C. Gómez-Campo (eds.), *Brassica Crops and Wild Allies*, Japan Scientific Societies Press, Tokyo, pp. 121-132.

Snogerup, S., M. Gustafsson and R. Von Bothmer (1990), "*Brassica* sect. *Brassica* (Brassicaeae). I. Taxonomy and variation", *Willdenowia*, Vol. 19, pp. 271-365.

Snow, A.A. and R.B. Jørgensen (1999), "Fitness costs associated with transgenic glufosinate tolerance introgressed from *Brassica napus* subsp. *oleifera* (oilseed rape) into weedy *Brassica rapa*", in: Lutmann, P.J.W. (ed.), *Gene Flow and Agriculture: Relevance for Transgenic Crops*, British Crop Protection Council Symposium Proceedings, Vol. 72, pp. 137-142.

Snow, A.A., B. Andersen and R.B. Jørgensen (1999), "Costs of transgenic herbicide resistance introgressed from *Brassica napus* into weedy *B. rapa*", *Molecular Ecology*, Vol. 8, No. 4, pp. 605-615, April, *http://dx.doi.org/10.1046/j.1365-294x.1999.00596.x*.

Snow, A.A., K.L. Uthus and T.M. Culley (2001), "Fitness of hybrids between weedy and cultivated radish: Implications for weed evolution", *Ecological Applications*, Vol. 11, No. 3, pp. 934-943.

Snowdon, R.J. (2007), "Cytogenetics and genome analysis in *Brassica* crops", *Chromosome Research*, Vol. 15, No. 1, pp. 85-95.

Snowdon, R.J. and W. Friedt (2004), "Molecular markers in *Brassica* oilseed breeding: Current status and future possibilities", *Plant Breeding*, Vol. 123, No. 1, pp. 1-8, February, *http://dx.doi.org/10.1111/j.1439-0523.2003.00968.x*.

Snowdon, R., W. Lühs and W. Friedt (2007), "Oilseed rape", Chapter 2, in: Kole, C. (ed.), *Genome Mapping and Molecular Breeding in Plants*, Vol. 2 Oilseeds, Springer-Verlag, Berlin, pp. 55-114.

Snowdon, R. et al. (2007), "*Brassica* oilseeds", in: Singh, R.J. (ed.), *Genetic Resources, Chromosome Engineering and Crop Improvement*, Vol. 4, CRC Press, Boca Raton, Florida, pp. 195-229.

Snowdon, R.J. et al. (2003), "Fishing for physical genome information-*Brassica* cytogenetics past, present and future", *Proceedings of the 11th International Rapeseed Congress*, Copenhagen, Vol. 1, pp. 116-119, *http://gcirc.org/fileadmin/documents/Proceedings/IRC2003Copen/Molecular_markers/BP3.15.pdf*.

Snowdon, R.J. et al. (2002), "Identifying the chromosomes of the A and C genome diploid *Brassica* species *B. rapa* and *B. oleracea* in their amphdiploid *B. napus*", *Theoretical and Applied Genetics*, Vol. 104, No. 4, pp. 533-538, March.

Snowdon, R.J. et al. (2000), "Development and characterisation of *Brassica napus-Sinapis arvensis* addition lines exhibiting resistance to *Leptosphaeria maculans*", *Theoretical and Applied Genetics*, Vol. 101, No. 7, pp. 1 008-1 014, November.

Sobrino-Vesperinas, E. (1988), "Obtainment of some new intergeneric hybrids between wild Brassiceae", *Candollea*, Vol. 43, pp. 499-504.

Song, G.K. et al. (1995), "Rapid genome change in synthetic polyploids of *Brassica* and its implications for polyploid evolution", *Proceedings of the National Academy of Sciences USA*, Vol. 92, No. 17, pp. 7 719-7 723.

Song, K. and T.C. Osborn (1992), "Polyphyletic origins of *Brassica napus*: New evidence based on organelle and nuclear RFLP analyses", *Genome*, Vol. 35, No. 6, pp. 992-1001, *http://dx.doi.org/10.1139/g92-152*.

Song, K.M., T.C. Osborn and P.H. Williams (1988a), "Brassica taxonomy based on nuclear restriction fragment length polymorphisms (RFLPs): 1. Genome evolution of diploid and amphidiploid species", *Theoretical and Applied Genetics*, Vol. 75, No. 5, pp. 784-794, May.

Song, K.M., T.C. Osborn and P.H. Williams (1988b), "Brassica taxonomy based on nuclear restriction fragment length polymorphisms (RFLPs): 2. Preliminary analysis of subspecies within *B. rapa* (syn. *campestris*) and *B. oleracea*", *Theoretical and Applied Genetics*, Vol. 76, No. 4, pp. 593-600, October, *http://dx.doi.org/10.1007/BF00260914*.

Song, K.M. et al. (1991), "A linkage map of *Brassica rapa* (syn. *campertris*) based on restriction fragment length polymorphisms", *Theoretical and Applied Genetics*, Vol. 82, No. 3, pp. 296-304, *http://dx.doi.org/10.1007/BF02190615*.

Sparrow, S.D., C.W. Knight and J.S. Conn (1990), "Canola seed survival over winter in the field in Alaska", *Canadian Journal of Plant Science*, Vol. 70, No. 3, pp. 799-807, *http://dx.doi.org/10.4141/cjps90-097*.

Spect, C.E. and A. Diederichsen (2001), "*Brassica*", in: Haelt, P. (ed.), *Mansfield's Encyclopedia of Agricultural and Horticultural Crops*, Vol. 3, pp. 1 453-2 456.

Squire, G.R. et al. (1999), "Gene flow at the landscape level", in: Lutmann, P.J.W. (ed.), *Gene Flow and Agriculture: Relevance for Transgenic Crops*, British Crop Protection Council Symposium Proceedings, Vol. 72, pp. 57-64.

Sridevi, O. and N. Saria (1996), "Reciprocal hybridization between *Sinapis alba* and *Brassica* species", *Cruciferae Newsletter Eucarpia*, No. 18.

Srivastava, A. et al. (2001), "AFPL based genetic diversity assessment among agronomically important natural and some newly synthesized lines of *Brassica juncea*", *Theoretical and Applied Genetics*, Vol. 104, pp. 1 092-1 098.

Staniland, B.K. et al. (2000), "Effectiveness of border areas in confining the spread of transgenic *Brassica napus* pollen", *Canadian Journal of Plant Science*, Vol. 80, No. 3, pp. 521-526, *http://dx.doi.org/10.4141/P99-117*.

Stanton, R., J. Pratley and D. Hudson (2002), "Sheep are potential vectors for the spread of canola (*Brassica napus*) seed", *Australian Journal of Experimental Agriculture*, Vol. 43, No. 6, pp. 535-538, *http://dx.doi.org/10.1071/EA02046*.

Stefansson, B.R. (1983), "The development of improved rapeseed cultivars", Chapter 6, in: Kramer, J.K.G. et al. (eds.), *High and Low Erucic Acid Rapeseed Oils*, Academic Press, Toronto, pp. 144-159.

Stefansson, B.R., F.W. Hougen and R.K. Downey (1961) "Note on the isolation of rape plants with seed oil free from erucic acid", *Canadian Journal of Plant Science*, Vol. 41, No. 1, pp. 218-219, *http://dx.doi.org/10.4141/cjps61-028*.

Stewart, C.N. Jr., M.D. Halfhill and S. Warwick (2002), "Gene flow and its consequences: *Brassica napus* (canola, oilseed rape) to wild relatives", *Proceedings of Scientific Methods Workshop: Ecological and Agronomic Consequences of Gene Flow from Transgenic Crops to Wild Relatives*, Ohio State University, Columbus, Ohio, 5-6 March, pp. 106-112, *www.biosci.ohio-state.edu/~asnowlab/Proceedings.pdf*.

Stiewe, G. and G. Röbbelen (1994), "Establishing cytoplasmic male sterility in *Brassica napus* by mitochondrial recombination with *B. tournefortii*", *Plant Breeding*, Vol. 113, No. 4, pp. 294-304, *http://dx.doi.org/10.1111/j.1439-0523.1994.tb00739.x*.

Stiewe, G. et al. (1995), "Natural and experimental evolution of CMS for rapeseed breeding", *Advanced Plant Breeding*, Vol. 18, pp. 59-76.

Stringam, G.R. and R.K. Downey (1982), "Effectiveness of isolation distance in seed production of rapeseed (*Brassica napus*)", *Agronomy Abstracts*, pp. 136-137.

Struss, D., U. Bellin and G. Röbbelen (1991), "Development of B-genome chromosome addition lines of *B. napus* using different interspecific *Brassica* hybrids", *Plant Breeding*, Vol. 106, No. 3, pp. 209-214, April, *http://dx.doi.org/10.1111/j.1439-0523.1991.tb00503.x*.

Struss, D., C.F. Quiros and G. Röbbelen (1992), "Mapping of molecular markers on *Brassica* B-genome chromosomes added to *Brassica napus*", *Plant Breeding*, Vol. 108, No. 4, pp. 320-323, May, *http://dx.doi.org/10.1111/j.1439-0523.1992.tb00137.x*.

Struss, D. et al. (1996), "Construction of *Brassica* B genome synteny groups based on chromosomes extracted from three different sources by phenotype, isozyme and molecular markers", *Theoretical and Applied Genetics*, Vol. 93, No. 7, pp. 1 026-1 032, November, *http://dx.doi.org/10.1007/BF00230120*.

Sun, W.C. et al. (1991), "*Brassica* and *Brassica*-related oilseed crops in Gansu, China", *Proceedings of the 8th International Rapeseed Congress*, Saskatoon, Canada, Vol. 4, pp. 1 130-1 135, *http://gcirc.org/fileadmin/documents/Proceedings/IRC1991vol4/P1-035.pdf*.

Sundberg, E. and K. Glimelius (1991), "Effects of parental ploidy level and genetic divergence on chromosome elimination and chloroplast segregation in somatic hybrids within Brassicaceae", *Theoretical and Applied Genetics*, Vol. 83, No. 1, pp. 81-88, November, *http://dx.doi.org/10.1007/BF00229229*.

Sweet, J.B. and R. Shepperson (1998), "The impact of releases of genetically modified tolerant oilseed rape in UK", *Acta Horticulturae*, Vol. 459, pp. 225-234.

Sweet, J.B. et al. (1999a), "Assisting the impact and consequences of the release and commercialization of genetically modified crops", in: Lutmann, P.J.W. (ed.), *Gene Flow and Agriculture: Relevance for Transgenic Crops*, Brighton Crop Protection Council Symposium Proceedings, No. 72, pp. 241-246.

Sweet, J.B. et al. (1999b), "Hybridisation and persistence in herbicide tolerant oilseed rape (*Brassica napus*)", *Proceedings of the 10th International Rapeseed Congress*, Canberra.

Sweet, J.B. et al. (1997), "The impact of releases of genetically modified herbicide tolerant oilseed rape in the UK", *Proceedings of the Brighton Crop Protection Conference: Weeds*, pp. 291-302.

Szabo, T. (1985), "Variability of flower, nectar, pollen and seed production in some Canadian canola varieties", *American Bee Journal*, Vol. 124, pp. 351-354.

Takahashi, O. (1987), "Utilization and seed production of hybrid vegetable varieties in Japan", in: Feistritzer, W.P. and A.F. Kelly (eds.), *Hybrid Seed Production of Selected Cereal, Oil and Vegetable Crops*, FAO Plant Production and Protection Paper, No. 82, Rome, pp. 314-328.

Takahata, Y. and K. Hinata (1986), "Consideration of the species relationships in subtribe Brassicinae (Cruficerae) in view of cluster analysis of morphological characters", *Plant Species Biology*, Vol. 1, No. 1, pp. 79-88, June, *http://dx.doi.org/10.1111/j.1442-1984.1986.tb00017.x*.

Takahata, Y. and K. Hinata (1983), "Studies on cytodemes in subtribe Brassicinae (Cruciferae)", *Tohoku Journal of Agricultural Research*, Vol. 33, No. 3/4, pp. 111-124.

Takahata, Y. and K. Hinata (1980), "A variation study of subtribe *Brassicinae* by principal component analysis", in: Tsunoda, S., K. Hinata and C. Gómez-Campo (eds.), *Brassica Crops and Wild Allies*, Japan Scientific Societies Press, Tokyo, pp. 33-49.

Takayama, S. and A. Isogai (2005), "Self-incmpatibility in plants", *Annual Review of Plant Biology*, Vol. 56, pp. 467-489.

Takayama, S. and A. Isogai (2003), "Molecular mechanism of self-recognition in *Brassica* self-incompatibility", *Journal of Experimental Botany*, Vol. 54, No. 380, pp. 149-156, January.

Takayama, S. et al. (2000), "The pollen determinant of self-incompatibility in *Braccica campestris*", *Nature*, Vol. 413, pp. 534-538.

Takeshita, M., M. Kato and S. Tokumasu (1980), "Application of ovule culture to the production of intergeneric or interspecific hybrids in *Brassica* and *Raphanus*", *Japanese Journal of Genetics*, Vol. 55, No. 5, pp. 373-387, *http://dx.doi.org/10.1266/jjg.55.373.*

Thalmann, C., R. Guadagnuolo and F. Felber (2001), "Search for spontaneous hybridization between oilseed rape (*Brassica napus* L.) and wild radish (*Raphanus raphanistrum* L.) in agricultural zones and evaluation of the genetic diversity of the wild species", *Botanica Helvetica*, Vol. 111, No. 2, pp. 107-119, *http://retro.seals.ch/digbib/view?pid=bhl-002:2001:111::223.*

The *B. rapa* Genome Sequencing Project Consortum (2011), "The genome of the mesopolyploid crop species *Brassica rapa*", *Nature Genetics*, Vol. 43, pp. 1 035-1 039, *http://dx.doi.org/10.1038/ng.919.*

The Multinational *Brassica* Genome Project (2015), brassica.info website, "Brassica genomes" page, Southern Cross Plant Science, Southern Cross University, United States, *www.brassica.info/resource/genome.php* (accessed 11 January 2016).

This, P. et al. (1990), "Dissection of the *Brassica nigra* genome by chromosome addition lines", *Plant Breeding*, Vol. 105, pp. 211-220.

Thomas, P. (1994), *Canola Growers Manual*, Canola Council of Canada.

Thompson, C.E. et al. (1999), "Regional patterns of gene flow and its consequences for GM oilseed rape", in: Lutmann, P.J.W. (ed.), *Gene Flow and Agriculture: Relevance for Transgenic Crops*, British Crop Protection Council Symposium Proceedings, Vol. 72, pp. 95-100.

Thompson, K.F. (1976), "Cabbages, kales etc. *Brassica oleracea* (Cruciferae)", in: Simmonds, N.W. (ed.), *Evolution of Crop Plants*, Longman, London, pp. 49-52.

Thompson, K.F. and J.P. Taylor (1966), "Non-linear dominance relationships between *S* alleles", *Heredity*, Vol. 21, pp. 345-362, *http://dx.doi.org/10.1038/hdy.1966.36.*

Timmons, A.M. et al. (1996), "Risks from transgenic crops", *Nature*, Vol. 380, No. 6 574, April.

Timmons, A.M. et al. (1995), "Assessing the risks of wind pollination from fields of *Brassica napus* subsp. *oleifera*", *Euphytica*, Vol. 85, No. 1, pp. 417-423, February.

Tonguc, M. and P.D. Griffiths (2004), "Transfer of powdery mildew resistance from *Brassica carinata* to *Brassica oleracea* through embryo rescue", *Plant Breeding*, Vol. 123, No. 6, pp. 587-589, December, *http://dx.doi.org/10.1111/j.1439-0523.2004.00987.x.*

Truco, M.J. and C.F. Quiros (1994), "Structure and organization of the B genome based on a linkage map in *Brassica nigra*", *Theoretical and Applied Genetics*, Vol. 89, No. 5, pp. 590-598, November, *http://dx.doi.org/10.1007/BF00222453.*

Truco, M.J. et al. (1996), "Inter- and intra-genomic homology of the *Brassica* genomes: Implications for their origin and evolution", *Theoretical and Applied Genetics*, Vol. 93, No. 8, pp. 1 225-1 233, December, *http://dx.doi.org/10.1007/BF00223454.*

Tsen, M. and S.H. Lee (1942), "A preliminary study of Chinese cultivated Brassicas", *Hortus Sinicus.*, Vol. 2, pp. 1-32.

Tsunoda, S. (1980), "Eco-physiology of wild and cultivated forms in *Brassica* and allied genera", in: Tsunoda, S., K. Hinata and C. Gómez-Campo (eds.), *Brassica Crops and Wild Allies*, Japan Scientific Societies Press, Tokyo, pp. 109-120.

Tsunoda, S. and S. Nishi (1968), "Origin, differentiation and breeding of cultivated *Brassica*", *Proceedings of the XII International Congress Genetics*, Vol. 2, pp. 77-88.

Uchimiya, H. and S.G. Wildman (1978), "Evolution of fraction I protein in relation to origin of amphidiploid Brassica species and other members of the Cruciferae", *Journal of Heredity*, Vol. 69, No. 5, pp. 299-303.

United Nations (1935), "Genome analysis in *Brassica* with special reference to the experimental formation of *B. napus* and peculiar mode of fertilization", *Japanese Journal of Botany*, Vol. 7, pp. 389-452.

United Nations (n.d.), UN comtrade, *http://comtrade.un.org/db* (accessed 23 May 2012).

United States Census Bureau (n.d.), Foreign Trade Statistics, *www.census.gov/foreign-trade* (accessed 23 May 2012).

USDA-ARS (2011), *Germplasm Resources Information Network* (GRIN) [Online database], United States Departement of Agriculture, Agricultural Research Service, *www.ars-grin.gov* (accessed October 2011).

USDA Foreign Agriculture Service (2015), "Production, supply and distribution online", Table 03: Major Vegetable Oils: World Supply and Distribution (Commodity View), *http://apps.fas.usda.gov/psdonline/psdHome.aspx* (accessed 11 January 2016).

Vašák, J.V. (2002), "The role and effects of glucosinolates of *Brassica* species – A review", *Rostlinná Výroba*, Vol. 48, No. 4, pp. 175-180, *www.agriculturejournals.cz/publicFiles/52985.pdf*.

Vasilenko, M. et al. (2003), "Obtaining and analysis of intergeneric somatic hybrids between *Brassica napus* and albino line of *Orychophragmus violaceus*", *Tristol. Genet.*, Vol. 37, No. 1, pp. 3-10 (in Russian).

Vaughan, J.G. (1977), "A multidisciplinary study of the taxonomy and origin of Brassica crops", *BioScience*, Vol. 27, No. 1, pp. 35-40, *http://dx.doi.org/10.2307/1297791*.

Vaughan, J.H. and J.M. Whitehouse (1971), "Seed structure and taxonomy of the Cruciferae", *Botanical Journal of the Linnean Society*, Vol. 64, pp. 383-409.

Vavilov, N.I. (1949), "The origin, variation, immunity and breeding of cultivated plants", *Chron. Bot.*, Vol. 13, pp. 1-364.

VDI Richtlinien (2007), "Monitoring the effects of genetically modified organisms, pollen monitoring, technical pollen sampling using pollen mass filter", VDI 4330 Blatt 3/Part 3, Association of German Engineers, *www.vdi.eu/guidelines/vdi_4330_blatt_3-monitoring_der_wirkungen_gentechnisch_veraenderter_organismen_gvo_pollenmonitoring_technische/*.

VDI Richtlinien (2006), "Monitoring the effects of genetically modified organisms, pollen monitoring, biological pollen sampling using bee colonies", VDI 4330 Blatt 4/Part 4, Association of German Engineers, *www.vdi.eu/guidelines/vdi_4330_blatt_4-monitoring_der_wirkungen_von_gentechnisch_veraenderten_organismen_gvo_pollenmonitoring/*.

Vera, C.L., D.I. McGregor and R.K. Downey (1987), "Detrimental effects of volunteer *Brassica* on production of certain cereals and oilseed crops", *Canadian Journal of Plant Science*, Vol. 67, No. 4, pp. 983-995, *http://dx.doi.org/10.4141/cjps87-135*.

Verkerk, R. et al. (2009), "Glucosinolates in *Brassica* vegetables: The influence of the food supply chain on intake, bioavailability and human health", *Molecular Nutrition & Food Research*, Vol. 53, Supplement 2, S219-S265, *http://dx.doi.org/10.1002/mnfr.200800065*.

Vijayakumar, C.H.M. et al. (1994), "An assay of basal branching F2 segregants from intra- and inter-specific crosses in *Brassica*", *Cruciferae Newsletter Eucarpia*, No. 16, pp. 25-26.

Von Bothmer, R., M. Gustafsson and S. Snogerup (1995), "*Brassica* sect. *Brassica* (*Brassicaceae*) II. Inter- and intraspecific crosses with cultivars of *B. oleracea*", *Gene Resource and Crop Evolution*, Vol. 42, pp. 165-179.

Von Ernst, D. et al. (1998), "Sicherheitsforschung zu Freisetzungsversuch in Roggenstein (Bayern)", *Bundesgesundheitsblatt*, Vol. 12, pp. 525-530 (in German).

Voss, A., R.J. Snowdon and W. Lühs (2000), "Intergeneric transfer of nematode resistance from *Raphanus sativus* into the *Brassica napus* genome", *ISHS Acta Horticulturae 539: III International Symposium on Brassicas and XII Crucifer Genetics Workshop*, pp. 129-134, *http://dx.doi.org/10.17660/ActaHortic.2000.539.16*.

Waddington, J. and K.E. Bowren (1978), "Effect of crop residue on production of barley, bromegrass and alfalfa in the greenhouse and in the field", *Canadian Journal of Plant Science*, Vol. 58, pp. 249-225.

Wang, H., C. Guan and C. Zhang (2007), "Studies on rapeseed production and cultivation science and technology in China", *Proceedings of the 12th International Rapeseed Congress*, Wuhan, China, Vol. III, pp. 2-7.

Wang, R., V.L. Ripley and G. Rakow (2007), "Pod shatter resistance evaluation in cultivars and breeding lines of *Brassica napus*, *B. juncea* and *Sinapis alba*", *Plant Breeding*, Vol. 126, No. 6, pp. 588-595, December, *http://dx.doi.org/10.1111/j.1439-0523.2007.01382.x*.

Wang, X.-W. (2010), "The genome of *Brassica rapa*", Plant and Animal Genomes XVIII Conference, San Diego, California, Abstract W104.

Wang, Y.P., K. Sonntag and E. Rudloff (2003), "Development of rapeseed with high erucic acid content by asymmetric somatic hybridization between *Brassica napus* and *Crambe abyssinica*", *Theoretical and Applied Genetics*, Vol. 106, No. 7, pp. 1 147-1 155, May.

Wang, Y.P. et al. (2004), "Production and characterization of asymmetric somatic hybrids between *Brassica napus* and *Crambe abyssinica*", *Votr. Pflanzenzüchtung*, Vol. 64, pp. 85-86.

Warwick, S.I. (2007), "Gene flow between GM crops and related species in Canada", in: Swanton, C. and R. Gulden (eds.), *The First Decade of Herbicide Resistant Crops in Canada*, Vol. 4 of Topics in Canadian Weed Science, Canadian Weed Science Society, Saint-Anne-de Bellevue, Quebec, Canada, pp. 101-113.

Warwick, S.I. and I.A. Al-Shehbaz (2006), "Brassicaceae: Chromosome Number Index and Database on CD-Rom", *Plant Systematics and Evolution*, Vol. 259, No. 2, pp. 237-248, July, *http://dx.doi.org/10.1007/s00606-006-0421-1*.

Warwick, S.I. and L.D. Black (1997), "Molecular phylogenies from theory to application in Brassica and allies (tribe Brassiceae, Cruciferae)", *Opera Botanica*, Vol. 132, pp. 159-168.

Warwick, S.I. and L.D. Black (1993), "Molecular relationships in the subtribe Brassicinae (Cruciferae, Tribe, Brassiceae)", *Canadian Journal of Botany*, Vol. 71, No. 7, pp. 906-918, *http://dx.doi.org/10.1139/b93-102*.

Warwick, S.I. and L.D. Black (1991), "Molecular systematics of *Brassica* and allied genera (subtribe Brassicinae, Brassiceae) – Chloroplast genome and cytodeme congruence", *Theoretical and Applied Genetics*, Vol. 82, No.1, pp. 839-850, *http://dx.doi.org/10.1007/BF00231281*.

Warwick, S.I. and J.C. Hall (2009), "Phylogeny of *Brassica* and wild relatives", in: Gupta, S.K. (ed.), *Biology and Breeding of Crucifers*, CRC Press, Boca Raton, Florida, pp. 19-36.

Warwick, S.I. and C.A. Sauder (2005), "Phylogeny of tribe Brassiceae (Brassicaceae) based on chloroplast restriction site polymorphisms and nuclear ribosomal internal transcribed spacer and chloroplast *trn*L intron sequences", *Canadian Journal of Botany*, Vol. 83, No. 5, pp. 467-483, *http://dx.doi.org/10.1139/b05-021*.

Warwick, S.I., H.J. Beckie and L.M. Hall (2009), "Gene flow, invasiveness, and ecological impact of genetically modified crops", *Annals of the New York Academy of Sciences*, Vol. 1 168, pp. 72-99, July, *http://dx.doi.org/10.1111/j.1749-6632.2009.04576.x*.

Warwick S.I., H.J. Beckie and E. Small (1999), "Transgenic crops: New weed problems for Canada?", *Phytoprotection*, Vol. 80, No. 2, pp. 71-84, *http://dx.doi.org/10.7202/706182ar*.

Warwick, S.I., L.D. Black and I. Aguinagalde (1992), "Molecular systematics of *Brassica* and allied genera (subtribe Brassicinae, Brassiceae) – Chloroplast DNA variation in the genus *Diplotaxis*", *Theoretical and Applied Genetics*, Vol. 83, No. 6, pp. 839-850, April.

Warwick, S.I., A. Francis and I.A. Al-Shehbaz (2006), "Brassicaceae: Species checklist and database on CD-Rom", *Plant Systematics and Evolution*, Vol. 259, No. 2, pp. 249-258, July, *http://dx.doi.org/10.1007/s00606-006-0422-0*.

Warwick, S.I., A. Francis and R.K. Gugel (2009), "Guide to wild germplasm: *Brassica* and allied crops (Tribe Brassiceae, Brassicaceae)", Third Edition, *www.brassica.info/info/publications/guide-wild-germplasm.php* (accessed 23 March 2010).

Warwick, S.I. et al. (2003), "Hybridization between transgenic *Brassica napus* L. and its wild relatives: *Brassica rapa* L., *Raphanus raphanistrum* L., *Sinapis arvensis* L., and *Erucastrum gallicum* (Willd.) O.E. Schulz", *Theoretical and Applied Genetics*, Vol. 107, No. 3, pp. 528-539, August.

Watanabe, M. and K. Hinata (1999), "Self-incompatibility", in: Gómez-Campo, C. (ed.), *Biology of Brassica Coenospecies*, Elsevier Science, pp. 149-183.

Weekes, R. et al. (2005), "Crop-to-crop gene flow using farm scale sites of oilseed rape (*Brassica napus*) in the UK", *Transgenic Research*, Vol. 14, No. 5, pp. 749-759, *http://dx.doi.org/10.1007/s11248-005-0943-2*.

Wei, W. and H. Darmency (2008), "Gene flow hampered by low seed size of hybrids between oilseed rape and five wild relatives", *Seed Science Research*, Vol. 18, No. 2, pp. 115-123.

Westcott, L. and D. Nelson (2001), "Canola pollination: An update", *Bee World*, Vol. 82, No. 3, pp. 115-129, *http://dx.doi.org/10.1080/0005772X.2001.11099514*.

Wikipedia contributors, (2008), "*Brassica oleracea*", Wikipedia, The Free Encyclopedia, *http://en.wikipedia.org/w/index.php?title=Brassica_oleracea&oldid=196822776* (accessed 12 March 2008).

Wikipedia contributors, (2011), "Computational phylogenetics", Wikipedia, The Free Encyclopedia, *http://en.wikipedia.org/wiki/Computational_phylogenetics* (accessed October 2011).

Wikipedia contributors, (2011), "Phylogenetics", Wikipedia, The Free Encyclopedia, *http://en.wikipedia.org/wiki/Phylogenetics* (accessed 27 October 2011).

Wikipedia contributors, (2011), "Triangle of U", Wikipedia, The Free Encyclopedia, *http://en.wikipedia.org/wiki/Triangle_of_U* (accessed 17 November 2011).

Wilkinson, M.J. et al. (2000), "A direct regional scale estimate of transgene movement from genetically modified oilseed rape to its wild progenitors", *Molecular Ecology*, Vol. 9, No. 7, pp. 983-991, July.

Wilkinson, M.J. et al. (1995), "Problems of risk assessment with genetically modified oilseed rape", *Proceedings of the Brighton Crop Protection Conference: Weeds*, Vol. 3, pp. 1 035-1 044.

Williams, I., A.P. Martin and R.P. White (1987), "The effect of insect pollination on plant development and seed production in winter oil-seed rape (*Brassica napus* L.)", *Journal of Agricultural Science*, Vol. 109, No. 1, pp. 135-139, August, *http://dx.doi.org/10.1017/S0021859600081077*.

Williams, I., A.P. Martin and R.P. White (1986), "The pollination requirements of oil-seed rape", *Journal of Agricultural Science*, Vol. 106, No. 1, pp. 27-30, February, *http://dx.doi.org/10.1017/S0021859600061670*.

Williamson, M.H. (1996), *Biological Invasions*, Chapman and Hall, London.

Winter, H. et al. (2003), "Transfer of new black leg resistance into oilseed rape", *Proceedings of the 11th International Rapeseed Congress*, Copenhagen, 6-10 July, Vol. 1, pp. 19-21.

Wu, X. et al. (2009), "Genetic diversity in oil and vegetable mustard (*Brassica juncea*) landraces by SRAP markers", *Genetic Resources and Crop Evolution*, Vol. 56, No. 7, pp. 1 011-1 022, *http://dx.doi.org/10.1007/s10722-009-9420-8*.

Yamagishi, H. and S. Nakagawa (2004), "Somatic hybrids between *Arabidopsis thaliana* and cabbage (*Brassica oleracea*)", *Journal of the Japanese Society Horticultural Science*, Vol. 73, No. 4, pp. 319-323, *http://doi.org/10.2503/jjshs.73.319*.

Yamagishi, H. et al. (2002), "Production of asymmetric hybrids between *Arabidopsis thaliana* and *Brassica napus* utilizing an efficient protoplast culture system", *Theoretical and Applied Genetics*, Vol. 104, Nos. 6-7, pp. 959-964, May.

Yang, Y.W. et al. (2002), "A study of the phylogeny of *Brassica rapa*, *B. nigra*, *Raphanus sativa* and their related genera using noncoding regions of chloroplast DNA", *Molecular Phylogenetics and Evolution*, Vol. 23, pp. 268-275.

Yang, Y.W. et al. (1999), "Molecular phylogenetic studies of *Brassica*, *Rorippa*, *Arabidopsis* and allied genera based on the internal transcribed spacer region of 18S-25S rDNA", *Molecular Phylogenetics and Evolution*, Vol. 13, No. 3, pp. 455-462, December.

Yarnell, S.H. (1956), "Cytogenetics of the vegetable crops: II. Crucifers", *Botanical Review*, Vol. 22, No. 2, pp. 81-166, February.

Yoshimura, Y. et al. (2006), "Transgenic oilseed rape along transportation routes and port of Vancouver in Western Canada", *Environmental Biosafety Research*, Vol. 5, No. 2, pp. 2-9, April-June.

Zhang, X. and S.R. Wessler (2004), "Genome-wide comparative analysis of the transposable elements in the related species *Arabidopsis thaliana* and *Brassica oleracea*", *Proccedings of the National Academy of Sciences USA*, Vol. 101, No. 15, pp. 5 589-5 594, *http://dx.doi.org/10.1073/pnas.0401243101*.

Zhu, J., D. Struss and G. Röbbelen (1993), "Studies on resistance to *Phoma lingam* in *Brassica napus-Brassica nigra* addition lines", *Plant Breeding*, Vol. 111, No. 3, pp. 192-197, October, *http://dx.doi.org/10.1111/j.1439-0523.1993.tb00629.x*.

Annex 3.A1
Common pathogens and pests

Table 3.A1.1. **Insect, mite and other Brassicacous crop pests and their regional distribution**

Order, genus and species	Common name	Regions affected
Coleoptera		
Acalymma vittatum (F.)	Striped cucumber beetle	North America
Agriotes lineatus (L.)	Lined click beetle	North America, Europe, Russian Federation
Baris laticollis Marsh.	Not reported	Europe
Ceutorhynchus assimillis Payk.	Cabbage seed weevil	North America, Europe
Ceutorhynchus hepaticus Gyll.	Not reported	Europe
Ceutorhynchus napi Gyll.	Rape stem weevil	Europe
Ceutorhynchus obstricus (Marsh.)	Cabbage seedpod weevil	Europe
Ceutorhynchus pallidactylus (Marsh.)	Cabbage stem weevil	Europe
Ceutorhynchus pleurostigma Marsh.	Turnip gall weevil	Europe, North Africa, Russian Federation
Ceutorhynchus rapae (Gyll.)	Cabbage curculio	North America, Europe, Russian Federation
Chaetocnema indica Weise	Not reported	India
Entomoscelis americana Brown	Red turnip beetle	Canada
Listroderes costirostris Schönh.	Vegetable weevil	United States, South America, Europe, Africa, Asia, Australia, New Zealand
Meligethes aeneus F.	Pollen, rape or blossom beetle	North America, Europe, North Africa, Russian Federation, China (People's Republic of)
Meligethes viridescens (F.)	Pollen or blossom beetle	North America, Europe
Phyllotreta aerea Allard	Leaf beetle	North America, Europe, North Africa, Russian Federation, India
Phyllotreta atra F.	Cabbage flea beetle	Russian Federation
Phyllotreta chotanica Duvivier	Striped flea beetle	India, South East Asia
Phyllotreta consobrina (Curtis.)	Turnip flea beetle	No distribution information found
Phyllotreta cruciferae (Goeze)	Crucifer flea beetle	North America, Europe, North Africa, Russian Federation, India
Phyllotreta flexuosa (Ill.)	Not reported	Thailand, Malaysia
Phyllotreta nemorum (L.)	Striped flea beetle	Europe
Phyllotreta striolata (F.)	Cabbage flea beetle	North America, Europe, Russian Federation, India, Asia
Phyllotreta undulata Kutschera	Lesser striped flea beetle	North America, Europe, Australia
Psylliodes chrysocephala L.	Cabbage stem flea beetle	Canada, Europe, North Africa, Russian Federation
Psylliodes punctulata Melsh.	Hop flea beetles	North America
Diptera		
Atherigona orientalis Schiner	Pepper fruit fly	United States, Central and South America, Africa, India, Asia, Australia
Chromatomyia horticola Gour.	Pea leaf miner	Europe, Africa, India, Asia
Contarinia nasturtii (Kief.)	Swede midge	North America, Europe
Dasineura brassicae (Winn.)	Brassica pod midge	Europe
Delia floralis (Fall.)	Turnip maggot	North America, Europe, Russian Federation, China (People's Republic of), Japan
Delia radicum (L.)	Cabbage root fly	North America, Europe, North Africa, Russian Federation, China (People's Republic of)
Liriomyza brassicae Riley	Serpentine leaf miner	Worldwide, except the Russian Federation
Liriomyza bryoniae Kltb.	Tomato leaf miner	Europe, Russian Federation, India, China (People's Republic of), Japan
Phytomyza horticola Gour.	Cruciferous leaf miner	Europe, India, Asia

Table 3.A1.1. **Insect, mite and other Brassicacous crop pests and their regional distribution** (*cont.*)

Order, genus and species	Common name	Regions affected
Phytomyza rufipes Meig.	Cabbage leaf miner	United States, Europe
Homoptera		
Brevicoryne brassicae (L.)	Cabbage aphid	Worldwide
Smynthurodes betae Westw.	Gall-forming aphid, bean root aphid	United States, Europe, Middle East, Australia
Hemiptera		
Aleyrodes proletella (L.)	Cabbage whitefly	Europe
Bagrada hilaris (Burm.)	Painted bug	India, Sri Lanka, Africa, Arabia
Bemisia tabaci (Genn.)	Tobacco whitefly	Worldwide
Eurydema olerace (L.)	Cabbage bug	Turkey, Russian Federation
Eurydema pulchrum (Westw.)	Small cabbage bug	India, Asia
Eurydema rugosum Mots.	Cabbage bug	Russian Federation, China (People's Republic of), Japan
Eurydema species	Orange stink or shield bugs	Europe, North Africa, Russian Federation, India, Asia, Australia
Eurydema ventralis Kolenati	Cabbage bug	Europe, Africa, Russian Federation
Lipaphis erysimi Kltb.	Mustard aphid	Worldwide
Lygus borealis (Kelton)	Not reported	Canada
Lygus elisus Van D.	Pale legume bug	North America
Lygus hesperus Knight	Western tarnished plant bug	North America
Lygus lineolaris (P. de B.)	Tarnished plant bug	North America
Lygus ruqulipennis Popp.	Bishop bug	Canada, Europe, Russian Federation
Murgantia histrionica (Hahn)	Harlequin bug	United States
Myzus persicae Sulz.	Spinach aphid or Green peach aphid	Worldwide
Nysius niger Baker	False chinch bug	India, North America, Caribbean
Nezara viridula (L.)	Green stink bug	Worldwide
Pemphigus populitransversus Riley	Poplar petiolegall aphid	United States
Pseudococcus calceolariae (Mask.)	Scarlet mealybug	United States, Central and South America, Europe, Africa, China (People's Republic of), Australia, New Zealand
Hymenoptera		
Athalia lugens (Klug)	Mustard sawfly	India
Athalia rosae (L.)	Turnip or cabbage leaf sawfly	Europe, Russian Federation, China (People's Republic of), Japan
Lepidoptera		
Acronicta rumicis (L.)	Knotgrass moth	Europe, Russian Federation, India, China (People's Republic of)
Agrotis exclamationis L.	Heart and dart moth	Europe, Russian Federation
Agrotis ipsilon (Hufn.)	Black cutworm	Worldwide
Agrotis orthogonia Morr.	Pale western cutworm	Canada
Agrotis segetum D. & S.	Turnip moth	Europe, Africa, India, China (People's Republic of), Japan
Argyrogramma signata (F.)	Green semi-looper	India, South East Asia
Ascia monuste (L.)	Gulf white cabbage worm	South America
Autographa californica Speyer	Alfalfa looper	North America, Malaysia
Autographa gamma (L.)	Silver Y moth	Europe, North Africa, India, Asia
Autographa nigrisigna (Wlk.)	Beet worm	Russian Federation, India, China (People's Republic of), Japan
Cacoecimorpha pronubana Hbn.	Carnation tortrix	United States, Europe, North Africa, Japan
Chrysodeixis agnata Stgr.	Three-spotted plusia	China (People's Republic of), Japan
Clepsis spectrana (Treit.)	Oblique-banded caterpillar	Europe, Canada

Table 3.A1.1. **Insect, mite and other Brassicacous crop pests and their regional distribution** (*cont.*)

Order, genus and species	Common name	Regions affected
Crocidolomia pavenana (F.)	Large cabbage-heart caterpillar	India, Africa, Asia, Australia
Cydia nigricana F.	Pea moth	Caribbean, Europe, Russian Federation
Diacrisia oblique Wlk.	Jute hairy caterpillar	India, Asia
Elasmopalpus lignosellus (Zell.)	Lesser cornstalk borer	United States, Central America, Thailand
Estigmene acraea Drury	Salt marsh caterpillar	United States, Central America
Euxoa ochrogaster (Gn.)	Red-backed cutworm	North America
'tEvergestis forficalis L.	Crucifer caterpillar	Europe, India, Japan
Evergestis rimosalis (Gn.)	Cross striped cabbageworm	North America
Hadula trifolii (Hufn.)	Clover cutworm	North America, Europe, Africa, Russian Federation, India, China (People's Republic of)
Helicoverpa armigera (Hbn.)	Cotton bollworm	Europe, Africa, India, Russian Federation, South East Asia, Australia, New Zealand
Hellula phidilealis (Wlk.)	Cabbage budworm	Central America
Hellula undalis (F.)	Cabbage webworm	Europe, Africa, India, Asia, Australia, New Zealand
Lacanobia oleracea (L.)	Bright-line brown-eye moth	Europe
Lacanobia suasa D. & S.	Not reported	Europe, Russian Federation
Loxostege sticticalis L.	Beet webworm	North America, Asia, Europe, Russian Federation
Mamestra brassicae (L.)	Cabbage moth	Europe, Russian Federation, India, Asia
Mamestra configurata Wlk.	Bertha armyworm	North and Central America
Noctua pronuba (L.)	Common yellow underwing moth	Europe
Ochropleura flammatra D. & S.	Indian cutworm	India
Peridroma saucia (Hbn.)	Pearly underwing moth	The Americas, Europe, India, China (People's Republic of), Japan
Pieris brassicae (L.)	Cabbage caterpillar	South America, Europe, Russian Federation, India, China (People's Republic of), Japan, Africa
Pieris canidia (Sparrman)	Small cabbage butterfly	China (People's Republic of), South East Asia
Pieris napi (L.)	Green-veined white butterfly	Europe, North Africa, Russian Federation, India, China (People's Republic of), Japan
Pieris rapae L.	Imported cabbageworm or cabbage white butterfly	North and Central America, Europe, North Africa, Russian Federation, India, Asia, Australia, New Zealand
Plutella xylostella L.	Diamondback moth	Worldwide
Pontia daplidice (L.)	Not reported	Russian Federation
Spodoptera exigua (Hbn.)	Beet armyworm	North and Central America, Europe, Africa, Russian Federation, India, Asia, Australia
Spodoptera frugiperda J. E. Smith	Fall armyworm	The Americas
Spodoptera littoralis (Bdv.)	Cotton leafworm	Africa, Middle East
Trichoplusia ni (Hbn.)	Cabbage looper	Worldwide, except Australia and New Zealand
Vanessa cardui L.	Painted lady butterfly	North America, Europe, Africa, Russian Federation, Australia
Xestia c-nigrum (L.)	Spotted cutworm	North and Central America, Europe, Russian Federation, India, Asia
Acari		
Halotydeus destructor (Tucker)	Redlegged earth mite	Australia, New Zealand, South Africa
Tyrophagus putrescentiae (Schr.)	Cereal mite	United States, Central and South America, Europe, Africa, India, China (People's Republic of)
Stylommatomophora		
Arion lusitanicus Mabille	Spanish slug	Europe
Deroceras reticulatum Müll	Grey field slug	North America, Europe, Russian Federation, Australia, New Zealand
Lissachatina fulica (Bowdich)	Giant African land snail	South America, Africa, India, South East Asia

Table 3.A1.1. **Insect, mite and other Brassicacous crop pests and their regional distribution** (*cont.*)

Order, genus and species	Common name	Regions affected
Thysanoptera		
Thrips tabaci Lind.	Onion thrips	Worldwide
Tylenchida		
Meloidogyne ethiopica Whitehead	Not reported	South America, Europe, Africa
Meloidogyne graminicola Golden & Birchfield	Rice root knot nematode	United States, South America, South Africa, India, South East Asia
Pratylenchus neglectus (Rensch) Filipjev & Stekhoven	California meadow nematode	Europe, Russian Federation, Pakistan

Source: Information drawn from CAB International Crop Protection Compendium; Bonnemaison (1965); Lamb (1989); Thomas (1994).

Table 3.A1.2. **Diseases of rapeseed = Canola (*B. napus* L. and *Brassica rapa* L. [= *B. campestris* L.])**

Common name(s)	Scientific name (and synonyms)
Bacterial black rot	*Xanthomonas campestris* pv. *campestris* (Pammel 1895) Dowson 1939 = *Xanthomonas campestris* pv. raphani = *Xanthomonas campestris* pv. aberrans
Bacterial leaf spot	*Xanthomonas campestris* pv. *armoraciae* (McCulloch 1929) Dye 1978
Bacterial pod rot*	*Pseudomonas syringae* pv. *maculicola* (McCulloch 1911) Young, Dye and Wilkie 1978) (Canada, United Kingdom)
Bacterial soft rot	*Erwinia carotovora* (Jones 1901) Bergey et al. 1923 *Pseudomonas marginalis* pv. *marginalis* (Brown 1918) Stevens 1925
Scab	*Streptomyces* spp. *Streptomyces scabiei* corrig. (ex Thaxter 1891) Lambert and Loria 1989 = *Streptomyces scabies* (Thaxter 1891) Waksman and Henrici 1948
Crown gall	*Agrobacterium tumefaciens* (Smith and Townsend 1907) Conn 1942
Fungal diseases	
Alternaria black spot = Dark pod spot (United Kingdom)	*Alternaria brassicae* (Berk.) Sacc. *A. brassicicola* (Schwein.) Wiltshire *A. japonica* H. Yoshii = *A. raphani* Groves and Skolko
Anthracnose	*Colletotrichum gloeosporioides* (Penz.) Penz. and Sacc. in Penz. *Glomerella cingulata* (Stoneman) Spauld. and H. Schrenk [teleomorph] *C. higginsianum* Sacc. in Higgins
Black leg = stem canker (United Kingdom)	*Leptosphaeria maculans* (Desmaz.) Ces. and De Not *Phoma lingam* (Tode: Fr.) Desmaz. [anamorph]
Black mold rot	*Rhizopus stolonifer* (Ehrenb.: Fr.) Vuill.
Black root	*Aphanomyces raphani* Kendrick
Brown girdling root rot*	*Rhizoctonia solani* Kühn (Canada) *Thanatephorus cucumeris* (A.B. Frank) Donk [teleomorph]
Cercospora leaf spot	*Cercospora brassicicola* Henn.
Clubroot	*Plasmodiophora brassicae* Woronin
Downy mildew	*Peronospora parasitica* (Pers.: Fr.)Fr.
Fusarium wilt	*Fusarium oxysporum* Schlechtend.: Fr. f. sp. *conglutinans* (Wollenweb.) W.C. Snyder and H.N. Hans
Gray mold	*Botrytis cinerea* Pers.: Fr. *Botryotinia fuckeliana* (de Bary) Whetzel [teleomorph]

Table 3.A1.2. **Diseases of rapeseed = Canola (*B. napus* L. and *Brassica rapa* L. [= *B. campestris* L.])** (*cont.*)

Common name(s)	Scientific name (and synonyms)
Head rot	*Rhizoctonia solani* Kühn
	Thanatephorus cucumeris (A.B. Frank) Donk [teleomorph]
Leaf spot*	*Alternaria alternata* (Fr.: Fr.) Keissl. (Canada)
	Ascochyta spp. (former USSR)
Light leaf spot	*Pyrenopeziza brassicae* Sutton and Rawlinson in Rawlinson et al.(1978)
	Cylindrosporium concentricum Grev. [anamorph]
Pod rot*	*Alternaria alternata* (Fr.: Fr.) Keissl. (Canada)
	Cladosporium sp.
Powdery mildew	*Erysiphe polygoni* DC.
	E. cruciferarum Opiz ex Junell.
Ring spot	*Mycosphaerella brassicicola* (Duby) Lindau in Engl. and Prantl
	Asteromella brassica (Chev.) Boerema and Van Kesteren [anamorph]
Root rot	*Alternaria alternata* (Fr.: Fr.) Keissl.
	Fusarium spp.
	Macrophomina phaseolina (Tassi) Goidanich
	Phymatotrichopsis omnivora (Duggar) Hennebert
	Phytophthora megasperma Drechs.
	Pythium debaryanum Auct. non R. Hesse
	P. irregulare Buisman
	Rhizoctonia solani Kühn
	Thanatephorus cucumeris (A.B. Frank) Donk [teleomorph]
	Sclerotium rolfsii Sacc.
	Athelia rolfsii (Curzi) Tu and Kimbrough [teleomorph]
Sclerotinia stem rot	*Sclerotinia sclerotiorum* (Lib.) de Bary
Seed rot, damping-off	*Alternaria* spp.
	Fusarium spp.
	Gliocladium roseum (Link) Bainier
	Nectria ochroleuca (Schwein.) Berk [teleomorph]
	Pythium spp.
	Rhizoctonia solani Kühn
	Thanatephorus cucumeris (A.B. Frank) Donk [teleomorph]
	Rhizopus stolonifer (Ehrenb.: Fr) Vuill.
	Sclerotium rolfsii Sacc.
Root gall smut*	*Urocystis brassicae* Mundkur (People's Republic of China, India)
Southern blight (leaf, root and seed rot)	*Sclerotium rolfsii* Sacc.
Verticillium wilt*	*Verticillium longisporum* (comb. Nov. Karapappa et al.) (Europe)
White blight*	*Rhizoctonia solani* Kühn (India)
	Thanatephorus cucumeris (A.B. Frank) Donk [teleomorph]
White leaf spot = grey stem (Canada)	*Pseudocercosporella capsellae* (Ellis and Everh.) Deighton
	= *Cercosporella brassicae* (Faitrey and Roum.) Höhn.
	Mycosphaerella capsellae (Inman and Sivansen) [teleomorph]
White rust = staghead	*Albugo candida (Pers.) Kunze*
	= A. cruciferarum (DC.) S.F. Gray
	(Peronospora sp. commonly present in staghead phase)
Yellows	*Fusarium oxysporum Schlechtend.: Fr.*
Nematodes, parasitic	
Cyst nematode	*Heterodera cruciferae* Franklin
	H. schachtii Schmidt
Lesion nematode	*Pratylenchus* spp.
	P. pratensis (de Man) Filipjev
Root-knot nematode	*Meloidogyne* spp.

Table 3.A1.2. **Diseases of rapeseed = Canola (*B. napus* L. and *Brassica rapa* L. [= *B. campestris* L.])** (*cont.*)

Common name(s)	Scientific name (and synonyms)
Viral diseases	
Crinkle*	genus *Carmovirus*, Turnip crinkle virus (TCV) (former 'Yugoslavia')
Mosaic	genus *Caulimovirus*, Cauliflower mosaic virus (CaMV)
	genus *Cucumovirus*, Cucumber mosaic virus* (CMV) (Hungary)
	genus *Comovirus*, Radish mosaic virus (RaMV)
	genus *Potyvirus*, Turnip mosaic virus (TuMV)
Yellows	genus *Luteovirus*, Beet western yellows virus (BWYV)
	genus *Cytorhabdovirus*, Broccoli necrotic yellows virus* (BNYV)
Phytoplasmal diseases	
Aster yellows and phyllody	Aster yellows phytoplasma
Miscellaneous diseases and disorders	
Autogenic necrosis	Genetic disorder
Black speck	Physiological
Sulfur deficiency	Sulfur deficiency
Tipburn	Calcium deficiency

Note: * Not known to occur naturally in the United States.

Source: Reproduced from the American Phytopathological Society listing of known *Brassica* pathogens and disorders (2001).

References

American Phytopathological Society (n.d.), listing of known *Brassica* pathogens and disorders at: *www.apsnet.org/publications/commonnames/Pages/Rapeseed.aspx* (last update 3 August 2001).

Bonnemaison, L. (1965), "Insect pests of crucifers and their control", *Annual Reviews of Entomology*, Vol. 10, pp. 233-256, January, *http://dx.doi.org/10.1146/annurev.en.10.010165.001313*.

Lamb, R.J. (1989), "Entomology of oilseed *Brassica* crops", *Annual Reviews of Entomology*, Vol. 34, pp. 211-229, January, *http://dx.doi.org/10.1146/annurev.en.34.010189.001235*.

Thomas, P. (1994), *Canola Growers Manual*, Canola Council of Canada.

Annex A. List of OECD consensus documents on environmental safety assessment, 1996-2015

	Consensus document	Lead country(ies)	Year of issue	Volume
Facilitating harmonisation	Designation of a Unique Identifier for Transgenic Plants (revised version) (guidance document)	Working Group	2006	Vol. 3
	Introduction to the OECD Biosafety Consensus Documents (*available in English and French)*	Working Group	2005	Vol. 1, 3, 4, 5 & 6
	Low Level Presence of Transgenic Plants in Seed and Grain Commodities: Environmental Risk/Safety Assessment, and Availability and Use of Information	Working Group	2013	Vol. 6
	Molecular Characterisation of Plants Derived from Modern Biotechnology	Canada	2010	Vol. 3
	Points to Consider for Consensus Documents on Biology of Cultivated Plants	Working Group	2006	Vol. 3
Traits	Crop Plants Made Virus Resistant through Coat Protein Gene-Mediated Protection	Task Group	1996	Vol. 1
	Genes and their Enzymes that Confer Tolerance to Glyphosate Herbicide	United States, Germany and Netherlands	1999	Vol. 1
	Genes and their Enzymes that Confer Tolerance to Phosphinothricin Herbicide	United States, Germany and Netherlands	1999	Vol. 1
	Herbicide Metabolism and the Residues in Glufosinate-Ammonium (Phosphinothricin)-Tolerant Transgenic Plants	Germany	2002	Vol. 1
	Transgenic Plants Expressing *Bacillus thuringiensis*-Derived Insect Control Protein	United States	2007	Vol. 3
Micro-organisms	***Information Used in the Assessment of Environmental Applications of Micro-organisms***			
	Acidithiobacillus	Canada	2006	Vol. 2
	Acinetobacter	Canada	2008	Vol. 4
	Baculovirus	Germany	2002	Vol. 2
	Pseudomonas	United Kingdom	1997	Vol. 2
	Guidance Documents on Biosafety Aspects of Bacteria			
	Horizontal Gene Transfer Between Bacteria	Germany	2010	Vol. 4
	Methods for Detection of Micro-organisms Introduced into the Environment: Bacteria	Netherlands	2004	Vol. 4
	Use of Information on Pathogenicity Factors: Bacteria	Netherlands and Canada	2011	Vol. 5
	Use of Taxonomy in Risk Assessment of Micro-organisms: Bacteria	Canada and United States	2003	Vol. 4
Biology of crops	Bananas and plantains (*Musa* spp.)	Spain	2009	Vol. 4
	Brassica crops (*Brassica* spp.)	Canada	2012	Vol. 5
	Cassava (*Manihot esculenta*)	Brazil, NEPAD-ABNE and ILSI-CERA	2014	Vol. 6
	Chili, hot and sweet peppers (*Capsicum annuum*)	Korea, Mexico and United States	2006	Vol. 1

	Consensus document	Lead country(ies)	Year of issue	Volume
Biology of crops (*cont.*)	Common bean (*Phaseolus vulgaris*)	Brazil and ILSI-CERA	2015	Vol. 6
	Cotton (*Gossypium* spp.)	Spain	2008	Vol. 4
	Cowpea (*Vigna unguiculata*)	Australia	2015	Vol. 6
	Maize (*Zea mays* subs. *mays*)	Mexico	2003	Vol. 1
	Squashes, pumpkins, zucchinis and gourds (*Cucurbita*)	Mexico and United States	2012	Vol. 5
	Oyster mushroom (*Pleurotus* spp.)	Korea	2005	Vol. 1
	Papaya (*Carica papaya*)	United States	2005	Vol. 1
	Potato (*Solanum tuberosum* subsp. *tuberosum*)	Netherlands and United Kingdom	1997	Vol. 1
	Rice (*Oryza sativa*)	Japan	1999	Vol. 1
	Oilseed rape (*Brassica napus*): ***replaced with* Brassica Crops *(2012) in Vol. 6***	Canada	1997	Vol. 1
	Sugar beet (*Beta vulgaris*)	Switzerland	2001	Vol. 1
	Sugarcane (*Saccharum* spp.)	Australia	2013	Vol. 6
	Sunflower (*Helianthus annus*)	France	2004	Vol. 1
	Wheat (*Triticum aestivum*)	Germany	1999	Vol. 1
	Soybean (*Glycine max*)	Canada	2000	Vol. 1
Biology of trees	***Timber trees***			
	Birch: European white birch (*Betula pendula*)	Finland	2003	Vol. 2
	Douglas fir (*Pseudotsuga menziesii*)	Canada	2008	Vol. 3
	Eucalyptus (*Eucalyptus* spp.)	Australia	2014	Vol. 6
	Larches : North American larches (*Larix lyalli, Larix occidentalis, Larix laricina*)	Canada	2007	Vol. 3
	Pines: Eastern white pine (*Pinus strobus*)	Canada	2002	Vol. 2
	Pines: Jack pine (*Pinus banksiana*)	Canada	2006	Vol. 3
	Pines: Lodgepole pine (*Pinus contorta*)	Canada	2008	Vol. 3
	Pines: White pine (*Pinus monticola*)	Canada	2008	Vol. 3
	Poplars (*Populus* spp.)	Canada	2000	Vol. 2
	Spruces: Black spruce (*Picea mariana*)	Canada	2010	Vol. 3
	Spruces: Norway spruce (*Picea abies*)	Norway	1999	Vol. 2
	Spruces: Sitka spruce (*Picea sitchensis*)	Canada	2002	Vol. 2
	Spruces: White spruce (*Picea glauca*)	Canada	1999	Vol. 2
	Fruit trees			
	Bananas and plantains (*Musa* spp.) [listed above in "Crops"]	Spain	2009	Vol. 4
	Papaya (*Carica papaya*) [listed above in "Crops"]	United States	2005	Vol. 1
	Stone fruits (*Prunus* spp.)	Austria	2002	Vol. 2

ORGANISATION FOR ECONOMIC CO-OPERATION AND DEVELOPMENT

The OECD is a unique forum where governments work together to address the economic, social and environmental challenges of globalisation. The OECD is also at the forefront of efforts to understand and to help governments respond to new developments and concerns, such as corporate governance, the information economy and the challenges of an ageing population. The Organisation provides a setting where governments can compare policy experiences, seek answers to common problems, identify good practice and work to co-ordinate domestic and international policies.

The OECD member countries are: Australia, Austria, Belgium, Canada, Chile, the Czech Republic, Denmark, Estonia, Finland, France, Germany, Greece, Hungary, Iceland, Ireland, Israel, Italy, Japan, Korea, Luxembourg, Mexico, the Netherlands, New Zealand, Norway, Poland, Portugal, the Slovak Republic, Slovenia, Spain, Sweden, Switzerland, Turkey, the United Kingdom and the United States. The European Union takes part in the work of the OECD.

OECD Publishing disseminates widely the results of the Organisation's statistics gathering and research on economic, social and environmental issues, as well as the conventions, guidelines and standards agreed by its members.

OECD PUBLISHING, 2, rue André-Pascal, 75775 PARIS CEDEX 16
(97 2011 17 1 P) ISBN 978-92-64-25044-4 – 2016

www.ingramcontent.com/pod-product-compliance
Ingram Content Group UK Ltd.
Pitfield, Milton Keynes, MK11 3LW, UK
UKHW051458290926
15471UKWH00008B/1276

9 789264 250444